THE WEATHER EXPERIMENT

ALSO BY PETER MOORE

Damn His Blood: Being a True and Detailed History
of the Most Barbarous and Inhumane Murder at Oddingley
and the Quick and Awful Retribution

THE WEATHER EXPERIMENT

The Pioneers Who Sought to See the Future

PETER MOORE

Farrar, Straus and Giroux
New York

Farrar, Straus and Giroux
18 West 18th Street, New York 10011

Printed in the United States of America
Originally published in 2015 by Chatto & Windus, an imprint of Vintage, Great Britain
Published in the United States by Farrar, Straus and Giroux
First American edition, 2015

Library of Congress Cataloging-in-Publication Data
Moore, Peter, 1983–
    The weather experiment : the pioneers who sought to see the future / Peter
Moore.
        pages   cm
    Includes bibliographical references and index.
    ISBN 978-0-86547-809-1 (hardcover) — ISBN 978-0-374-71127-6 (e-book)
    1. Meteorologists—Biography.   2. Climatologists—Biography.
    3. Meteorology—History—19th century.   4. Weather forecasting—History—
19th century.   I. Title.

QC858.A2 M66 2015
551.5092'2—dc23

                                                          2015011212

www.fsgbooks.com
www.twitter.com/fsgbooks • www.facebook.com/fsgbooks

1   3   5   7   9   10   8   6   4   2

To my mother

A fool, you know, is a man who never tried an experiment in his life.

ERASMUS DARWIN TO RICHARD LOVELL EDGEWORTH

We see nothing truly till we understand it.

JOHN CONSTABLE

The Weather Clerk himself, after a life of practice, as well as study, values the 'forecasts' more and more as a scientific ground on which to hoist the drum.

ADMIRAL ROBERT FITZROY

Contents

Author's Note

In the nineteenth century temperature was measured in Fahrenheit. Ice melted at 32°F (0°C), water boiled at 212°F (100°C), and the temperature of the human body was 98°F (37°C). In Britain and the United States barometers measured atmospheric pressure in inches of mercury. The length of a column of mercury at mean sea level would be just beneath 30 inches (1013 millibars). Thirty and a half inches of mercury was typical high pressure. Twenty-nine and a half inches was typical low pressure.

I have used original weights and measures throughout. Most will be familiar to readers, except perhaps a fathom, which is approximately 6 feet or 1.83 metres, and an Irish Mile which is 1.27 statute miles.

Göttingen Mean Time was adopted as a standard for simultaneous magnetic observations in the 1830s and was used on occasion in the 1840s for meteorology before it was replaced by Greenwich Mean Time.

Illustrations

Unless otherwise stated, images come from the author's private collection.

Illustrations in the text

p. 11: Richard Lovell Edgeworth's telegraph, 1830.

p. 17: Claude Chappe's telegraph, 1794 (© The British Library Board, 2394.f.3 p33).

p. 48: Cloud formations above a rural landscape, with a key to the types. Engraving by E. Radcliffe (© Wellcome Library, London).

p. 51: Sketch from *Researches about Atmospheric Phaenomena*, by Thomas Forster, 1816.

p. 71: Engraving from *A Voyage Towards the South Pole, Performed in the Years 1822–1824*, by James Weddell (© The British Library Board, G2558 between p34–35).

p. 73: Map of South America by Robert Wilkinson, 1813.

p. 95: A combined thermometer, hygrometer, and barometer. Engraving after B. Martin (© Wellcome Library, London).

p. 98: 'The distressed situation of His Majesty's Ship *Egmont*' in the West Indies hurricane of 1780 (PX8417 © National Maritime Museum, Greenwich, London).

p. 156: Samuel Morse making his experiments with telegraph transmission (© Wellcome Library, London).

p. 164: Samuel Morse's telegraphic language (© The British Library Board, 1398.f.24 p27).

p. 169: Cecilia Glaisher's snow crystal designs, 1855 (© The Fitzwilliam Museum, Cambridge).

p. 180: London from Blackheath, *Illustrated London News*, 1846 (with thanks to the Gladstone Library for permission to reprint).

p. 214: Cecilia Glaisher's snow crystal designs, 1855 (© The Fitzwilliam Museum, Cambridge).

p. 237: *Royal Charter* Wreck, *Illustrated London News*, 1859 (with thanks to the Gladstone Library for permission to reprint).

pp. 240–41: Synoptic chart of the Royal Charter Gale, 1859, from *The Weather Book*, by Robert FitzRoy, 1863.

p. 243: Cone signals, *Illustrated London News*, 1860 (with thanks to the Gladstone Library for permission to reprint).

p. 253: James Glaisher's instruments deck in the hot-air balloon, from *Travels in the Air*, by James Glaisher et al, 1871.

p. 257: The Pigeons, from *Travels in the Air*, 1871.

p. 264: Flight profile of a balloon ascent in 1862, from *Travels in the Air*, 1871.

p. 324: Air temperature observed in the balloon ascent and descent in 1864, from *Travels in the Air*, 1871.

Plate Section

p. 1: Portrait of Robert FitzRoy, *c.* 1831 (© Science Photo Library).

p. 2: 'Study of Cirrus Clouds' by John Constable, *c.* 1821/2 (© Getty Images); *Spring: East Bergholt Common*, by John Constable, *c.* 1814, 1821 or 1829 (© Victoria and Albert Museum, London).

p. 3: Mount Sarmiento, Tierra del Fuego, showing the survey ship HMS *Beagle* (PW6229 © National Maritime Museum, Greenwich, London); the Arctic Council planning a search for Sir John Franklin, by Stephen Pearce, 1851 (© De Agostini Picture Library/Bridgeman Images).

p. 4: A meeting of the Royal Society at Somerset House (© The Royal Society); *Men of Progress*, a group portrait of the great American inventors of the Victorian Age, by Christian Schussele, 1862 (© Smithsonian Institution, Washington, D.C./Bridgeman Images).

p. 5: (clockwise from top left) Robert FitzRoy, *c.* 1864 (© Royal Astronomical Society/Science Photo Library); James Pollard Espy, by Thomas Sully, 1849 (© Photo National Portrait Gallery/Smithsonian/Art Resource/Scala, Florence); Sir Francis Galton by Gustav Graef, 1882 (© National Portrait Gallery, London); James Glaisher *c.* 1860s.

p. 6: Mirage and Luminous Aureola from *Travels in the Air*, 1871.

p. 7: Isothermal Chart, or View of Climates & Production from William C. Woodbridge, school atlas, published by Oliver D. Cooke & Co., Hartford, 1823 (Graphic Arts Division, Department of Rare Books and Special Collections © Princeton University Library); Chart of the Christmas Day snowstorm in 1836, by Elias Loomis, 1859 (© The David Goldsmith Collection).

p. 8: Tropical and polar air current, from *The Weather Book*, 1863.

Acknowledgements

While writing this book I was awarded a residency at Gladstone's Library at Hawarden in Flintshire. For a month in early spring 2014 I lived in the library, an exquisite Gothic Revival building just across the River Dee into Wales. I worked every day on a creaking first-floor gallery surrounded by 32,000 of Gladstone's own annotated books – famously transported to the library in a wheelbarrow by the Grand Old Man himself. For me there could have scarcely been a better place to dream, read or write about Victorian society. To Peter Francis, for the award and several conversations about weather and religion, I am enormously grateful. For help navigating the collections, which yielded several undiscovered FitzRoy letters, and for permission to republish illustrations, I'm indebted to Louisa Yates and Gary Butler. Thanks are also due to the indomitable trio of Siân Morgan, Phillip Clement and Ceri Williams, as well as the library's always cheerful, always helpful, other employees.

For permission to consult and quote sources I would like to acknowledge the British Library, the National Library of Ireland, the Royal Society (Keith Moore), the Huntington Library (Vanessa Wilkie), the Beinecke Rare Books Library at Yale (Sandra Markham and Anne Marie Menta), the Wellcome Library and the National Meteorological Library and Archives in Exeter. For access to his set of Elias Loomis' weather charts of the 1836 storm and for his enthusiasm for the subject, I'm grateful to my intrepid friend David Goldsmith, whose own weather project is eagerly anticipated.

I would also like to thank Sheila Newman, Jo and Ben in Wexford, Dame Julia Slingo, David Whiting, Julie Wheelwright, and, for all her sagacity and good humour, Sarah my fellow Dolphinite. Dr Christopher Prior of the University of Southampton has once again endured an early draft and returned with valuable advice. Professor John Thornes of the University of Birmingham was also an enormous help. In an age when cloud spotting has been elevated in popularity to the levels of two centuries ago, and people are again faced with the dilemma of distinguishing altocumulus from stratocumulus, his book on John Constable's skies provides a superb introduction to meteorological science. Ever since a serendipitous meeting in 2012 John Thornes has played a valued part in this project – both as a mine of information and a source of encouragement. While he has rescued me from the odd meteorological blunder, any that remain are my own responsibility.

I'm fortunate to have the support and terrific expertise of all at Peters Fraser & Dunlop. Particularly my literary agent, Annabel Merullo, but also Rachel Mills, Laura Williams, Marilia Savvides, Kim Méridja, Silvia Molteni and James Carroll. At Farrar, Straus and Giroux I'd like to thank Mitzi Angel, Jeff Seroy, Stephen Weil, Daniel del Valle and Will Wolfslau, as well as Katja Scholz at Mareverlag.

The greatest debt is due to Juliet Brooke, my wonderful editor at Chatto & Windus, who has nurtured this book from the beginning and challenged me at all the right times. I am also grateful to Clara Farmer, Susannah Otter, Kate Bland and Mikaela Pedlow, and to Kris Potter for the fabulous jacket design of Chatto's edition.

A special thanks to Claire for tolerating a home crammed with nineteenth-century journals and other weather paraphernalia. Alongside everything else, her unquenchable optimism and editorial instincts have remained prized commodities. When London becomes too much I am lucky to have a Staffordshire hideaway with a quiet desk and a familiar bed to escape to, not forgetting a father who often seems as encyclopaedic as Francis Beaufort.

My mother, raised on the raw east Yorkshire coast, still watches the sky with a knowing eye. She would often call me up from the comfort of the sofa in my childhood to savour a red sunset or to glimpse a curious cloud. This book, dedicated to her, is my riposte.

THE WEATHER EXPERIMENT

The Weather Experiment

We are never far from a weather forecast. An average Briton on an average day might encounter five or six of them, broadcast, printed, tweeted, passed on second hand. You can wake in the morning to the sparkling enthusiasm of a breakfast weather presenter, and be lulled off at night to the mantric rhythms of Radio 4's shipping forecast and its signature tune, 'Sailing By'.

Whatever the medium, the weather forecast is an ingrained component of modern life, its ever-evolving projection of what the atmosphere has in store always at hand. As a rule forecasters are clean-cut, bright-eyed, full of empathy and concern when something grim is on the way. The kind tenor of their broadcasts, the sharp suits, the mild manners, the crisply delivered chunks of meteorological caution, can fool you into thinking they are paragons of conservatism. The reality is completely different. These forecasters are in fact the product of one of the most notorious and daring scientific experiments of the nineteenth century.

It is a strange thought. Such is the ubiquity of the forecast today it is difficult to imagine an age before it existed. The bright, breezy afternoon of 24 November 1703, for example, when the Great Storm – the most intense ever to strike England – was hurtling wolf-like towards the west coast. Hardly anyone could have foreseen what was about to happen. Winds stripped lead from church roofs, windmills were blown with such force that they spun into flame like giant Catherine wheels. Cattle and sheep were swept over hedges. Ships were blown across the North Sea from Harwich to Sweden, and others were driven on to the Goodwin Sands where an estimated 2,000 were swallowed up by the waves. No final tally was recorded but it is thought 10,000 died in a handful of hours. For Daniel Defoe it was a disaster far worse than the Great Fire of London.

For all Defoe knew, another storm might come at any time. A century and a half more would pass before the first storm warnings and weather forecasts were issued in the 1860s. This time lag reflects the complexity of the problem: the daunting task of decoding the atmosphere, and co-ordinating a response. That this ambition was achieved at all is testament to the industry and intellect of a remarkable group of individuals who lived between the years 1800 and 1870. They came from eclectic backgrounds – they were sailors, artists, chemists, inventors, astronomers, hydrographers, businessmen, mathematicians and adventurers. They coined radical theories, invented instruments, established networks and convinced governments that they had a moral duty to protect their citizens. Stretching across seventy years this book tells their story. It examines how they laid the foundations for the meteorological science of today, and gave us the ability to glimpse into the future.

In 1800 the weather remained a mystery. Standing on the quarterdeck of the *Victory* at Trafalgar, Horatio Nelson had no scientific way of measuring the winds. Rising up in his hydrogen balloon, the daredevil aeronaut Vincenzo Lunardi could not have explained why the sky appeared blue. A young J.M.W. Turner, who was forging a reputation as a landscape artist, had no vocabulary to describe the clouds he painted, nor could he explain how they stayed suspended in the air. Thomas Jefferson, founding father, President of the United States and keen weather diarist, had no way of knowing how high the earth's atmosphere extended over his mountaintop home at Monticello in Virginia. And Mary Shelley, who would write so vividly of a storm on the night of Victor Frankenstein's wedding, had no scientific understanding of what a storm actually was, how it functioned or where it came from.

Various theories attempted to fill the void. Some believed the weather was cyclical, that temperatures of one year would be repeated in another somewhere along the line. Others thought that weather was governed by the orbit of the moon or the planets, the pulse of the sun, the soil of the earth or the electricity in the sky. 'The powers of reason have been bewildered in the inextricable labyrinth of causes and effects,' wrote one frustrated theorist in 1823.[2] For most, weather was a divine force, mood music conducted by God sent to foreshadow change or to punish sin. As Psalm 19 asserted, 'The heavens declare

the glory of God; the skies proclaim the work of his hands.'[1] Helpless
in the face of nature, Christians rang church bells when storms
approached hoping these would drive bad weather away. Often the
bells were blessed by the clergy. The Director of the Paris Observatory
François Arago jotted down the words of a typical benediction: 'whenever it rings may it drive far off the malign influences of evil spirits,
whirlwinds, thunderbolts, and the devastations which they cause, the
calamities of hurricanes and tempests'.[3]

It was only right. The sky was God's wilderness, a place apart, an
impenetrable barrier between God's divine kingdom and the tainted
world below. Many still referred to this space as 'the heavens', a catch-
all expression that encompassed clouds, rainbows, meteorites and stars.
It was a suitably vague and deferential term for such an uncertain,
quicksilver space: a place that was at once both incredibly close and
impossibly distant.

Weather-watchers had no linguistic framework to scientifically
explain what they saw. 'Our Language is exceeding scanty & barren
of words to use & express ye various notions I have of Weather &c,
I tire myself with Pumping for apt terms & similes to illustrate my
Thoughts', wrote a Worcestershire diarist in 1703. Striving to describe
what was happening up above, he wrote of skies:

> loaded & varnished with Bulging, dull swelling Bas-Releive clouds
> bloated & pendulous. I style them ubera caeli fecunda: sky-
> cubbies or udders cloudy; they enclosed & stufft ye whole visible
> Hemisphere in colour like Lead-vapours or a tall Fresco ceiling,
> or marble veined grotto.[4]

The writer's attempt to impose order on nature was suggestive of
a time to come. The catalytic moment came in 1735 when Carl Linnaeus' *Systema Naturae* was published. The book provided 'observing
gentlemen', as Gilbert White would later call them, with a simple
method for dividing all the variety of nature into neat groups. In time,
Linnaeus' ambition became an Enlightenment ideal. Everything –
plants, animals, rocks, diseases – was to be studied, grouped, assigned
logical Latin names, made sensible.

Yet the sky escaped this. A hundred years after the Worcestershire
weather diarist felt let down by the 'scanty & barren' meteorological
language, there was still no set vocabulary to describe atmospheric

processes. The sky was the last part of nature to be classified: a relic of the arcane, chaotic world that had existed before Newton and the Scientific Revolution. Those scattered few who did track air temperature and pressure, such as Jefferson at Monticello or Gilbert White at Selborne, not only lacked a standardised language but they also had no outlet, no forum to share their research. Rooted in one place, able to see perhaps ten or twenty miles to the horizon, they had a flavour of their own weather, but no conception of what was going on beyond their little scientific fiefdoms. Of fronts, cyclones, cumulus clouds, lapse rates or radiation flows, they knew nothing.

It was 1800 before change finally came. In intellectual circles there was an increasing use of the word 'atmosphere', a compound Greek term signifying a surrounding vapour. The linguistic shift reflected a changing intellectual stance. Unlike the heavens, the atmosphere was as deserving of rational analysis as a human heart, the corolla of a flower or a sandstone rock. The discovery of the core gases hydrogen, oxygen and nitrogen by Cavendish, Priestley and Rutherford had given fresh personality to the air that floated around people's heads. Poets and philosophers began to envisage gases flowing like rivers in the sky: streams of wind, avalanches of cloud, torrents of moisture. This was a new world to explore, as real in the enlightened imagination as the deserts of Africa or mountains of Asia.

Luke Howard, who became internationally renowned for his work on clouds at the beginning of the nineteenth century, caught the zeitgeist in a rallying passage:

> The *sky* too belongs to the Landscape: – the ocean of air in which we live and move, with its continents and islands of cloud, its tides and currents of constant and variable winds, is a component part of the great globe: and those regions in which the bolt of heaven is forged, and the fructifying rain condensed, – where the cold hail concretes in the summer cloud, – and from whence large masses of stone and metal have descended at times upon the earth, can never be to the zealous Naturalist a subject of tame and unfeeling contemplation.[5]

People were looking at the skies in new ways. In 1803 Howard published his *Essay On the Modifications of Clouds*, allotting scientific names to clouds for the first time. A few years later Francis Beaufort sketched out his idea for a quantified wind scale. In 1823 John Frederic Daniell's *Meteorological Essays* would appear, a work that revived interest in the subject. By the 1830s meteorological articles and reports were filling scientific journals, meteorological societies were being formed and networks of weather-watchers were being established. Inspired, people tracked the atmosphere as never before. They took readings at home, at sea, on mountaintops and in balloons. For John Ruskin, a wide-eyed undergraduate at Christ Church, Oxford, meteorology was no longer the poor relation. It had become 'the young Hercules', 'full of the soul of the beautiful'.[6]

More achievements followed: the first synoptic maps, the earliest weather reports, better understanding of dew, snow crystals, hail and storms. Increasingly, the question was what to do with all this knowledge. Should meteorologists continue until they ground out atmospheric laws – the laws that controlled the weather – just as Newton had revealed the laws that governed the tides? Or should they turn what they had learnt to some practical use? In his essay, 'Remarks on the Present State of Meteorological Science', John Ruskin set out a manifesto:

It is for [the meteorologist] to trace the path of the tempest round the globe, – to point out the place whence it arose, – to foretell the time of its decline, – to follow the hours around the earth, as she 'spins beneath her pyramids of night,' – to feel the pulses of the ocean, – to pursue the course of its currents and its changes, – to measure the power, direction, and duration of mysterious and invisible influences, and to assign constant and regular periods to the seed-time and harvest, cold and heat, summer and winter, and day and night, which we know shall not cease, till the universe be no more.[7]

There was bound to be a philosophical collision. If weather was a capricious wonder of Nature, then to pursue it across land and sea and accurately record its movements was a difficult enough task. But to predict what it might do was a step too far. When one MP suggested to the Commons in 1854 that it might soon be possible to know the weather in London twenty-four hours in advance, the House roared with laughter.

It took a further seven years, the amassing of ledgers of data and the application of a new term, 'forecast', before national weather predictions were first officially issued, in 1861. But even then the project was fraught with difficulty. It was only two years since Charles Darwin had plunged the Church into an existential crisis with the publication of *On the Origin of Species*. Now science with its forecasts was threatening to explain the future just as the theory of evolution had explained the past.

By a quirk of history the man behind these forecasts, Robert FitzRoy, had been Darwin's captain on the *Beagle*'s famous voyage thirty years earlier. Today we are familiar with Darwin's story, from prospective country parson to revolutionary evolutionary theorist. We are less familiar, though, with FitzRoy's. Once a dashing star of the Royal Navy, a thoroughbred member of the British establishment, a champion of humanitarian causes, FitzRoy's career would veer off in an unexpected direction after he began his weather work in the 1850s.

A complex and contradictory character, full of fight and fire, FitzRoy is often misremembered today merely as Darwin's captain. He was much more than this. From his earliest footloose days in Tierra del Fuego to his later, equally colourful, career in Whitehall he strove to understand the weather. Alongside his contemporaries FitzRoy stands out. High-minded and morally driven, he was impatient to use his science for greater good. It was a stance that endeared him to the public but brought him enemies and left him exposed to accusations of recklessness, megalomania and vanity.

FitzRoy believed that he was simply moving with the times. By the 1850s meteorologists were no longer isolated figures. They were increasingly clustered into networks, their data shared over the wires of a bewildering new technology: the telegraph. Conceived a century before as a plaything, by the 1860s the telegraph had developed from its origin as an optical device to become fully electrified. It was the machine that made weather forecasting possible.

The invention of the telegraph, the evolution of meteorological theory, the contributions of those behind the progress – Beaufort, Constable, Redfield, Espy, Reid, Glaisher, Loomis – add up to something more. They unite to form what I have come to think of as a generational experiment: a quest to prove that earth's atmosphere was not chaotic beyond comprehension, that it could be studied, understood and, ultimately, predicted. Like a scientific experiment this story splits into its component parts – seeing, contesting, experimenting and, most importantly of all, believing.

The action blows across territory like a keen spring breeze. It travels from the Irish Midlands to the vales of Suffolk, from New York City to Tierra del Fuego at the tip of South America. Whether in the bristling beauty of a frosty winter's dawn, the dewy meadows, the fading blues, pinks and oranges of a summer's evening, or the aftermath of a transatlantic hurricane, those who sought the truth did so with growing confidence that they had the ability to find it.

Dawn

It is approaching dawn on a summer morning. The atmosphere is cool, clear and still. Above, the sky is filled with stars.

For hours the grass of a meadow has been cooling. Infrared radiation is being released from the earth into the air. With no incoming solar energy to replenish this heat, temperatures have been falling: 21°C, 18°C, 16°C. Each individual blade is losing its store of energy as the ground grows colder. The radiation flows at different rates from different species – white clover, sweet vernal, spear thistle, meadow foxtail and dandelion – depending on the plant's size, surface area and distinct radiative power.

The air above the meadow is damp: there have been some sharp showers over the last few days. This vapour is contained in the air and as it cools the relative humidity rises. Soon the air is saturated with water. The temperature has reached a tipping point: the dew point. From now on the vapour will begin to condense. If temperatures drop far enough there will be a dawn mist or fog, but it is not cold enough for this today and there is no breeze to mix the molecules together. Instead water droplets begin to form on the blades of grass, slowly, imperceptibly at first. As time passes these drops swell. Soon they become visible flecks clinging to the tips and stems of every blade.

As the sun breaks across the horizon the meadow is struck by the first rays of ambient light of a new day. The impression from afar is that there is a milky film stretched over the meadow. But what we really see is the eye's merged impression of the dawn sunlight rebounding off millions and millions of droplets that have formed this day's unique harvest of dew.

The dew does not wet the blades of grass as raindrops do. Instead it is balanced on microscopic hairs, separated from the stem by a thin membrane. It hangs like a jewel on a neck, twinkling and sparkling, acting like a lens for the morning sunlight. If you study a dewdrop with just one eye, and change your angle of incidence with the sun, you can see it shine with iridescence: blue, then green, yellow, orange, red. It is a rainbow in miniature.

Walking across the dewy meadow in the morning you might notice another effect. Your shadow is long, cast by the dawn sun for several metres. A dazzling white light shines like a halo around your head. This is *Heiligenschein* – German for 'light of the holy one'. It is generated by the dew. Each imperceptible droplet is concentrating the sunlight on the plant behind it, just as the eye of a human focuses light on to the retina.

The intensified light bounces immediately back through the globe of water towards our eye. We should see a greenish light, but our eyes' colour receptors are overpowered and instead we perceive a white halo around our heads, shining in the golden sunlight of a summer's dawn.

PART ONE

Seeing

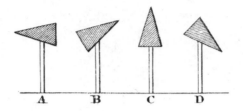

CHAPTER I

Writing in the Air

At a quarter to eight on a breezy spring morning in 1804, Francis Beaufort of the Irish Telegraph Corps came racing up the broad upper slopes of Croghan Hill, his militiamen close on his tail. He was instantly at work. He jammed 'at least nine ounces' of tobacco leaves into a lead pipe, pulled out a match, held it close and let it catch. The flare ignited, smoke coiled into the morning air. In seconds Beaufort and his men were engulfed by the thick, earthy aroma of tobacco. In a letter to his sister Fanny written two days later Beaufort declared proudly that his flare 'made the hollow between the little moat and the summit of the hill look like the crater of Mount Vesuvius in an eruption'.[1]

Beaufort was a short man, not much over five feet. On mornings like this his men might catch sight of the sabre scars on his arms, reminders of his days at sea. Now they had a moment to rest at the top of Croghan Hill, a whale-backed elevation that sloped out of the Bog of Allen in the Irish Midlands, and watch the smoke rise. This was part of a predetermined plan, Beaufort's way of signalling their location to Richard Lovell Edgeworth, the Chief Telegrapher, who was lodged nine miles away in the hamlet of Kilrainey.

Beaufort had woken late that morning. Anxious that he was going to miss Edgeworth, he had set off on a fifteen-minute scramble up the hill. In his letter to Fanny he noted that he had almost broken his neck on the way. His sister would have recognised the description. This childlike, superabundance of energy characterised all he did. Even his letters home rang with exclamation marks, or skipped from one sentence to the next in a chain of breathless dashes.

But this was just one side of Beaufort, a passionate edge reserved

for an intimate few. Outwardly he was a practical man. He had a
tidy, scrupulous mind that had served him well during his ten years
of service in the Royal Navy. Now he was bringing this experience
to bear overseeing the construction of Ireland's first ever optical
telegraph line. This was the brainchild of Beaufort's brother-in-law
Richard Lovell Edgeworth. It consisted of a chain of hilltop stations
out of which a tall pole rose fifteen feet into the air. To the top of
this pole was fastened a large isosceles triangle that could be turned
around like the hands of a clock to any one of eight distinct positions.
The whirling of the triangle corresponded with a vocabulary that
Edgeworth had invented so that words or phrases could be transmitted along the line, with one station mimicking the movements
of another.

It was an exciting project with stations planned to link Dublin on
the east coast with Galway on the west. If the machines worked, and
Edgeworth was certain that they would, it should be possible to
transmit messages between the two places in minutes – a dazzling
thought. For the past six months, Beaufort's job had been to lead his
militia from one location to another: sourcing the raw materials,
building guardhouses and stations and training men to understand the
telegraph code. Little by little they had progressed, and throughout
the winter and spring months of 1804 telegraph stations had been
appearing on the landscape, looking to the untrained eye like diminu-
tive windmills.

It was difficult work but Beaufort loved being out in the open
air. That morning on Croghan Hill, with his flare burned out and
his militiamen 'cold and tired', he decided to enjoy the view. He
discharged his men and remained on the top alone. Beaufort was
used to analysing the atmosphere and watching its subtle shifts.
Now in his solitude he gazed out across one of the most exhilarating
views in Ireland. Many were drawn to the top of Croghan Hill to
enjoy the panoramic vista and on that spring morning the terrain
of his homeland stretched out in miniature beneath him. Far to his
east were the Wicklow Mountains, rising and falling, on the horizon.
Closer were the deep brown hues of the bogs, a treeless barren
country, notoriously perilous for those on foot, but cast in glossy
morning light even they glowed with lustre. To the north the shifting
patterns of the sky were reflected in the shallow water of Lough

Ennell, a locale that, tradition held, had inspired Jonathan Swift's miniature kingdom Lilliput a century before.

In his letter to his sister Fanny, Beaufort wrote:

> It was a most great and sublime scene . . . the honesty of my situation – the anxiety about the heights – and the awful magnificence of the still, silent and half visible world, far above which I seemed to be elevated – the brilliancy of the moon and the rapidity with which the clouds flew overhead (from my nearly being in them) – kept my thoughts sufficiently and delightfully employed.[2]

Beaufort's letter is steeped in the language and passions of his age. He responds to the view more like a Romantic poet than a military man. His eyes are attuned to the 'brilliancy' and, paradoxically, 'awful magnificence' of the world around him. The racing, pulsing atmosphere has brought on a giddiness, a tightening of fibres, a sensation that he is eager to convey to his sister. This was a typical response. Beaufort basks on the hill, overawed by an atmosphere that seems beyond comprehension. Like others of his time – he was born within four years of Southey, Coleridge and Wordsworth – he remained in thrall to the philosophy set down by Edmund Burke half a century before: the Sublime, that exquisite combination of terror and bliss, and its effect on the soul.

Edgeworth's optical telegraph had been commissioned as a part of the Irish response to Napoleon Bonaparte's gathering armies on the French coast. A prototype had been displayed and tested before Lord Hardwicke, the Lord Lieutenant in Dublin, and was now the source of considerable excitement. Should the French attack, as seemed likely at the start of 1804, it would give the government a way of communicating the news across the country and perhaps even the chance to raise the militia in response.

For years before, governments had been forced to rely on hilltop beacons or flares – like that burned by Beaufort on Croghan Hill – to spread news of invasions. Other methods had been tried too: church bells, trumpets, cannon fire, carrier pigeons, drums and torchlight had all been used with varying degrees of success to send simple dispatches of life or death, peace or war. But each method had been hampered by its own specific difficulty. Well into the eighteenth century written letters remained the most reliable way of transmitting complex messages across distance. But even in urgent cases these could only travel as fast as the gallop of a horse. More often they crawled between the cities and towns, towns and villages, bearing tidings of events that had long since happened.

So slow was the trickle of information that people remained almost completely ignorant of anything that happened beyond their own personal sphere. Reports, for example, of Captain Cook's murder in Hawaii in 1779 took eleven months to reach England. A decade later, in July 1789, ten days passed before Parson Woodforde heard the news in Norfolk of the Storming of the Bastille in Paris. Matters had improved gradually throughout the eighteenth century with the expansion of the turnpike network in England. This offered a steady and straight surface for the red, maroon and black mail coaches that rolled sluggishly along them (at 7 mph on a good day). But in Ireland with its dangerously rutted roads, overgrown bridleways and meandering country lanes, a letter written in Dublin would commonly take a week to arrive at its destination in Galway.

The emergence of the optical *télégraph* in France promised to revolutionise this. News of its invention had spread through British society, in August 1794, when a design for the device was discovered in the pocket of a prisoner in Germany. Newspapers seized on the story, equally exciting and terrifying, that their revolutionary enemy had invented a machine that allowed them to communicate across hundreds of miles in a flash. The *télégraph* was the invention of Claude Chappe, a bright and determined engineer, displaced clergyman and member of the Société Philomatique in Paris. Forced from his clerical living at the start of the Revolution, Chappe had turned his mind to invention and, with the help of his brothers, conceived the idea of a machine capable of sending messages with clarity, speed and confidentiality. After several prototypes he settled on a design that, the *Annual Register* observed, imitated the form of the human body. The *télégraph* was fifteen feet tall and had two adjustable arms that were fixed to an upright pole. 'Were two men to make signs to each other at a distance,' the *Register* had explained, 'too great for seeing the ordinary motions as made by dumb people, they would move their arms as Monsieur Chappe moves his telegraph.'[3]

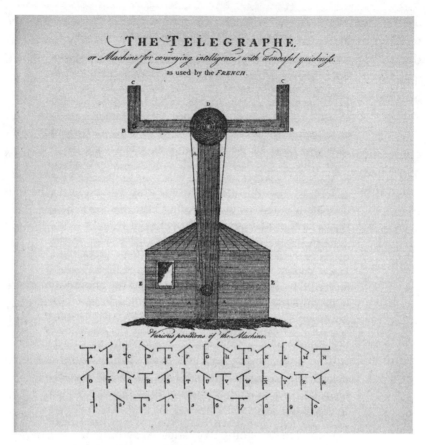

Mounted at stations twenty miles apart so that messages could be relayed with great speed, Chappe's telegraph was truly ground-breaking. Eager to show its potential, enterprising businessmen held telegraphic demonstrations in London's theatres. The British actor and writer Charles Dibdin seized the moment to produce a ballad, which thundered:

> If you'll only just promise you'll none of you laugh
> I'll be after explaining the French Telegraph!
> A machine that's endow'd with such wonderful pow'r
> It writes, reads, and sends news fifty miles in an hour[4]

The abrupt appearance of this device had shattered existing notions of velocity. The very word *télégraph* – a fusion of the Greek *tele* and *graph* that literally meant 'far-writer' – became a fashionable euphemism for speed, efficiency and confidentiality. That a gentleman might divulge intimate conversation to another in a distant place without being exposed was a tantalising prospect. That William Pitt the Younger in Downing Street might be able to converse, harry, jostle or intrigue with the Lord Lieutenant in Ireland over his nightly bottles of wine or direct faraway battles from the solitude of his study was an idea that had set imaginations ablaze.

A decade had passed since Chappe's first *télégraph* line had been completed in northern France and since then a rich variety of designs had been created and tested across Europe. There were shutter telegraphs that winked and blinked, others that waved and whirled. If Edgeworth's device was to work then, like the French, the Irish would theoretically be able to send encoded messages at the speed of light. It would be a visual nerve that stretched the breadth of the country.

To oversee his plans Edgeworth had turned to a talented acquaintance Francis Beaufort. Now aged thirty, Beaufort's life had already followed a colourful course. He had travelled to the far corners of the world, survived a shipwreck, served King George a dozen times in battle and discovered the allure of exploration, a thrill that would come to dominate his life. For Beaufort, at a loose end back in Ireland, the country of his birth, Edgeworth's telegraph project was a chance to plough his talents into a scientific, patriotic pursuit. They both believed that their Irish telegraph was going to change everything. It was a perfect match.

From boyhood Francis Beaufort had been marked for his talent. Quick-witted and naturally curious, in the 1780s he had filled notebooks with formulas and theorems in an immaculate copperplate hand. An entry from one book kept by his father gives a revealing snapshot of Francis at the age of fourteen. It was penned on a winter's night in Dublin. Long after dark he had lain awake at the family home in Mecklenburg Street, staring into the sky. His attention had been caught by an intriguing circle that ringed the moon. It shone with subtle brilliance. On a scrap of parchment entitled an 'Observation of Francis Beaufort' he recorded what he saw.

On the 12th Decr. 1788 at a little after 11 o'clock I saw a circle around the moon at a distance of about 8´ or 9´ the breadth of it was a semi [diameter] of the moon it consisted of three shades, the internal one that next the ☾ was a lightish purple next that a light red, and next a greenish yellow.[5]

Beaufort is awed by the lunar halo, a sight that he had most likely never experienced before. Rather than letting the moment escape he bottles it for future reference like a botanist with an unidentified specimen. He jots down the time of his observation, and adds quantitative detail to his description, preserving a picture of the scene. This was characteristic of Beaufort. It shows his natural desire to capture and record. It reflects his flair for empirical study: to watch, analyse and distil subjects into a legible form.

It was an early sign of Beaufort's organising mind. Another was a cryptic code invented for himself and his elder brother William – a combination of Greek letters, astronomical symbols and twirling lines – to enable the two to communicate in secret about risqué or forbidden topics like sex and religion. Knowing their father would notice the code, Francis once appealed to him, 'never take ill of my writing things [to] William in a concealed hand or manner, for let me assure you 'tis only little jokes or trifles between us.'[6]

Francis' father could hardly be annoyed, as it was just the kind of behaviour he would have indulged in himself. Reverend Daniel Augustus Beaufort was the guiding force in Francis' young life. Daniel – known affectionately as 'DAB' to his friends – was no ordinary man. Among his many and varied accomplishments was a beautifully accurate map of Ireland. This was his crowning achievement but he was also a classicist, gentleman farmer, architect, hobbyist philosopher and all-round society man who played a role in establishing the Royal Irish Academy. Talented though he was, Daniel Beaufort was constantly hindered by his Micawberish propensity for debt. The state of the family finances meant the Beauforts could never live the cosy life enjoyed by many clergy families. Always in the shadow of the bailiff, they lived a cat-and-mouse existence, fleeing from one place to another. During the first sixteen years of Francis' life they uprooted on six occasions: from Navan in County Meath – Francis' birthplace – to Chepstow in England, then Cheltenham, Dublin, London and, finally, Collon in County Louth in 1789.

Francis' education suffered as a result. He managed just a short spell at a marine academy in the 1780s in Dublin but for much of his boyhood he was schooled at home. His father's contacts did, however, bear fruit in 1788 when he was entered for a spell of private tuition with Dr Henry Usher, Professor of Astronomy at Trinity College Dublin. The classes came at a perfect moment in Francis' intellectual development. At dusk he would set out from his home in Mecklenburg Street, down past the legal chambers on Marlborough Street to the noise and bustle of Bachelors Walk and Ormond Quay on the north bank of the Liffey. He would pass the hospital, Royal Square and cut through the market gardens and open countryside of Phoenix Park and the winding road up to Castleknock, away from the dirt, damp and lamps of the city to the clear air and open skies above the newly founded Dunsink Observatory.

Dunsink, four miles outside the city's bounds and 275 feet above sea level, was the finest setting for celestial observation in Ireland. The splendour of the observation house with its lofty dome reflected the importance of astronomy at a time when the limits of the universe were being redrawn by astronomers like William Herschel whose recent discovery of Uranus had delighted the scientific world. At Dunsink Francis received tuition and was given the use of powerful telescopes, charts of the sky and sextants. He learnt to sweep the heavens for stars like Sirius or Polaris, for comets, and to calculate longitude and latitude by celestial observation.

Dr Usher's classes caught Francis on the cusp of adulthood and they would prove invaluable to him in his chosen career at sea. Later he said his heart had been set on becoming a sailor from the age of five and, after a decade of waiting, in 1789 he left Dublin with his father for London where he was 'torn from the paternal wing and launched on the boisterous ocean'. Through DAB's connections, Francis had secured a berth on board an East Indiaman. It was the beginning of a seafaring career in the golden age of sail.

Despite being the 'guinea pig' of the crew, within three weeks Francis had assumed responsibility for the midday latitude measurements. And when not on deck he would spend hours in the crow's nest, watching the world revolving about him. It was a world rich with new words – eddies, fathoms, hawseholes, furling, reefing – and ideas. Sailors still lived in fear of Davy Jones, the spirit of the sea. For luck they carried the caul of a baby or the feather of a wren. They told stories of sirens – sea nymphs who charmed with their melodious voices – and Aeolus, who kept the winds locked up in a mountain and 'loosed them at his pleasure; to afford a passage to the mariner, or to ruin him by a storm'.[7]

After an initial bout of seasickness passed, Francis thrived. By the time his ship, the *Vansittart*, had reached Batavia in the Dutch East Indies, he had grown so confident in his observations that, using a borrowed sextant, he was able to amend a previously accepted and erroneous calculation of the city's latitude by three miles. 'I am so conceited as to think my Lat. is much nearer to the mark as I have got so many [observations] and none of them disagree more than 20° from each other,' he noted.

From the calm of Batavia's Observatory, Francis could not have anticipated the turn his naval career was about to take. Just days out of harbour the *Vansittart* would strike a shoal in the Gaspar Straits and sink, along with its enormous treasure of £90,000 – more than three times as much as George III had paid for Buckingham House (later Palace) twenty-nine years earlier. Francis survived the wreck and miraculously avoided all subsequent perils in an ocean awash with Malay pirates. It marked the beginning of a roving period in his life. Having made passage back to Britain Francis joined the Royal Navy shortly before the outbreak of war with Revolutionary France. He went on to serve at the Glorious First of June and in a handful of other skirmishes against marauding Spanish and French vessels in the Mediterranean and Atlantic.

Francis' earliest surviving weather diaries stem from this time. A journal kept during his year on HMS *Latona* in 1791 shows his devotion to recording the passing weather in the conventional language of the day – moderate, light airs, clear or cloudy.[8] By the time he was serving on HMS *Aquilon* in 1792 his journal had expanded to eight columns that detailed the day of the week, the date of the month, the wind, course, distance, latitude, longitude, and 'place where bearing taken'. Beaufort was remarkable for his skill in observation – he would later dub it a 'hobby or insanity' – and also for the breadth of his knowledge. He sifted books, gathering facts like a magpie, building up a vast floating library that included works by the poets Pope and Dryden, Edward Gibbons' *Decline and Fall of the Roman Empire*, Tobias Smollett's magnificent picaresque novels *Roderick Random* and *Peregrine Pickle* and Adam Smith's *Wealth of Nations*. He read in English, French and Latin and added notes in Greek and Italian.

Throughout the 1790s Francis was borne upwards through the ranks. Trusted as both an expert navigator and a leader of men, by the turn

of the century he had been gazetted an officer. As first lieutenant of
the battle-hardened *Phaeton* Beaufort criss-crossed the Mediterranean
hunting for prizes. It was then, on a warm October afternoon in the
Mediterranean, that Lieutenant Beaufort's fortunes turned. Storming
a Spanish brig off the town of Fuengirola, he nearly lost his life.
Clambering onto the quarterdeck of the prize with his infantrymen,
he was shot at point-blank range with a musket and set upon by a
man with a sabre. The second of the two blows Beaufort took to his
head might well have killed him, he later reasoned, but for the cushion
of a silk handkerchief folded in his hat.

His injuries – nineteen in total, he recorded with typical exactitude –
were enough to curtail his promising career. He lost all feeling in three
fingers, a shotgun slug lay embedded in his left lung and other shrapnel
had torn holes into an arm – 'into one of which I could just cram an
inkbottle'. These would heal in time, but during the year he was
convalescing in Gibraltar and Lisbon, the Revolutionary War drew to
a close. By the time he arrived back at the Admiralty in London in
1802 the Navy was busy paying off officers of surplus ships. Without
influence, Beaufort was advised by Earl St Vincent, First Lord of the
Admiralty, to quietly accept his promotion to commander, along with
the traditional half-pay pension of £45 12s. 6d, and leave for home.

Years later when he had become a venerated man of science,
Beaufort would appreciate the value of his spell back home in Ireland.
At the time, though, he hated it. Stripped of his naval duties for several
years Beaufort was essentially rudderless. Approaching his thirtieth
birthday – approximately halfway through life by the standards of the
age – he had little money, no home or wife and he had been forced
to languish under his parents' roof at Collon in County Louth.

Prone to depression – Beaufort dubbed it his 'Blue Devils' – he
spent months dogged by low spirits, hankering after nothing more
than a secluded farm and a successful conclusion to his long-running
courtship of Charlotte, Edgeworth's beautiful daughter. He gained
neither. In one of a series of long, introspective letters to Charlotte,
he complained, 'I have been mending spectacles, and filing and riveting,
and fixing up pulleys, and growling all day, and almost swore in my
mouth that I would immediately mount my nag and be off to seek
my fortune in some happier clime where my ire would have time to
evaporate.' He rounded off with a dispiriting coda, 'I fancy I shall
think better of it and continue to vegetate here.'[9]

In the autumn of 1803 Edgeworth saved him from this miserable, listless existence when he offered him a job on his telegraphic project. For Beaufort it was the ideal challenge. Not only was it the prospect of steady work, but it was an opportunity to work at close quarters with a man who had lived and learned and schemed with some of the most accomplished minds of his time.

In 1803 Richard Lovell Edgeworth was fifty-nine years old and known as one of Ireland's most enlightened men. He was both a man of literature and a man of science, blessed with formidable creative intelligence and versatility. During his long career he had gained friends such as Sir Joseph Banks, Thomas Day, Erasmus Darwin, Matthew Boulton and Thomas Beddoes. To Daniel Beaufort, Richard Lovell Edgeworth was 'my learned and ingenious friend'.[10] It was a soubriquet that stuck: The Ingenious Mr Edgeworth.

The Beaufort and Edgeworth families had been drawn together while Francis was away at sea in the 1790s. Sharing an educated, inquisitive outlook, they were a good fit, and with Edgeworth's recent marriage to Francis' sister Fanny – she became his third wife – ties had been strengthened. As Maria, Edgeworth's novelist daughter, would write, 'It seldom happens, that, when two large families are connected by marriage, they suit each other in every branch of the connexion . . . but, happily, all the individuals of the two families, though of various talents, ages, and characters, did from their first acquaintance coalesce.' And the most firm of all the friendships that blossomed was that between her father and Francis.[11]

Edgeworth was almost exactly thirty years Beaufort's senior and was full of advice. He taught Francis scientific skills and an enquiring attitude – a way to think and test a problem. This was the continuation of a pattern. For many years in his youth Edgeworth had lived in England where he had joined a philosophic dining society based in the Midlands around Lichfield and Birmingham, later remembered by Joseph Priestley as 'The Lunar Society'.

Other leading members were the engineer James Watt, the potter Josiah Wedgwood, the industrialist Matthew Boulton and the medical doctor Erasmus Darwin. These men formed the nucleus of a group of supremely gifted thinkers who applied their imagination to the problems of the age. There was no limit to their interests. They discussed Watt's plan for a steam engine, Priestley's ideas on photosynthesis, Wedgwood's invention of his iconic Jasperware and Darwin's ideas about the transmutation of species, forerunner of his grandson's theory of evolution by natural selection. For Josiah Wedgwood 'they were living in an age of miracles in which anything could be achieved'.[12]

The Lunar Society did not operate to set rules and there was no membership list. It was the union of a group of friends who enjoyed and were stimulated by each other's company. They were not bound by modern conceptions of science. The word science itself, derived from the Latin *scientia* (knowledge), had a much broader meaning than it does today. It encompassed all theoretical knowledge, from nature, rhetoric, religion and language. Nor had the descriptive term 'scientist' been coined. Instead they considered themselves *philosophers*, sages who speculated 'on the nature of things and truth'. They called their pursuits 'philosophic endeavours' and working in the Midlands they were free to nurture their identities as they could not have in the cloistered academic environments of London, Oxford or Cambridge.

The Society met about once a month, usually at Boulton's Birmingham home at full moon so they could find their way home through the dark streets. These meetings were jovial affairs. They experimented together, sometimes seriously, sometimes playfully. They examined gases, metals, rocks, animals. They invented clocks, speaking machines, carriages, wind vanes, barometers. Erasmus Darwin – the most ambitious of them all – once conceived of a horizontal-axis windmill for Wedgwood that he claimed would be three times more powerful than the conventional vertical ones. Sometimes Darwin let his imagination wander to even grander visions. One plan was to unite the navies of the world to tug icebergs from the poles to the equator, where he thought they would equalise the global temperature.

It was the late 1760s when Edgeworth had met Darwin and his fellow 'Lunaticks'. He was then a footloose young man of twenty-three, free from the hereditary bind of his estate in Ireland and just down from Corpus Christi College in Oxford. With nothing to occupy his time, Edgeworth was enjoying a carefree stay in England, flitting between the boisterous West End clubs of Georgian London and his domestic retreat in the picturesque village of Hare Hatch in the Berkshire countryside. Here, in his own laconic description, he had begun to 'amuse himself with mechanics'. Among his inventions were a highly efficient turnip cutter, a one-wheeled chaise, a beautifully precise clock, a velocipede – the ancestor of a bicycle – and a sail-propelled carriage that travelled along at phenomenal speed, to the terror of the neighbourhood.

It was in this whimsical phase of Edgeworth's career that he had first conceived an idea for a machine to transmit intelligence across space. One night in the summer of 1767 Edgeworth had joined Sir Francis Delaval and his friends, the 'turf club', at Ranelagh Gardens in Chelsea. The talk was of an imminent horse race at Newmarket. Two of the finest horses in the country were to compete. They were, Edgeworth remarked, 'in every respect as nearly equal as possible'. Desperate to profit from the vast bets being placed, one of them, Lord March, had told his friends he would await the result at the Turf Coffee House and that he planned to 'station fleet horses on the road, to bring me the earliest intelligence of the event of the race, and I shall manage my bets accordingly.'[13]

Hearing this, Edgeworth asked March what time he expected to know the winner; March replied at about nine in the evening. 'I asserted,' Edgeworth remembered,

> that I should be able to name the winning horse at four o'clock in the afternoon. Lord March heard my assertion with so much incredulity, as to urge me to defend myself; and at length I offered to lay five hundred pounds, that I would, in London, name the winning horse at Newmarket at five o'clock in the evening of the day when the great match was to be run.[14]

In a superb surge of bravado the wager spiralled. 'Sir Francis having looked at me for encouragement, offered to lay five hundred pounds on my side; Lord Eglintoun did the same; Shaftoe and somebody else took up their bets; and the next day we were to meet at the Turf Coffee House, to put our bets in writing.'

The wager, on which a small fortune now rested, seemed fixed firmly against Edgeworth. The roads of eighteenth-century England were notoriously bad. In many places they were in a worse condition than when the Romans had left them 1,300 years before, and the chance of a message being transmitted across the sixty-six-mile route in anything less than five hours seemed impossible. The Green Dragon Coach that ran a twice-weekly service between London and Newmarket took the best part of a day. Yet Edgeworth had a different plan. Lord March's boast had reminded him of two monographs: one by the seventeenth-century polymath Robert Hooke and another by an old Bishop of Chester, John Wilkins, who had produced a tract called *Mercury: or the Secret and Swift Messenger*.

The argument of both books was that an idea might be transmitted across distances by a set of predetermined signals – a simple, optical vocabulary agreed on by the two parties. Hooke, an incisive and prolific experimenter, had invented a system that used deal boards of various shapes – squares, triangles, octagons – to represent letters of the alphabet: these were displayed in a large frame and viewed through a telescope. In a paper read to the Royal Society in 1684 Hooke had championed his idea, describing the distance between stations, the use of a telescope to view signals and the substitution of the deal boards at night for a series of lanterns. His idea, he revealed, had been tested across the Thames and with good operation he thought the same character might be seen in Paris the minute after it was shown in London.

Eighty years later Edgeworth planned to use a similar method to transmit results of the Newmarket race. He sketched his plan.

> After we went home, I explained to Sir Francis Delaval the means I proposed to use. I had early been acquainted with Wilkins's 'Secret and Swift Messenger'; I had also read in Hooke's Works of a scheme of this sort, and I had determined to employ a telegraph nearly resembling that which I have since published. The machinery I knew could be prepared in a few days.
>
> Sir Francis immediately perceived the feasibility of my scheme, and indeed its certainty of success. It was summer time, and by employing a sufficient number of persons, we could place our machines so near as to be almost out of the power of the weather.[15]

Days before the race and with Edgeworth's plans developing, the parties met again at the Turf Coffee House. 'I offered to double my bet, so did Sir Francis,' Edgeworth recalled. 'The gentlemen on the opposite side were willing to accept my offer; but before I would conclude my wager, I thought it fair to state to Lord March, that I did not depend on the fleetness or strength of horses to carry the desired intelligence, but upon other means.' Edgeworth's candour undid him. Lord March thanked him for his honesty and, now suspicious, decided it was better to quietly withdraw. 'My friends blamed me extremely for giving up such an advantageous speculation,' Edgeworth later conceded.[16]

Edgeworth may have lost his chance of considerable bounty, but he left the Turf Coffee House with the nucleus of an idea. He spent the next weeks in partnership with Delaval, experimenting with his signalling device. Edgeworth installed four prototype machines across London: at Delaval's town house in Downing Street, one in Great Russell Street in Bloomsbury, one in Piccadilly, and one far away in the village of Hampstead. Edgeworth does not disclose how his invention worked, although a later source indicates that it involved lamps. He merely concluded that 'this nocturnal telegraph answered well, but was too expensive for common use'. Nevertheless, the image of Edgeworth and the roguish Delaval transmitting surreptitious messages by night across a bustling, unsuspecting Georgian London is an enticing one. What advantage, if any, the friends gained from their invention is not documented. What is clear, however, is that Edgeworth had manufactured a device quite unlike anything else.

It was about this time that Edgeworth met Erasmus Darwin on a visit to Lichfield. Darwin was won over, thrilled by Edgeworth's mastery of mechanics and showman-like talent for practical demonstration. He wrote excitedly to Boulton,

Dear Boulton

I have got with me a mechanical Friend, Mr Edgeworth from Oxfordshire – The greatest Conjurer I ever saw – G-d send fair Weather, and pray come to my assistance . . .

E Darwin

He has the principles of Nature in his Palm, and moulds them as He pleases.

Can take away Polarity or give it to the Needle by rubbing it thrice on the Palm of his Hand

And can see through two solid Oak Boards without Glasses, wonderful! astonishing! diabolical!!!

Pray tell Dr Small He must come to see these Miracles[17]

It was the beginning of a friendship that would endure for the rest of their lives. When Edgeworth moved back to Ireland in the 1780s they continued to correspond regularly about their latest preoccupations. By now Edgeworth had given up on his telegraph, there being no obvious market for one. Instead he concentrated on his educational philosophy and an improved system of land management. Darwin's interests remained eclectic as ever and in the 1780s he had begun to turn to meteorology.

Erasmus Darwin's latest passion grew out of the contemporary belief that climate had a profound influence on human health. As a practising doctor, he resolved to keep an eye on the weather. He had a pointing-device installed on the ceiling of his study that was connected to a weather vane on his roof, so that he would always know the direction of the wind. To measure wind speed he ran a tube up his chimney that was connected at the top to a windmill sail. When the wind blew the contraption twirled around like a modern anemometer and Darwin counted the number of revolutions with a series of cogged wheels.

These instruments gave Darwin an unusually strong grip on the changing weather. He kept observations and turned them into theories in an attempt to explain global wind patterns. Even for Darwin, this was ambitious stuff. For centuries meteorology had been characterised by mystery and superstition. While many sciences – geology, botany, physics and chemistry – had flowered under enlightened analysis, meteorology had barely progressed from its classical conception as a science of 'meteors'. To the modern mind, 'meteor' conjures up images of glowing balls of extraterrestrial rock. But in classical meteorology 'meteor' related to any event that happened in the so-called sublunar zone, the uncertain territory between the earth and the moon. In his *Dictionary of the English Language*, in 1755, Dr Johnson defined meteors as, 'Any bodies in the air or sky that are of a flux and transitory nature.'[18] It was a suitably roving definition for a variety of phenomena that ranged from a shooting star to the appearance of a rainbow; the onset of a storm or a solar halo, a flash of lightning or a gust of wind.

For almost two millennia meteorological thought had been based on the ideas set out by Aristotle in his treatise *Meteorologica*, produced during the fourth century BC. Like others, Erasmus Darwin had been brought up on Aristotle. For Aristotle the meteoric zone was filled by two principal and distinct agents that he termed exhalations. These lay at the root of all meteoric activity. The first was warm and dry, generated by the light of the sun falling on the earth's surface. The other was cool and wet and was the result of sunlight mingling with the water of the oceans, rivers and lakes. The hot and dry exhalation would rise into the fiery sphere of the globe where it produced shooting stars, comets and the Milky Way. The cooler and moist exhalation clung to ground: clouds, dew, rain and snow were the result. Winds were distinct hot and dry rivers of air, fast-swimming vapour, while thunder was either the sudden escape of a hot exhalation caught in the condensation of a cloud or the violent collision of a wind and a cloud, the force of the impact generating lightning.

Aristotle's *Meteorologica* was the product of observation rather than experimentation. He drew on the prior work of Hippocrates, Democritus and Empedocles to explain rainbows, halos, rain, clouds, hail, snow and dew. In doing so he founded the subject of meteorology, or, literally translated, 'the study of things on high'. Today Aristotle's treatise forms a delightful, eccentric read. It is monumental in scope and ambition, and entirely wrong on almost every count. Even so, *Meteorologica* dominated intellectual thought for centuries. The concept of meteors and exhalations would have been familiar to such disparate individuals as Galileo, Descartes, Cook, Newton, Columbus and Shakespeare. The poet laureate John Dryden would still be advancing Aristotle's ideas in verse in the seventeenth century: 'Then flaming meteors, hung in air, were seen, And thunders rattled through a sky serene.'[19]

By Erasmus Darwin's time, this old edifice was beginning to crack. Instruments like the barometer and thermometer had been in general use for about a century, enabling people to study the weather in ways unknown to Aristotle. Edgeworth's telegraph, too, had potential use as a meteorological instrument to warn of coming storms. Such an idea had already been mooted in France by a member of the Assembly, Gilbert Romme, in 1793, who wrote of 'the possibility for predicting storms and for giving warnings to sailors and farmers'.[20] Surprisingly, this idea never occurred to Edgeworth. Instead when news of Chappe's telegraph broke in the newspapers Edgeworth's instinct was to resurrect his old design as part of a defensive shield to protect the Irish from an expected French attack. Darwin felt the same. In 1795 he urged Edgeworth to place telegraphs around the Irish coast, like 'Friar Bacon's wall of brass round England'.[21] It was a typically playful note from Darwin. A few years later, in April 1802, he would be halfway through another letter to Edgeworth, telling him of his latest schemes, when he suddenly collapsed. His death brought to an end their thirty-five years of friendship. It was in the months following Darwin's death that Beaufort returned to Ireland and a new partnership was born.

Edgeworth enjoyed Beaufort's vitality and aptitude while Beaufort drew on Edgeworth's many years of experience. Edgeworth gave Beaufort a template for thinking, a way of attacking a scientific question. Science was the great tool to thrill, to simplify, to improve, to progress; as Wedgwood had once said, 'to unleash wonders upon the world'. The mantle was passed down. Darwin, Edgeworth, Beaufort – one of the formidable lineages in British science.

Edgeworth and Beaufort formally began work on the Irish telegraph on 4 November 1803 after a successful trial near Dublin. 'Yesterday I tried the telegraphs at Castleknock before Lord & Lady Hardwick, Lady Hosgall, Mr and Mrs Wickham (of whom I wish to see more) & a variety of Lords and Ladies – Everything including wind & weather succeeded beyond my most sanguine expectations,' Edgeworth had written home.[22]

Edgeworth had waited for this moment for decades. He had been horrified in 1794 to discover that his old idea had been independently invented and perfected by Chappe in France. Soon after news of Chappe's telegraph was made public an anonymous letter had been published in London's *Morning Post*, pointing out that Edgeworth had designed the same machine years before. Although Edgeworth denied writing the letter that did not stop him from quoting it whenever he had the opportunity. It was all, though, to no avail. He had to wait almost a decade for his chance to rectify the situation and it was in 1803, with the resurgence of Napoleon's grand armies across the Channel, that he finally got his chance.

The opening years of the nineteenth century were some of the rockiest in British history. There had long been fears that Revolutionary France was targeting the Wicklow and Cork coastlines for invasion, the French military command having pinpointed them as Britain's weak underbelly. Only a terrible stretch of winter gales had prevented a force of 16,000 from landing at Bantry Bay in December 1796 and now the likelihood of a repeated attempt was growing.

By 1803 Napoleon had amassed an army of 200,000 at Boulogne, where they awaited a fine spell of weather to execute their crossing. The preparations had continued in public and fear across Britain was mounting. The Bayeux Tapestry had been carried on a celebratory tour of the French coast and at the Paris Mint a die for an invasion medallion had been cast with the words 'Struck in London 1804'. The British press, obsessed with Napoleon, had relayed accounts of his movements to their readership. In July a typical dispatch from *The Thunderer* informed readers, 'The First Consul reached Calais at 5 o'clock on Friday afternoon. His entry, as might be expected, was in a grand style of parade. He rode on a small iron grey horse of great beauty . . . The whole place resounded with *Viva Bonaparte!*'[23]

Time was of the essence so from the beginning Edgeworth and Beaufort adopted distinct roles. As Edgeworth politicked in Dublin, Beaufort was put in charge of the practical execution of the project: establishing the stations – often on elevated forts, isolated hills or churches – sourcing the raw materials for the towers, guardhouses and stations; drilling the corps of 'telegraph men' and teaching them the telegraphic vocabulary, the importance of accurately transcribing signals and how to react in an attack. Each station was manned by a captain and two or three militiamen who were under the command of 'confidential' officers in Dublin, Athlone and Galway. Edgeworth had assigned himself the role of overseer of the whole enterprise, 'an all seeing eye'.[24]

Beaufort was a man of action. He travelled the breadth of the country on his little grey colt, riding the sodden lowland lanes and climbing isolated hills, scouring the land for suitable elevations: sites that were both accessible and commanded clear views of up to twenty miles. He was constantly on the move. In November he was with Edgeworth in Dublin. In January he was in Galway. Two months later he was camping out near Athlone. An account book kept by Edgeworth reveals that most stations were built within a fortnight as the line stretched westward. The project also gave his daughter Maria, always keen-eyed, the opportunity to observe Beaufort close up. She was impressed. In December she noticed how he had invented a lexical system, 'a set of words of command' that helped the militia learn their drills more effectively. The system had 'some analogy with their soldier language & consequently was easily learned by rote & not easily forgotten',[25] she noted. It was a sign of what was to come.

Away from the fireside at Edgeworthstown in Longford it was hard, demanding work that stretched through the Irish winter. Beaufort was compelled to lead men deep into rural areas where revolutionary sentiment against the British government had long festered. There they encountered more than political hostility. Their strange equipment drew a suspicious response from the peasantry. 'Telegraphs, of course, became the subject of conversation,' he wrote to his sister Fanny.

> No wit of mine could make anyone comprehend them. But when I told them that I was just going up the hill to converse with a man at Kilrainey, and some ladies in the county of Longford who were 26 miles distant – they almost shrieked. They climbed on to the table and said it must be by the black art. I made her look through the first telescope she ever saw, I held it out in one hand and she saw the candle in her eye. I turned it and held it by the other and it appeared as far off as Phillipstown . . . In short these good people were so simple, so ignorant, so enclosed and so much entertained that I instantly got the hero of my tale.[26]

By April a string of stations had been completed, spanning sixty Irish miles from the Royal Hospital in Dublin to Athlone. Progress had been fast but it had not come without difficulty. Ireland's notoriously wet climate was a constant frustration for Beaufort. Conditions were often raw or blustery with storms gambolling in off the Atlantic, opaque clouds hanging low overhead, sucking the definition from objects and the saturation from colours, making it impossible to see stations fifteen or twenty miles away. In the mornings low black skies, fogs and mists would create an atmosphere hazy with humidity, often reducing visibility to below a hundred yards. Once Beaufort wrote to Edgeworth, 'None but telegraphic minds can feel the delights of a fine day, and a clear day, after unceasing storms, fogs and deluges.'[27]

Such weather slowed the process of selecting stations and, even worse, plunged the entire idea of visual communication into doubt. Although Edgeworth maintained that his telegraph would work on all but one day in a hundred – a fact he claimed to have established in his own weather diary – Beaufort remained quietly sceptical.

There were other problems too. When scoping out stations Beaufort encountered a backlash of 'nimbyism'. One resident who found out that a station was to be built near his garden told Beaufort 'he would look upon it as a very great grievance whatever others might make of it, – the men coming down to his house to ask for a light to make a fire or for to beg a drink of buttermilk!' 'The race of Levites is not yet extinct,'[28] Beaufort grumbled. A further problem stemmed from the aptitude and attitude of the telegraph corps, a unit raised from local militias. These were particularly unsuitable for the work which required a high degree of accuracy. Often the telegraphers were illiterate, and those who were not spent too much time composing letters of complaint.

By late spring 1804 Beaufort was writing fretfully to his father about 'the casual smoke of Dublin and the storm which always blows on the top of Caston Tower'.[29] Although on fine days a message might travel well enough, on others scrambled, incomprehensible messes would return, leaving Beaufort and Edgeworth's device looking less like a marvel than a jabbering infant. It was becoming obvious that any invention that depended on bright blue skies in Ireland was not assured of a long, prosperous future. In Edgeworthstown there was talk of 'provoking disappointments', unseasonably violent storms, 'men tired & quitting their stations'.[30] Beaufort's composure was slipping too. After a bad test communication he dashed off a fiery letter to Edgeworth, 'Good Heavens, what does this all mean? Is it that after eight months' practice not one word out of the whole should come right? I am almost out of my senses.'[31]

By the end of June the line was completed and on Monday 2 July Lord Hardwicke and a group of influential politicians were invited to a gala opening. As written up in *Freeman's Journal* the following week, it was an unqualified success. Edgeworth managed to send a complete message to Beaufort, who was stationed 130 miles away, in just seven minutes. In five minutes more Beaufort replied in acknowledgement. 'The velocity of this mode of communication is astonishing,' *Freeman's Journal* reported. 'Intelligence sent from Dublin at eleven o'clock, for instance by the sun dial at the Royal Hospital may be received at seven minutes and a half after eleven by the dial in Galway.'[32]

But for all the outward confidence, politicians were becoming nervous. The telegraph was growing more expensive, the militia proving troublesome, and the number of successful transmissions was worrying. Within a week news had reached Edgeworthstown that the government had lost faith in Edgeworth's plans. He was to be replaced as chief telegrapher with immediate effect and responsibility passed to the Army. It was the last that history would hear of Edgeworth's long-cherished optical telegraph.

The government's duplicity stung Beaufort, who chalked it up as another of life's cruel blows. Writing to his brother William he raged, 'At present I am nothing, I have nothing, I expect nothing, I am doing nothing and have nothing decidedly in view.'[33] In his anger, though, Beaufort overlooked what he would later come to appreciate. His time working with Edgeworth had been an invaluable education. Edgeworth had proven helpful in another way too. Over the last two years he had been using his influence on Beaufort's behalf, and it was through Edgeworth's contacts that he eventually found his way back into the Navy.

By July 1805 Beaufort had swapped the green hills of Ireland for the chaotic bustle of the Royal Dockyard at Deptford, the British naval base near London. There he had been promoted to his first command. His ship was the *Woolwich*, a fifth-rate, 44-gun vessel, once a keen and fleet fighting machine but now converted into a storeship. It was a bittersweet moment for Beaufort, the joy of promotion tempered by the ignominy of his ungainly command.

He felt the shame keenly. As far as he was concerned the Admiralty had simply thrown a drowning man an anvil. He took to his journals in despair:

> To a storeship! Good Heavens! It is for the command of a storeship that I have spilled my blood, sacrificed the prime of my life, dragged out a tedious economy in foreign climates, wasted my best hours in professional studies . . . For a storeship, for the honour of carrying new anchors abroad and old anchors home!

For a ship more lumbered than a Dover packet, more weakly
manned than a Yankee carrier – four fourths of her arms and
ammunition on shore, three feet deeper than her trim, and with
jury masts and sail! – So that she can neither fight nor run; in
short, for a ship where neither ambition, promotion or riches . . .
can be obtained![34]

Matters were made worse by the fact that the war against Napoleon
had entered a critical phase. The French Grande Armée was just a
short distance away across the Channel and Beaufort was forced to
endure the misery of reading newspaper accounts of Nelson's pursuit
of Vice Admiral Villeneuve across the Atlantic. On Thursday 7
November 1805, while anchored at Spithead in the Solent, he heard
news of Nelson's spectacular and tragic victory over the combined
fleet at Trafalgar. It was a blithe and bucolic setting, too gentle a place
to hear such magnificent news. History was happening elsewhere.

But as so often when an active mind like Beaufort's is left idle, it
wandered. In August 1805 he was thinking about the storage of the
Woolwich's hold, estimating how the spread of the ballast could best
benefit the speed of the ship. By the turn of the year another problem
was occupying him. His weather diaries were growing increasingly
detailed. Sometimes they included four different records of wind direction in a day and they always began with an estimation of speed – 'fresh
breeze', 'moderate breeze', 'squally', or 'light airs'.

By using this terminology Beaufort was simply reflecting naval
culture. Throughout the eighteenth century there had been no
successful scientific explanation of wind. It was a rushing stream of
vapour: each wind distinct from the next. The best that could be done
was to note down its characteristics in lively prose. A dashing breeze
off Southampton, a galloping gale near the Goodwin Sands, a sudden
gust, a Shakespearian tempest off Plymouth. Each different wind
presented a creative challenge.

Beaufort realised the limitations of this practice. Descriptive
records might leave behind a vivid picture of the scene, but the
data was not scientific. This problem had been satirised by Daniel
Defoe in his panoramic account of the Great Storm of November
1703. In an opening chapter of The Storm he had lamented the
differing perception of winds and weather among English and
foreign sailors.

Such Winds as in those Days wou'd have pass'd for Storms, are called only a *Fresh-gale*, or *Blowing hard*. If it blows enough to fright a South Country Sailor, we laugh at it: and if our Sailors bald Terms were set down in a Table of Degrees, it will explain what we mean.

Stark calm.	*A Top-sail Gale.*
Calm weather.	*Blows fresh.*
Little Wind.	*A hard Gale of Wind.*
A fine Breeze.	*A Fret of Wind.*
A small Gale.	*A Storm.*
A fresh Gale.	*A Tempest.*[35]

English vessels, Defoe argued, were so superior that this distorted the sailors' impression of the wind. 'If the *Japoneses*, the *East Indians*, and such like Navigators, were to come with their thin Cockle-shell Barks and Calico Sails; if *Cleopatra's* Fleet, or *Caesar's* great Ships with which he fought the Battle of Actium, were to come upon our Seas, there hardly comes a March or September in twenty Years but would blow them to Pieces, and then the poor Remnant that got Home, would go and talk of a terrible Country where there's nothing but Storms and Tempests.'[36]

Proud and patriotic, Defoe was nonetheless making an important point about the relative merits of subjective or objective record-keeping. And in the century between Defoe's book and Beaufort's appointment to the *Woolwich* there had been several attempts to establish a clear, quantifiable wind scale. The lighthouse engineer John Smeaton, the naval hydrographer Alexander Dalrymple and a Dutch surveyor, Jan Noppen, had all devised schemes, though none of them had come to be universally applied.

This was the problem that Beaufort was grappling with at the beginning of 1806. All British commanders were required to keep a ship's log. Why not then make a virtue of necessity? On 13 January 1806 he sat down to write an entry that would enshrine his name in history.

It read:

Hereafter I shall estimate the force of the wind according to the following statement, as nothing can convey a more uncertain

idea of wind and weather than the old expressions of moderate
and cloud, etc. etc.

0	Calm	7	Gentle steady gale
1	Faint breeze just not a calm	8	Moderate gale
2	Light air	9	Brisk gale
3	Light breeze	10	Fresh gale
4	Gentle breeze	11	Hard gale
5	Moderate breeze	12	Hard gale with heavy gusts
6	Fresh breeze	13	Storm[37]

Beaufort did not stop at that. Having set down the beginnings of
his quantitative wind scale he continued: if the strength of the wind
was to be measured by a series of numbers, then, he felt, the letters
of the alphabet could be employed to describe the state of the atmosphere. It was a logical extension. In the same journal entry he
continued to write out a code of twenty-nine distinct symbols – either
single letters of the alphabet or a combination of two letters – to
describe different types of weather: blue skies (*b*), sultry (*s*), hazy (*h*),
damp air (*dp*), foggy (*fg*), rain (*r*), squally (*sq*), thunder (*t*) and so on.
All of these and twenty-one other states of atmosphere were to be
recorded in a column, with each type of weather separated by a
comma. It would allow him to record not just wind direction, speed,
barometric pressure, temperature and time, but also, in a flexible and
simple framework, all the many fluctations of the atmosphere
around him.

Beaufort put his method to the test. It was Monday 13 January 1806,
four days after Nelson's state funeral in Westminster Abbey, ten days
before William Pitt the Younger would die and three days after the
surrender of the Dutch to the British at Cape Town. In Woolwich on
the Thames, the iconic imperial waterway, where Beaufort lay at
anchor, the wind was blowing from the north with forces between 4
(a gentle breeze) and 10 (a fresh gale). The skies above were blue (*b*),
the atmosphere squally (*sq*).

Beaufort's idea was beautifully simple. Historians have long examined the origins of the system that would go on to become a world-
famous, crucial building block of modern meteorology. It is true that
it evolved from previous ideas, work of men like Dalrymple and
Smeaton. But it was also influenced by the time he spent with

Edgeworth on the telegraph in Ireland. It was there that Beaufort wrestled with the art of communication. Working daily with the telegraphic vocabulary, he was forced to form his own series of easily understood signals. Beaufort had a natural aptitude for scientific observation but it was the time he spent with Edgeworth that gave his raw skill shape and purpose.

Beaufort's wind scale is often considered the solution to an old and frustrating problem, but in reality it was more of a beginning. Embittered and angry, among the din and drama of Woolwich dockyard, without even knowing it Beaufort had started a process that would develop as the years passed. Hunched over his journal in solitude Beaufort may have felt like the most powerless man in Britain. Yet in the way he planned, observed and recorded he had set a template for the experiments and achievements that followed. He had caught a glimpse of the future. A long process of civilising the skies had begun.

CHAPTER 2

Nature Caught in the Very Act

Beaufort would have to wait for four more years before his chance came. As commander of HMS *Woolwich* he was charged with delivering official dispatches, sending provisions and herding merchant vessels as they sailed around the trade routes of the globe. Over the years that followed he dropped anchor in Madeira, the Cape of Good Hope, Tristan da Cunha, Madras, St Helena and Montevideo. At length he gained admirers for his aptitude – on one occasion he impressed the old hydrographer Alexander Dalrymple with a beautifully executed sketch of the seafront at Montevideo. In 1809 he was given a better ship, the *Blossom*, an 18-gun fighting sloop, and sent to escort a convoy up the St Lawrence River to Quebec. A year later he received his long-awaited promotion to captain. It was cause for celebration and nowhere was the news greeted with more enthusiasm than in Edgeworthstown. 'There are no reasonable bounds to my joy and exultation,' wrote Edgeworth. He gave Beaufort's old telegraph corps the day off to toast his triumph, which they did in suitable style.[1]

Cheerful letters had flowed to and fro between Beaufort and Edgeworth since he had left Ireland. In December 1809 Beaufort wrote from the *Blossom* to tell Edgeworth that he had discovered a fascinating waterfall in Canada. The water dropped from 200 feet with such force that it had scooped a hollow in the rocks beneath. Beaufort had studied the waterfall closely, especially the spray that separated from the central curtain of water. He saw that this was so fine and light that it almost seemed to escape the pull of gravity and float in the air. It reminded him of a squall at sea and he wondered whether the force of falling water was enough to rouse a breeze. Might this be air rushing in to fill a vacuum?[2]

Beaufort had touched on what he described as a 'well known,

but most curious fact'. Winds were particularly volatile in a storm, not just in their strength but in their direction. It was a puzzle that, Beaufort acknowledged, was unlikely to be solved soon. If only, he moaned to Edgeworth, the Admiralty did not fritter away the data captured in ships' logs. 'There are at present 1000 King's vessels employed, from each of them there are from 2 to 8 logbooks deposited every year in the Navy Office,' he pointed out. 'These logbooks give the wind and weather every hour, and numbers of them must have been spread at the same time over a great extent of ocean – what better data could a patient meteorological philosopher desire?'[3]

But for the moment Beaufort had other concerns. Soon after his promotion to captain he was plucked out of the whole British fleet to complete a hydrographic survey of the southern coasts of Asia Minor. For Beaufort it was the perfect challenge. For eighteen months he skimmed the Mediterranean coastline of modern-day Syria and Turkey in HMS *Frederickssteen*. Later described with beautiful understatement as 'an errand', it plunged Beaufort deep into a lost world of antiquity that many had imagined but few had ever seen. All went well until a bloody skirmish with a band of Turks in 1812. Beaufort was left badly wounded and for a time he feared for his life. Though he recovered, the incident brought an end to his active career. Soon he was back in England, convalescing. His return marked the beginning of a happy period in his life. In London he married and began his long retirement.

In December 1813 Beaufort travelled to Edgeworthstown – the 'Seat of the Muses' – for a reunion with his now elderly mentor. A decade had passed since they had worked together on the telegraph, and Edgeworth was thrilled to see him. Ever scheming on Beaufort's behalf, Edgeworth had news. He had been in communication with Sir Joseph Banks, the President of the Royal Society, and had taken the opportunity to ask whether, in light of his surveying work in the Mediterranean, Beaufort might be eligible for election to the elite scientific club. Banks had replied enthusiastically, telling Edgeworth his recommendation was all any candidate should need. Edgeworth surprised Beaufort with the news during their Christmas reunion. Beaufort was overjoyed. So long overlooked, so convinced his talents were to be wasted, it was a present like no other. He returned to Britain with new-found optimism. He was becoming a man of substance, with avenues of possibility opening up before him.[4]

Beaufort arrived in London in January 1814 to a city in the grip of extreme weather. A Christmas fog had preceded a heavy snowfall. For weeks afterwards temperatures had languished below freezing, leaving the streets clotted with snow and the parks transformed into winter wonderlands, the Serpentine an ice rink. The Thames had frozen too between Westminster and Blackfriars, its surface like 'a solid rock of adamant', and Londoners had seized the opportunity for fun and games. Swing boats, bookstalls and scuttling booths were set up on the ice and thousands had flocked to see the prospect of Londoners walking on water. There were tents and booths decorated with streamers, bunting, flags and gaudy signs – one of which advertised the 'City of Moscow' – while everywhere, 'there was a plentiful store of those favourite luxuries, *gin, beer and gingerbread*'.[5]

The frost fair of February 1814 was the latest in a series of unusual weather events. May 1811 had seen astonishing thunderstorms over the capital; the spring and summer of 1812 had been the coldest since 1799. In May 1813 a double rainbow had appeared over the East End, 'with the most brilliant bow',[6] that had shone for forty minutes, and for three winters noxious fogs had filled the capital's streets. Unable to explain events, the scientific community had been attacked by an exasperated commentator in *Nicholson's Journal* who pointed out that nothing was so 'striking a proof of the little progress hitherto made in meteorology, than the difficulty of proposing a legitimate explanation of a phenomenon so common and familiar as a thick fog during winter'.[7]

One man who kept an unusually close eye on the weather was the landscape artist John Constable. Just a mile north of the frozen Thames, in Charlotte Street in February 1814, he, like many Londoners, was enduring the cold by his fireside. The festivities passed with barely a mention in his letters, just a miserable admission to Maria Bicknell, his fiancée, that he had 'thoughtlessly put on a great many clothes which were very damp', thereby 'fixing' a cough on his chest 'that laughs at present at every attempt to dislodge it'.[8]

Dark-eyed, with a thinning sweep of chestnut hair and bristling side whiskers, Constable was thirty-seven years old, just two years younger than Beaufort. As he crept towards middle age he had become a more pensive character than the vigorous romantic who had once ranged through the lanes and meadows around East Bergholt, his village home on the Suffolk–Essex border. Although he loved English weather a frost fair was not the sort of thing to interest him. He would not be remembered for winter scenes. Instead his modest reputation at the Royal Academy had come from his tender renderings of midsummer tranquillity in the English shires.

Like Beaufort's, Constable's career had yet to amount to anything much. Although he had advanced enough to make friends of Benjamin West, Sir Thomas Lawrence and, most notably of all, to sit next to J.M.W. Turner at the previous year's Royal Academy annual dinner, few outside the close-knit artistic community centred on the Academy had heard his name. A decade later with his fortunes finally turning he would look back on 1814 as a time of personal crisis: 'long I tottered on the threshold and floundered in the path,' he remembered wistfully, 'and there was never any young man nearer being lost than myself.'⁹

But to those who had noticed, Constable was revealing himself as an artist with a difference. Over the last decade he had been quietly staking his territory, his own distinctly personal style of landscape painting that blended rural charm and natural realism. For years he had been spending much of his favourite spring and summer months back in Suffolk, sketching the expansive views across the fertile Dedham Vale and Stour Valley. He wanted to capture the rustic, the everyday: a comely world of drowsing lanes, meandering streams, crooked half-timber cottages. Like a farmer bringing his produce to market he would return to his London home loaded with ideas to be transformed into pieces for the Academy over the winter, ready for the summer exhibitions.

What really marked Constable out was his commitment to accuracy. His aspiration to catch the telling details of country life can be traced back to the beginning of his career in 1802. Then a young artist in London trying to find his way, he had become disillusioned with the fashion for epic history painting and, particularly for overblown, idealised landscapes or 'works of fancy'. In exasperation he had written to his childhood friend in East Bergholt, John Dunthorne, outlining his ambitions. Constable admitted that he had wasted years trying 'to make my performances look like the works of other men' and vowed,

'I shall return to Bergholt, where I shall endeavour to get a pure and unaffected manner of representing the scenes'. Such a style would set him apart. '*There is room enough for a natural painter*. The great vice of the present day is *bravura*, an attempt to do something beyond the truth. Fashion always had, and will have, its day; but truth in all things only will last, and can only have just claims on posterity.'[10]

This letter, lucid and fizzing with intention, later came to be viewed as the genesis of his career. To Constable's mind, to be a natural painter meant to be faithful in the portrayal of nature. This he could achieve only by years of patient empirical study of people, animals, flora and trees. Already his devotion to such fine details had become his personal trademark: the stag-headed oak, a donkey browsing in the hedgerow, the family dog, his father's wagon, the cattle that would always be Suffolk bred. All these were painted with such attention that they appeared, one of his biographers noted, 'like minor characters in a novel'.[11]

Constable's fascination with the natural world extended to the sky and atmosphere. He understood that this was the prism through which all his landscapes were seen. The gaps in the clouds or the elevation of the sun generated the quality of chiaroscuro that he treasured – the strong contrast between light and shade that he defined with a flourish as 'that power which creates spaces; we find it everywhere and at all times in nature; opposition, union, light, shade, reflection and refraction'. A dark, lowering sky would strip the clarity from objects and saturation from colours, while a bright sky at midday would destroy shadows and texture. To fill his landscapes with chiaroscuro, Constable liked to depict what later would become known as cumulus clouds because they gave the maximum leeway to fill the ground below with alternating patches of dark and light. This search for light became synonymous with his style. Once when he heard a lady dismiss a picture as 'ugly' he responded, 'Madam, I never saw an ugly thing in my life, but light, shade, or perspective would always make it beautiful.'[12]

During the frozen days of January and February 1814 he had been at work on a new composition, *Landscape Ploughing Scene in Suffolk (A Summerland)*, based on a drawing he had made in his 1813 sketchbook. In complete contrast to the thronging icy streets of Fitzrovia, it was a summer scene: the viewer looks out from an elevation over a little wood and the sloping fields outside East Bergholt towards two lowly hills. Two churches and a windmill are visible in the distance; in the foreground two ploughmen are at work guiding swing ploughs, drawn by horses, along the furrows. The sky is threatening and uncertain. The composition is a mixture of cool and warm tones. The temperature in the ploughing field is hot enough for a labourer's dog to laze on a blanket in the sunshine.

While painting this sky Constable ran into difficulty. On 22 February he had voiced his frustrations to Dunthorne. 'I must try and warm the picture a little more if I can. But it will be difficult as 'tis now all of a piece – it is bleak and looks as if there would be a shower of sleet, and that you know is too much the case with my things.'[13] This was a typical problem. Harmonising landscape and atmosphere was one of the great technical challenges for artists. To sidestep the issue most landscapists simply painted bland, unobtrusive skies, which had the advantage of not distracting from the action below.

To Constable this was a flagrant dodge, but his principles brought him problems. The sky was an awkward subject and clouds the most difficult of all. Seen from a specific point below it was difficult to gauge their dimensions and, gusting along on the breeze, they rarely stayed still long enough to study them properly. George Harvey, a mathematician, noted:

> The same cloud which to one spectator may be glowing with light, to another may be enveloped with shadow. That which appears to be its summit may be only a portion of its anterior edge; while that which seems to be its lower bed, may really be a portion of its posterior border . . . The young observer must indeed apply, with perseverance and caution, to the masses of Cloud that float in the aerial regions above him.[14]

The problem with this painting troubled Constable for some weeks. But by the time the exhibition came round in May *Landscape Ploughing Scene in Suffolk* was entered as planned. There it was seen and admired by a collector called John Allnutt, a wine merchant from Clapham. In what Constable's great friend and later biographer Charles Leslie would later christen 'an extraordinary event', Allnutt surprised everyone by buying the picture.[15]

This was so extraordinary because Constable had never sold a landscape before to anyone beyond his own personal sphere. He was shocked and delighted. In later years he would thank Allnutt – lauding him as the man who had given him confidence to continue. Constable's joy might well have been tempered had he known the truth at the time. When Allnutt received the landscape at his Clapham home he decided 'I did not quite like the effect of the sky.' Perhaps it was the coolness, the shower of sleet or the threat of rain. Perhaps it was simply the sky's prominence in the composition. Allnutt did not say, nor did he tell Constable. He simply decided to 'obliterate' his sky and have it repainted with an idealised replacement by a different artist.[16]

In 1814 John Constable wasn't the only Londoner occupied by the skies. As John Allnutt went about dismantling his landscape in Clapham, elsewhere a talented young scholar called Thomas Forster was preparing a second edition of his highly popular work, *Researches About Atmospheric Phaenomena*. At only twenty-four Forster had already made his mark on British scientific life. Raised by free-spirited parents, a child of Rousseau, from boyhood Forster had felt an acute connection with the natural world. A vegetarian, animal lover and stargazer, while others of his age had run wild or gone to war Forster had attended Joseph Banks' *conversaziones* and lectures at the Linnean Society, becoming a member in 1811. He had written several articles for the *Philosophical Magazine* and *The Pamphleteer*, and had produced two splendidly esoteric monographs, *The Action of Spirituous Liquors of the Human Stomach* and *Observations on the Brumal Retreat of the Swallow*, which he had published under the jaunty pseudonym 'Philochelidon'.

These works set the tone for a long and roving intellectual career, but for the moment Forster's prevailing passion was for meteorology. It was something of a family hobby. Both his grandfather and father had kept weather journals, which meant Forster had an unbroken chain of observations to draw on that stretched back to January 1767. In an age before standardised record-keeping began this was an invaluable resource, and over the years Forster's own meteorological records had continued to evolve. Raised with the old Aristotelian ideas of competing vapours, at just fifteen, on 13 August 1805, Forster had been alert enough to notice 'a very unusual exhalation from an elm tree at Clapton, in the parish of Hackney'.[17] Like a young Beaufort, Forster had sketched the scene. It was early evening, between six and seven o'clock. It was a warm day, the air was clear and a wind blowing from the south-east. Forster had watched as 'a column of darkish vapour appeared to arise from the top of an elm tree at some distance: it looked about two or three feet high'. Forster observed the strange exhalation for half an hour as it came and went, came and went.

Forster remembered this for years and would mention it a decade on when *Researches About Atmospheric Phaenomena* was published in 1812. The book was a passionate synthesis of accepted meteorological theory, weather wisdom and other snippets from Forster's journal. It sold well. Years later Arago at the Paris Observatory would still be quoting from it. In his preface Forster set out the allure of the subject. 'The atmosphere and its phaenomena are everywhere,' he wrote, 'thunder rolls, and rainbows glitter in all conceivable situations, and we may view them whether it may be our lot to dwell in the frozen countries of polar ice, in the mild climate of the temperate zone, or in the parched region which lay more immediately under the path of the sun.'[18]

This was a clarion call for meteorology at a moment that can be seen as its rebirth as a modern science. From the start of the eighteenth century Aristotle's ideas had been losing ground to a new generation of rational thinkers. The most popular belief was that the moon or planets were the driving forces behind weather, but other theories had abounded too. Some had argued that sulphurous soils in a region interacted with the air to create storms, others that weather ran in repetitive cycles over several years. All of the theories had seemed more plausible than the notion of exhalations and meteors. A telling blow to Aristotle's views had come in the 1750s when John Pringle, physician and future President of the Royal Society, demonstrated that meteorites were not forged in the sublunar zone but came from a mysterious extraterrestrial space. At almost the same time news of Benjamin Franklin's fabled kite experiment in Philadelphia – where he sensationally managed to draw sparks out of a thunderous sky – spread throughout Europe. If Pringle's work had undermined Aristotle, then Franklin's split it in two, destroying in one stormy Pennsylvanian night the ancient belief that lightning was produced by colliding winds.

A craze for electrical meteorology had followed. A young John Barrow, who would later become a powerful secretary at the Admiralty, was fascinated by Franklin's experiment. At home in the north of England, he built a kite of his own and flew it in a thunderstorm. All was going smoothly until an old woman came up to see what he was about. 'It was too tempting an opportunity not to give her a *shock*,' Barrow remembered wickedly.[19] Just as Galvini later demonstrated that the muscles of the human body were animated by electrical currents, philosophers began to envisage an atmosphere of potent fluid, one reaction sparking the next so that the air whirled in perpetual motion. All of a sudden explanations were recast. The aurora borealis was explained as the flow of electrical fluid between positive and negative clouds. Shooting stars became the flashings of distant electricity.

These were the discoveries Thomas Forster was raised with at the beginning of the nineteenth century, an exhilarating time when the atmosphere seemed at last ready to yield its ancient secrets. Already in the new century a ground-breaking idea had been advanced in Britain. It had come from an unlikely source. An unknown chemist and Quaker called Luke Howard, a man who had never appeared in print or public to any significant extent, had stood up at an academic gathering in London in December 1802 and set out a comprehensive system for classifying different types of cloud.

Until then clouds had remained beyond the realm of science. Most people simply regarded them as ornaments of the sky or airy noth-ings. Describing them had always been a challenge. Shape-shifting, mercurial, sleek or vast, they had roamed wild without definition. In a bid to explain what they saw people had been forced to rely on their literary powers. As with the wind, the success of the description relied on the talent of the writer. A notable attempt to capture the character, shape and texture of different clouds was made by a Worcestershire weather diarist in 1703. Soaring to daring poetic heights, he likened the clouds to hair combs, cobwebs, spun wool, palm branches, foxes' tails, crêpe or raw, finished silk, and a hundred other everyday objects.[20]

Luke Howard's *Essay on the Modification of Clouds* – presented to the Askesian Society in 1802 – would bring an end to this. It is now rightly considered one of the most significant contributions in nineteenth-century meteorology. 'If Clouds were the mere result of the condensation of Vapour in the masses of atmosphere which they occupy, if their variations were produced by the movements of the atmosphere alone,' he began his paper, 'then indeed might the study of them be deemed a useless pursuit of shadows, an attempt to define forms which, being the sport of winds, must be ever varying, and therefore not to be defined.'[21]

E. Radcliffe

CLOUDS.

1 Cirrus — 2 Cirro Cumulus — 3 Cirro Stratus — 4 Cumulo Stratus.
5 Cumulus — 6 Nimbus — 7 Stratus.

But this was not the case with clouds, Howard argued. Schooled in the classics and influenced by Carl Linnaeus' system for organising plant species and types, Howard had conceived the idea of a similar framework for clouds. For years he had observed the skies around his London home and by 1802 he had managed to distil their myriad fluctuations into a simple classification system. Howard had pinpointed seven modifications and gave each a Latin name. The three principal types were *cirrus* for the wispy, or 'flexuous' clouds that formed in the high atmosphere; *cumulus* for the familiar, mid-level 'convex or conical heaps, increasing upwards from a horizontal base'; and *stratus*, 'A widely extended, continuous horizontal sheet, increasing from below upward'. To these he added two 'intermediate modifications': *cirro-cumulus* and *cirro-stratus*; and two compound modifications: *cumulo-stratus* and *cumulo-cirro-stratus*. This final modification was more colloquially known as the *nimbus* or rain cloud.

From its modest beginnings as an occasional paper of a little-known philosophical society, Howard's cloud classification theory and his reputation as a pioneering meteorologist spread rapidly. After the paper's appearance in the *Philosophical Magazine* his system was soon adopted all across Europe, with Howard even congratulated by the German writer Goethe who claimed his classification system had emerged like a lighthouse out of a fog, helping to impose order on chaotic nature. In a few years Howard was famous. His success lay in his use of Latin – still the intellectual lingua franca – something that gave him an unassailable advantage over a rival (and strikingly similar) system devised at almost the same time by the French naturalist Jean-Baptiste Lamarck.

Forster was among the generation of thinkers that came of age in the era of Howard's triumph, and Howard's influence suffuses *Researches About Atmospheric Phaenomena*. Characteristically, Forster wanted to improve Howard's terminology. Believing the Latin prohibitive and elitist, he set out to produce terms of his own, with easily remembered names, among them *curlcloud* for cirrus, *stackencloud* for cumulus and *fallcloud* for stratus. Indeed the first third of Forster's book is essentially a personal meditation on Howard, 'whose theory of the formation and destruction of clouds appears', Forster writes, somewhat begrudgingly, 'as far as I am capable of judging, to be extremely accurate in most particulars'.

Like Aristotle many centuries before him, Forster creates brief weather narratives. He describes in intimate detail the formation of a cumulus from a minute speck to a bulging giant. He characterises the cirrus as a mercurial presence, stretched out in the heavens like a dog before a fire with its long fibrous body and pointed tail, sometimes visible for days, sometimes gone in a glance. Of stratus clouds he writes of a fine mist creeping 'along the valleys of a summer's evening', striking in their whiteness and, when viewed at a distance by moonlight, 'very fanciful in their appearance'. High above the stratus came the cirrostratus, which Forster likened to a shoal of fish. 'Sometimes the whole sky is so mottled with it as to give the idea of the back of a mackerel,' he wrote. The cirrocumulus reminded him of a flock of sheep, while the cirrostratus clouds 'swelling somewhat in the middle, and seen below a more thin and extensive sheet of cloud, give the idea of the back of a great dolphin rising out of the ocean'.[22]

Forster's vision of the sky was held together by his belief in the binding force of electricity. High above the others the cirrus, with their tails drawn out by an electric current, were, he argued, the conductors of everything. Beneath, the other clouds swapped their positive and negative charges in an endless electrical interaction. Occasionally the modifications would lock together to create an opaque, rain-bearing nimbus.

Forster projected this vision of an electrical theatre, the clouds sometimes discordant like characters in a farce, or united like musicians in an orchestra. And to enliven the second edition of his work he added a series of illustrative plates to show clouds *in situ* above the landscape. The art was crude. The engravings were based on his own weather sketches. The curling tips of two distant cirri look more like the curved rails of a sleigh than any cloud, while the exaggerated shape of a cumulonimbus appears disturbingly to the modern eye like the mushroom cloud of an atomic explosion. Inelegant as they were, Forster's publisher, Robert Baldwin of Paternoster Row, deemed them an attractive selling point. On the book's release he wrote up a promotional paragraph for the newspapers.

This edition will contain a series of plates illustrative of Mr Howard's nomenclature of the clouds and other atmospheric phaenomena. The want of such a nomenclature has rendered all descriptions of atmospheric appearances unintelligible or indistinct. It is hoped therefore that this attempt to lay down some general rules for observers may be found extremely advantageous to the painter and engraver.[23]

The second half of the Georgian age was a wide-eyed, watchful time. People were on the alert for a moment of personal enlightenment or a chance discovery. For the many who believed that nature, after the Bible, was God's second book, the quest to understand the natural world, to peek into the celestial machine, became a spiritual experience. A careful eye and inquisitive mind might unravel a divine wonder that would bring them, for a moment, closer to God. Joseph Priestley had demonstrated how easily these wonders might be found on a memorable occasion in 1767. Walking in the streets of Leeds, he had noticed a distinctive gas being emitted from a brewery. He had bottled the gas, carried it home and released it into a flask of water. He was delighted with the fizzing, tongue-tingling result. He named it soda water and it soon became a popular, refreshing drink that was marketed as a health remedy.

This idea of bottling nature was expressed in other ways, too. Just a year after Priestley's success with soda water a clergyman, William Gilpin, began work on his own pet project: the picturesque. To Gilpin the picturesque was an aesthetic ideal, a refined form of the Sublime that Edmund Burke had so fashionably defined a decade earlier. Gilpin's idea of the picturesque was rather loose, 'such objects as are proper subjects for painting',[24] but nonetheless his

writings on the subject were influential. By the 1780s and 1790s when
Constable was a boy in Suffolk the idea of picturesque walking
tours had taken hold. In 1794 Gilpin expanded his philosophy further.
He urged those embarking on picturesque tours to sketch as they
went, so that when they returned home they could continue to
enjoy the sights of their tours. These sketches were an early incar-
nation of the holiday snap.

Gilpin distilled his ideas on sketching into the paper, *Essay on the
Art of Sketching Landscape*. He included notes on composition and
method, outlining how anyone could watch nature, copying what they
saw in a sketching pad in black lead or Indian ink as a fleeting memento.
For him it was a rewarding pastime but it was not without its limits. 'The
art of painting,' Gilpin argued, 'cannot give the richness of nature . . .
generally an attempt at the highest finishing would end in stiffness'.[25] To
Gilpin's mind this ideal of high finishing – or perfect realistic detail –
'belongs only to a master, who can give expressive touches'. He went
on, 'Painting is both a science and an art; and if so very few attain
perfection, who spend a life-time on it, what can be expected from
those, who spend only their leisure?'[26]

Gilpin's work was widely read. It found its way into the hands of
Sir Joshua Reynolds, first President of the Royal Academy. 'The essay
has lain upon my table; and I think no day has passed without my
looking at it, reading a little at a time,'[27] he wrote, congratulating
Gilpin. Hundreds of others followed Gilpin's instructions, descending
on picturesque points in Britain – Symonds Yat in the Forest of Dean,
or the barren rugged terrain beside Hadrian's Wall – on the lookout
for beauty. Gilpin reminded readers that each suitable object they
found, every tree or wood or hill or stream, is 'varied a second time
by combination; and almost as much, a third time, by different lights,
shades, and other aerial effects'.[28]

Constable owned a copy of Gilpin's *Essays* and it was an early influence on his method – particularly his fondness for outdoor sketching.
An even more powerful influence was his hero, the seventeenth-
century French painter Claude Lorrain. One hundred and fifty years
before Constable, Claude had made his own efforts to unpick nature
while living in Rome, lying on his back in the meadows in the early
mornings watching the rising of the sun and the colouring of the sky,
seeing it change from black to red to gold to blue. The key, Constable
understood, was to watch carefully, closely. It was a meditative

experience, and one that Constable repeated over and over again, piling up more studies as the years passed.

The sheer volume of Constable's (often undated) preparatory sketches has left a lifetime of dilemmas for his cataloguists today. One in particular is an oil entitled *Spring: East Bergholt Common*. Painted on an oak panel, on the back of an earlier night scene, it has been variously estimated as dating from 1821 or 1829, although one source suggests it may be a preliminary sketch for the *Ploughing Scene in Suffolk* of 1814.

Whatever the case, *Spring: East Bergholt Common* is one of the most beautiful and eloquent of all his oils. It fits an exaggerated landscape format, 19cm by 36cm, and features a relatively simple composition: a ploughman driving his horses across a field with one of East Bergholt's windmills standing just behind. The windmill anchors the picture's theme: the interaction of human society and nature. The sky is the cold frosty blue of a March morning. The atmosphere is forbidding and uncertain, filled with lively cumulus. The wind seems strong enough to chill the ploughman's back and spin the windmill's sails. Beyond the cold browns and greens of the fields there's the faint outline of a church spire in the background and a hint of human life beneath an elm tree by the left margin. 'Constable could never consent to patch up the verdure of nature to obtain warmth,' his biographer Charles Leslie later wrote. *Spring: East Bergholt Common* is the very type of picture that would later prompt the Swiss painter Henry Fuseli to write of Constable 'he makes me call for my great-coat and umbrella'.

The sketch was etched into a mezzotint for Constable's *English Landscape Scenery* later in his career. Here Constable would add his own description of the scene:

This plate may perhaps give some idea of one of those bright and animated days of the early year, when all nature bears so exhilarated an aspect; when at noon large garish clouds, surcharged with hail or sleet, sweep with their broad cool shadows the fields, the woods and hills; and by the contrast of their depths and bloom enhance the value of the vivid greens and yellows, so peculiar to this season; heightening also their brightness, and by their motion causing that playful change always so much desired by the painter.[29]

Constable knew days like this well. As a young man he had worked as a miller for his father in East Bergholt, operating the very mill depicted in this sketch. Leslie later noted that Constable had left a physical trace of his presence on the landscape, carving *John Constable 1792* 'very accurately and neatly' into one of the windmill's timbers. Constable himself would tell friends that it was during his time as a miller that 'he made his earliest studies and most useful observations on atmospheric effects'.

The art of windmilling is almost completely lost today but in Constable's time it was a common practice that required mental alertness. 'By a wind-miller every change of the sky is watched with a particular interest,' Leslie wrote in his biography. The East Bergholt mill was a typical post windmill, so named because it was mounted on a tall upright post that allowed the miller to twist its body to catch the wind. Constable would have worked inside, operating the gearing system, rotating the sails and applying the brake when the wind blew too strongly. This was a real danger. A windmill was a relatively delicate instrument compared with the force of wind it was intended to harness. If the sails were not furled in a storm its arms could spin around at an ever-quickening pace, propelled by its own force as well as by that of the wind. The friction of the movement had been known to set mills alight, as had attempts to belatedly jam the brake wheel. The miller, therefore, had to be attuned to coming weather, judging the subtle shifts in the cloud base, the dimming of the ambient light or the quickening of the breeze. In eighteen months of labour Constable – known in East Bergholt village as 'the Handsome Miller' – became a natural weather-watcher. Later his brother Abram declared, 'When I look at a mill painted by John, I see that it will go *round*, which is not always the case with those by other artists.'

In solitude Constable was free to study the moods of the season, to watch clouds sweep over Bergholt Common and out towards the coast. He might have noticed 'wind shear' when conflicting winds blew clouds in opposite directions at different altitudes in the atmosphere. He must have watched the small scudding clouds that accompanied rain-bearing cumulus. These 'opaque patches', which moved closer to the earth on a faster current of wind, Constable called *messenger clouds* that 'always portend bad weather'. He continued:

They float midway in what may be termed the lanes of the clouds; and from being so situated, are almost uniformly in shadow, receiving a reflected light only, from the clear blue sky immediately above them. In passing over the bright parts of the large clouds they appear as darks; but in passing the shadowed parts, they assume a grey, a pale, or a lurid hue.[30]

These messenger clouds were typical of the scientific detail that Constable liked to absorb into his art, a fact which gives his works a curious extra dimension today. At one of his lectures, given in 1836 at the Royal Institution, he would argue that 'Painting is a science and should be pursued as an inquiry into the laws of nature. Why then may not landscape painting be considered as a branch of natural philosophy, of which paintings are but the experiments?' He displayed his analytical abilities by dissecting a 'small evening winter piece' by Salomon van Ruysdael. Typically the piece included a windmill. This picture', he said,

represents an approaching thaw. The ground is covered with snow, and the trees are still white; but there are two windmills near the centre; one has the sails furled, and is turned in the position from which the wind blew when the mill left off work; the other has the canvas on the poles, and is turned another way, which indicates a change in the wind; the clouds are opening in that direction, which appears by the glow of the sky to be the South (the sun's winter habitation in our hemisphere), and this change will produce a thaw before the morning. The concurrence of these circumstances shows that Ruisdael *understood* what he was painting.[31]

Few painters could match such analytical precision. Constable would once tell Leslie that the composition of a picture resembled 'a sum in arithmetic; take away or add the smallest item and it must be wrong'.

After 1816 Constable began to replace East Bergholt as the main subject of his art. Now he was married, he had less flexibility to travel to the Stour Valley on research trips and although rural Suffolk remained an inspiration, his horizons widened. He resolved to commemorate the official opening of Waterloo Bridge in June 1817, a grandiose state occasion in contrast to the sleepy East Bergholt fields. And two years later he would find a new geographic challenge when he moved his family from central London to a rented Georgian cottage in Lower Terrace, Hampstead. The move was the catalyst for his greatest atmospheric experiments yet.

Hampstead was a natural and increasingly fashionable route away from the bustle and smoke of the capital's streets. Just four miles north-west of London, and a shilling coach-fare from Tottenham Court Road, Holborn or the Bank, Hampstead village was set on the slopes of a steep hill and had what would later be called a Pickwickian English quality with cobbled lanes and coaching inns. It also had something of a reputation as an artistic enclave. George Romney the celebrated portrait painter had spent his last years in Hampstead and in 1819, the year of Constable's arrival, John Keats would draw inspiration from the same bucolic surroundings to write his 'Ode to a Nightingale'.

Hampstead's artistic heart would have appealed to Constable as would its reputation for healthy air and salubrious springs (or 'wells'). Such was the high estimation of Hampstead water that it was ferried down to Charing Cross, Bloomsbury, Temple Bar and Fleet Street on a special daily shuttle service and sold at threepence per flask. Having such an elixir close at hand was another benefit for Constable, whose move was sparked partly out of concern for his wife Maria's fragile health. Away from the city Constable would encounter a place renowned for its microclimate with cleaner, fresher air.

The move to Hampstead was a liberation for Constable. He had lived in central London for nearly two decades without fully embracing city life. 'In London nothing is to be seen, worth seeing, in the natural way,' he would say.[32] Hampstead was an almost perfect compromise. It allowed him to stay close to the vital community – the lifeblood for his career – that gathered about the Academy, while enabling him to stray into his favoured wild places. Hampstead village was nestled beside Hampstead Heath, a semi-wilderness 'remarkable for the prodigious extent of view over the city of London and the adjoining counties'. The Heath was a blend of steep undulating mounds, shallow ponds and isolated cottages.

In October 1819 he produced his earliest dated oil sketch of the Heath, *Branch Hill Pond, Hampstead*. At this time he had lived in the area for just a few months, and his first impression of the Heath is fresh, keen. He paints the earth in fawns, bronze and tea greens, all built up with a vigorous impasto. A rider stands by as his horse drinks in the waters of the pond. But once again it is the sky that sets the tone. It is leaden with menace, the colour of a two-day-old bruise, broken with crepuscular rays. Looking west, towards Harrow in the distance, rain is falling heavily.

Over the next two years Constable produced studies of elm trees, sandbanks and rural cottages, all emblematic of the local landscape. But overshadowing these sketches in volume and scope would be a completely fresh series of oils, dedicated to capturing the sky. Painted on small sheets of paper, each taking about an hour, these studies came at a quickening pace. Between October 1820 and October 1822 he painted more than a hundred, recording the sky at all hours of the day, from all directions and in all weather conditions. Constable began by finding an open viewpoint. There he would sit with his paintbox resting on his knees, thick sturdy paper fastened to the inside lid. Then he would paint, applying layers of purples, lead whites, red lead, black and iron earth pigments to create the illusion of the clouds. Freezing nature at a glance.

On Tuesday 17 October 1820 Constable painted a sunset scene, with a line of elms and the scrub of the Heath straddling the foreground. Behind and above this landscape the sky is warm and dappled with dark, frail cumulus, scudding clouds. He annotated the sketch: 'Hampd. 17 October 1820 Stormy Sunset. Wind. W'.

The level of detail in the annotation is significant. He records the location, date, and the weather conditions. Rather than the stereotypical wandering, artistic mind, this seems more the behaviour of a man of science. Constable had long been in the habit of annotating his sketches. Fifteen years before he had written, 'Nov 4. 1805 'noon, very fine day, the Stour', on the back of one East Bergholt sketch. He had continued to do this from time to time, ever since. But during his 'skying' phase between 1820 and 1822, this commitment to scientifically annotating his sketches became more purposeful.

18 Oct 1820 – 4 to 5 1/2 . . . wind north
Hampstead July 14 1821 6 to 7 p.m. N.W. breeze strong
'5 oclock afternoon: August 1821 very fine bright & wind after rain slightly in the morning
Sepr. 10. 1821, Noon, gentle Wind at West, very sultry after a heavy shower with thunder, accumulated thunder clouds passing

slowly away to the south East. very bright & hot. all the foliage
sparkling [?with the] and wet[33]

The annotations, even without the accompanying sketches,
preserve a picture of the scene. As his project develops you can sense
Constable striving to depict the atmosphere truthfully. His notes
become longer, more descriptive. He tests himself with thundery
afternoons, glowing sunsets, breezy mornings, stormy sunsets, sudden
showers and clouds of every type and tint during every hour of the
day. Painted two years after Turner filled a sketchbook with images
of clouds and skies and vivid sunsets, the annotations, the rooting in
a historical moment, make Constable's skies stand apart. In bursts
over two years Constable's annotated sketches continued to come.
They reached a pitch in early September 1822:

> Sept. 5. 1822. looking S.E. noon. Wind very brisk. & effect bright
> & fresh. Clouds moving very fast. with occasional very bright
> openings to the blue.
> Sept. 6th. 1822. looking S.E. – 12 to 1 oclock, fresh and bright, be-
> tween showers – much the look of rain all the morning, but
> very fine and grand all the afternoon and evening.

The sketches invite historical analysis. In a recent research project
John Thornes, Professor of Applied Meteorology at the University of
Birmingham, compared each of Constable's sky sketches with contemporary weather records made by Luke Howard, *Cowe's Meteorological
Register*, the *Philosophical Magazine* and the Greenwich Observatory.
Of thirty-six of Constable's dated studies Thornes concluded that nine
agreed excellently with the sources, fifteen very well, eleven well and
only one poorly. Furthermore, using the same methods he was able
to speculate 'with some certainty' on dates for a series of undated
studies.

Scholars have long debated why Constable painted these skies.
Drawn together they comprise hundreds of hours of work that he
never showed any interest in exhibiting. Was it simply evidence of
his love for nature? His move from the narrow streets of central
London to the 'atmospheric laboratory' of the Heath? The artistic
challenge they posed? A long-held love for *plein-air* sketching? A reference point for future works? On the back of a study on 5 September

1821, for example, he scribbled: 'Very appropriate for the coast at Osmington'.

It is telling that these sketches emerged at the same time as Constable was progressing to working on six-foot canvases. Unrecognised by the Academy for years, he had determined to paint on a larger scale: a scale impossible to ignore. This cast his skies, already prominent, into an even greater focus. Now he had the freedom and challenge of depicting each whorl and ridge of a cumulus, each shade of the midday sunlight. As he was working on his sky series, between 1820 and 1822, he embarked on his most ambitious painting yet – a work that, like his cloud studies, he gave a descriptive title: *Landscape: Noon*. Later this would be rechristened *The Hay Wain* and would be judged a masterpiece for its scope, narrative power and skyscape. Robert Hunt in the *Examiner* lauded a sky, 'which for noble volume of cloud and clear light we have never at any time seen exceeded except by Nature'.[34]

But while Hunt was enthusiastic, others were still not convinced. In September 1821 Constable received a letter from his friend and patron, John Fisher, telling him that a 'grand critical party' had sat in judgement on one of his compositions, and that they had found 'objections' to his sky.

The letter reached Constable as he was in the midst of his period of skying in September 1821. It was the kind of criticism that he had lived with for many years and, a month later, he replied in one of his most revealing letters. He thanked Fisher for fighting his battles and revealed how many times friends or tutors had advised him 'to consider my <u>Skey</u> – as a <u>White Sheet</u> "<u>drawn behind the Objects</u>"'.

Cirtainly if the Skey is *obtrusive* – (as mine are) it is bad. but if they are *evaded* (as mine are not) it is worse, they must stand and alwa[y]s shall with me make an effectual part of the composition. it will be difficult to name a class of Landscape – in which the skey is not the "<u>key note</u>" – <u>the standard of "Scale"</u> – and the chief *Organ of Sentiment'* – You may conceive then what a "*White Sheet*" would do for me. impressed as I am with these notions, and they cannot be Erroneous. the 'skey' is the "*source of light* in nature – and governs every thing – Even our common observations on the weather of every day – are suggested by them but it does not occur to us – their difficulty in painting both as to composition and Execution is very great.[35]

The letter stands as a lucid, compelling argument on a matter that long troubled its author. It shapes our image of Constable on the Heath into that of a man on a determined quest to understand and master a skill, in his own way like Newton experimenting with a prism in his Trinity rooms or Franklin flying his kite in a Philadelphia storm. Painting with grace, skill and lyrical intensity on Hampstead Heath Constable emerges as a philosopher, a student of nature, searching for truth.

An inventory of his library taken after his death proves Constable's enlightened credentials. He owned monographs on chemistry, fishes and biology, five volumes of Cuvier's *The Animal Kingdom* and Gilbert White's *Natural History of Selborne* – a book that he received as a present from his friend Fisher at the beginning of his skying period in March 1821. His ambition to be a natural painter had led him on many paths of investigation. 'In such an age as this painting should be *understood* not looked on with blind wonder, nor considered only as a poetic aspiration, but as a pursuit, *legitimate, scientific and mechanical*,'[36] Constable argued, a philosophy he refashioned with beautiful laconic grace in his Royal Institution lectures: 'We see nothing truly till we understand it.'[37]

'We see nothing truly till we understand it' – this is a perfect mantra not just for Constable but for his time. It encapsulates a growing attitude, an evolution of Burke's theory of the Sublime. Burke's philosophy drew on the psychological power of sublime objects: cliffs, crashing waterfalls, thunderous clouds. These objects were overwhelming, just as Beaufort had been overwhelmed by the moving atmosphere at the summit of Croghan Hill. They were chaotic, hectic, inexplicable, and evoked simultaneous reactions of pleasure and terror. The drug of choice for much of the Georgian Age, sublime sights, Burke wrote, 'always produce a delight' when they do not 'press too close'.[38]

Twenty years later on Hampstead Heath, the situation was much the same but the response to it was entirely different. Unlike Beaufort, Constable was not overawed. Instead he remained dispassionate, watchful, curious. It was reflective of nineteenth-century empiricism rather than the old thrill-seeking of the past. As Constable said, his aim was to be able to see nature truthfully, and to do this he had two advantages. The first was his temperament, a natural propensity to observe and record. A second was the civilising force of science. Among the titles in Constable's library was a second edition of Forster's *Researches about Atmospheric Phaenomena*. It is a little link of great significance that binds Constable to the meteorological enlightenment of the nineteenth century.

It is unknown just when Constable came across a copy of Forster's book. Today it remains in the possession of the Constable family, simply marked with the inscription 'Constable, 6/- Published at 10/6 scarce'. The artist's only direct reference to the book comes in 1836, in a letter written to a friend (but not a relative) called George Constable:

> Any observations on clouds and skies are on scraps and bits of paper, and I have never yet put them together so as to form a lecture, which I shall do, and probably deliver at Hampstead next summer . . . If you want anything more about the atmosphere, and I can help you, write to me . . . Forster's is the best book – he is far from right, still he has the merit of breaking much ground.[39]

There are strong hints, though, that Constable purchased Forster's work much earlier. His copy was a second edition, of 1815, that Baldwin had specifically pitched 'to the painter and engraver'. This remained on sale until 1823 when it was replaced by a subsequent edition, so it seems likely that Constable acquired the book between 1815 and 1823. A further clue that he had absorbed Forster's book by the time of his skying period comes with the annotation of one of his sketches as '*cirrus*'.[40] The cirrus sketch is one of the most beautiful of Constable's skies. The cirrus are delicate, gossamer-like in the distant atmosphere. Had Constable not read Howard's papers or Forster's *Researches*, there is little chance he would have been able to give this sketch such an accurate meteorological title.

Better still, Constable's copy of Forster's *Researches* was annotated, a fact that has left to history a unique trace of his meteorological understanding. An annotated book has a special attraction: the two competing voices; the hurried marginalia, scribbled in a fleeting moment of agreement or disgust. In annotated books both thinkers are alive. Reading these works years later it feels as if you are eavesdropping on a conversation that happened long ago.

Constable picked out sections, drawing thin lines of his pencil beside memorable quotes. Quotes are picked out in the preface and throughout the first chapter: the allure of meteorology, its accessibility, the reawakening of interest in the subject. As Forster moves through his analysis of Howard's cloud classification theory, Constable expresses his opinions. He is 'doubtful' of Forster's description of cirrus clouds; he disagrees with some of Forster's comments about the stratus. In a section explaining cumulus clouds he underscores the sentence, 'It is commonly of a dense structure, forming the lower atmosphere and moving along in the current of wind which is next to the earth.' Next to the section on cirrostratus Constable writes, 'heat, wind[s?], electricity moisture', and beside Forster's description of the towering cumulostratus, Constable has written 'it is only a mushroom in thunder'.[41]

The annotations stretch on, Constable underlining and emphasising his thoughts on clouds becoming lighter and darker, the falling of rain, the appearance of dew and the varieties of cumulus on a summer's day. Throughout, Constable has the confidence to challenge Forster's opinions when he does not agree, which he often does not.

Constable's copy of *Researches* cements him as a man of his time. It transforms his cloud sketches from beautifully observed and executed visions of nature into scientific studies. Rather than just being the vigilant witness, Constable also becomes the talented student: reading and watching, thinking and painting, sketching and experimenting. This cross-pollination of ideas, from Howard to Forster, Forster to Constable, is a hallmark of the meteorological reawakening at the beginning of the nineteenth century, a time when the atmosphere and the weather at last seemed within reach.

Constable's cloud sketches would not be seen until much later in the century when the majority were left by his daughter in a bequest to the Victoria and Albert Museum. But the landscapes that they would inform were. The skies of these works – *The Hay Wain* in 1821, *The Leaping Horse* in 1824, *The Chain Pier in Brighton* in 1827, *Vale of Dedham* in 1828, *Old Sarum* in 1829 or *Salisbury Cathedral from the Meadows* in 1831 – bought by the Tate in 2013 for £23.1 million – would go on to dazzle viewers in the nineteenth century. Today they are treasured as glories of British fine art.

Constable's attitude to painting skies came directly from the intellectual climate of his day – a culture which urged students to observe, distil and record. This attitude unites Constable and Forster. Though they probably never met, their shared aim is evident. Through their different mediums they endeavoured to create truthful records of passing weather. Forster glancing up from his desk at a sky brushed with electrified cirrus. Constable sat on the Heath looking into a pale blue atmosphere, cumulus clouds carried on a current of wind that blows in gusts over his paintbox. In the distance the skies have darkened and rain is falling. A storm is on its way.

CHAPTER 3

Rain, Wind and the Wondrous Cold

Captain Phillip King of HMS *Adventure* watched the pampero from the shore of Maldonado Bay. Although the day had been hot and sultry, no one had expected the storm that came or the chaos it brought. The *Adventure* had lain at anchor in the bay with her crew camped on the beach. The day had been set aside for repairs and resupplying stores of water, oranges and meat as King waited for his companion vessel HMS *Beagle* to arrive from Rio de Janeiro. It was late afternoon on Friday 30 January 1829. The day had passed as usual. Supplies were being ferried across the breakwater. A French frigate, *L'Aréthuse,* was drifting along the coastline. The sky was 'gathered and unsettled'. A little after five o'clock King had glanced at his barometer. It had been reading a steady 30" of mercury, but suddenly it had dropped to 29.50. There was a fluttering of the flags. King felt the wind veer. Black clouds were overrunning the sky. All too quickly, the pampero was upon them.[1]

The winds roared with raw violence. A squall filled the bay. From the shore King saw the *Adventure* pitched broadside, tugging at its anchors. Tents went scurrying down the beach. One of the *Adventure's* boats, fully laden at oar in the bay, was driven up a beach. Another was 'shaken to atoms'.[2] 'The spray,' King wrote, 'was carried up by whirlwinds, threatening complete destruction to everything that opposed them.' Anxious about his men and the fate of *L'Aréthuse,* King kept his eyes fixed on the scene before him. He did not glance along the coast to the east. If he had, even for a fleeting second, he might have seen the faint outline of HMS *Beagle* on the horizon, fighting for her life.

The pampero had hit as the *Beagle* and her new commander Robert FitzRoy were beating west down Rio de la Plata. They were within sight of Maldonado Bay where FitzRoy had been told to meet King and receive his orders. Until now the long leg from Rio had gone smoothly, the *Beagle* skimming south along the Atlantic coast under a summer sun. The *Beagle* was a dextrous bark of just ninety feet, tiny in comparison to the monstrous ships of the line of the Napoleonic Wars. But the *Beagle* had not been built with fighting in mind. Instead she was flexible: she had the ability to hug shorelines and could be beached for running repairs – advantages that made her ideal for surveying work.

FitzRoy had kept his eye on the weather that afternoon. At 1 p.m. a steady breeze had been blowing from the NNE, an hour later this had freshened, but by 3 p.m. the wind had died away to a near calm. Sailors had an unusual affinity with weather. Knowing how to catch a breeze or to shorten sail was their core skill. A good judge could shave days off a voyage. But that day the weather had been difficult to read. By half past four it was 'gathered and unsettled'. At 5 p.m. the sky was overcast and FitzRoy noticed thunder and lightning to the south-west. At 5.20 he had trimmed the topgallant sails, jib and spanker, and hoisted the mainsail in their place. There was something else too. Like King, FitzRoy noticed that his barometer had dipped. In the last half-hour it had fallen: 29.90, 29,80, 29.60. It meant only one thing. But before FitzRoy could act, just after 5.40 p.m., a 'tremendous and sudden squall' struck the ship.[3]

The crew, midway through hoisting the mainsail, had it ripped out of their hands. The lines and yards were carried overboard, as was a boy called Thomas Anderson, blown into the waters from thirty feet. Suddenly the *Beagle*'s prow was twisting in the sea, waves crashing into her bows. For a dramatic few minutes all was confusion. It was a pampero. Everyone on board had heard of these viciously powerful south-westerly gales that rose with mounting force over the Argentine Pampas and exploded across La Plata. They could whip up a terrifying squall in minutes. Brief, ferocious, deadly, they were the meteorological equivalent of a piranha attack.

FitzRoy saw lightning, rain and hail almost at once. The fragile bark, one of a class ridiculed as coffin brigs by naval men, was helpless in the face of the assault. Her topmasts and jib boom were sheared off along with a handful of spars. At one terrifying moment she was pitched back on her beam ends in the rolling sea within a few degrees of capsizing. It was only when FitzRoy cut away the best bower and small bower anchors that she was brought to the wind and righted.

By 6 p.m., just fifteen minutes after the pampero hit, the worst had passed – but not before a second seaman, Charles Rosenberg, had been lost overboard.

This incident, on his very first cruise as a commander, would haunt FitzRoy. Charged with a ship and the lives of its crew he had come within a whisker of being wrecked within six weeks of his appointment. Thirty years later he retained a vivid picture of the scene; and of the 'two fine fellows' who were blown from aloft. '[They] swam hard for their lives,'[4] FitzRoy recalled, 'but were immediately overwhelmed by the sea.' Had he struck sail earlier, perhaps the crew would have been sheltered on deck. In the age of sail a commander might make a hundred decisions a day: tacking, turning, reefing, furling, sending down yards, leading drills. Each order was important, and in bad seas the margins between speed, safety and disaster were perilously thin.

It was a reality FitzRoy understood. Commanders were selected for their character: they were men of judgement who acted with conviction. FitzRoy himself would later write,

Those who never run any risk; who only sail when the wind is fair; who heave to when approaching land, though perhaps a day's sail distant; and who even delay the performance of urgent duties until they can be done easily and quite safely; are, doubtless, extremely prudent persons: but rather unlike those officers whose names will never be forgotten while England has a navy.[5]

But FitzRoy felt the responsibility. He particularly dwelt on his failure to heed the barometer. 'Signs in the sky, barometric evidence, and temperatures shewed what was coming,' he would remember, 'but want of faith of such indications, and the impatience of a very young commander in sight of his admiral's flagship induced disregard, and *too late* an attempt to shorten sail sufficiently.'[6]

Two days later, on 1 February, the battered and broken *Beagle* was finally able to join King and the *Adventure* in Maldonado, where she underwent repairs. Nothing else is said of FitzRoy's misjudgement. A pampero was the kind of misfortune that could befall a commander in the southern oceans. The loss of the two sailors was considered one of the perils of the sometimes brutal, sometimes serene, life at sea.

★   ★   ★

Aristocratic, dashing to the eye and fiercely capable, FitzRoy was among the Navy's finest young hopes. He was born on 5 July 1805 into a family with impeccable Tory credentials. His father, Lord Charles, was an army general and later the MP for Bury St Edmunds. His mother Lady Frances Stewart was the half-sister of the statesman Lord Castlereagh. His grandfather, the Duke of Grafton, had served as prime minister and through his paternal blood FitzRoy could trace his lineage right back to King Charles II. It was a pedigree few could match. Although FitzRoy's mother died when he was five his child-hood was seemingly a happy one, spent at the family's fine Palladian mansion in Northamptonshire. From their Midland base the FitzRoys maintained links to society life in London and the shires. During Robert's childhood Castlereagh had been advancing through the polit-ical ranks and, best of all, twice in the Regency years the family scooped the ultimate social prize when their horses, Whalebone and then Whisker, romped home at the Derby.

But it was the sea and not politics or society that had caught FitzRoy's imagination. One of the few surviving stories of his early years is a playful account of his maiden voyage. It took place on a large pond at his family's estate. Always one to seize an opportunity, FitzRoy commandeered a laundry tub from the kitchen during the servants' dinner hour. He hauled the tub to the pond, loaded it with bricks for ballast, and launched his craft. Like a Cook or Bligh in miniature FitzRoy progressed valiantly from one side to the other, propelled by a long pole. Reaching his destination he ruined his triumph by overbalancing, capsizing and tumbling right in. Soaked through, he was fished out by the gardener.[7]

FitzRoy's naval career recovered from this inauspicious beginning. After brief stretches at Rottingdean and Harrow he rounded off his education at the Royal Naval College in Portsmouth. Here he blos-somed. He swept through the three-year course, a heady mix of clas-sics, mathematics, Newtonian philosophy, navigation, languages, fencing, dancing, painting, gunnery and drawing, and emerged the star pupil. Within a year and a half he was serving his midshipman apprenticeship on the oceans of the world. It was a practical education and served him well.

After five years he returned to Portsmouth, by now a hardy tar, to sit his lieutenant's examination – a notorious test of skill and a frightful ordeal. FitzRoy was forced to stand before a panel chaired by Sir William Hoste – veteran of the Napoleonic Wars – and suffer hours of close questioning. It was an intellectual Battle of Trafalgar, nautical scenarios lobbed from all angles like missiles. 'All the questions were worked correctly,'[8] FitzRoy wrote. 'Many of them by three methods (one being by algebra and spherical trigonometry) – the other two practical ways, suitable to quick or to rigorous calculation.' Out of twenty-six entrants, FitzRoy came first. He left the college with an unprecedented double of full marks and the Gold Medal.

FitzRoy's education did not end there. On being gazetted a lieutenant he stocked his cabin with an incredible 400 volumes, outdoing even Beaufort. The books fed a mind that was more disciplined than brilliant. His success was grounded in determination and hard work more than raw talent. While at sea he found time to study Latin, Greek, French, Italian and Spanish. He kept pace with the latest scientific news, and developed a particular relish for phrenology, the fashionable subject of the day. An odd, boisterous science, phrenology sought to expose links between the size and shape of the human head and the character of the individual. It burned brightly for some years because it fed into the contemporary mania for categorising everything from animals to plants to clouds. One of phrenology's loudest British champions was Thomas Forster, who was credited with coining the word. FitzRoy subscribed to the idea completely; it was just the subject to appeal to a young mind bent on order.

In contrast to frivolous novels and dusty philosophical books by men like Burke and Gilpin, science was considered wholesome and pure: a proper pursuit for those of refined tastes. In the 1820s one of the great ways to dazzle the opposite sex was to talk of the stars or the planets. Science offered a way of making sense of an over-complex and cluttered world, and phrenology promised to bring order to the most complicated subject of all: the human mind. FitzRoy followed a simple mental checklist – the head and temple, nose and chin – to diagnose the personality of the person before him. Were they bold or timid? Clever or stupid? Lazy or fizzing with life?

Phrenology fuelled the outlook of a man who was already climbing the professional ladder with noticeable speed. Most in the Navy had already heard of FitzRoy, the handsome aristocrat with the fierce work ethic and brooding temper. The Admiralty liked what they saw. His intellectual prowess and aristocratic heritage was a potent mix. Some speculated that he had a fortune of £20,000. He was marked down as cool-headed, shrewd, undaunted, true: equal to the challenge ahead.

Captain King had established Maldonado Bay as the fair-weather launch pad for the Admiralty's survey of the southern end of the South American continent. Since 1826 the *Adventure* and *Beagle* had navigated the wild Patagonian coasts and channels, adding detail to the few threadbare charts that existed. They sought to map an area that mirrored the Scottish coast in its geography, sweeping in shallow coves on the east, and fractured and splintered into thousands of tiny islands on the west. One hundred and sixty miles from the southern-most point of the continent, the survey had stretched through the fabled Strait of Magellan, a jagged vein of water that cleaved the Atlantic and Pacific, offering merchants bound for Tahiti, the Sandwich Islands, New South Wales or Van Diemen's Land an alternative route to the infamous Cape Horn. Between Strait and Cape they had explored the outer fringes of Tierra del Fuego, a mountainous, inhospitable archipelago thinly populated with guanacos, foxes, condors, kingfishers, pumas and primitive Fuegian tribes.

The South American survey had been commissioned by Viscount Melville, Lord of the Admiralty, in 1825. It was part of an invigorated naval policy that included Arctic exploration and fresh attempts to locate the North-West Passage. Newly free from a generation of war, the Admiralty had decided to turn Britain's naval mastery and surplus of ships to different ends. Still poorly charted, South America had a powerful allure. Though Britain's possessions in the region were restricted to the Falkland Islands, the government had plans to extend its interests. As a Christian nation that thought itself uniquely blessed by God, it was Britain's duty to civilise the continent and to liberate its untapped mineral deposits.

In the 1760s John Byron, or 'Foul Weather Jack', had returned from Tierra del Fuego with enticing tales of a landscape filled with 'the finest trees I ever saw'. 'I make no doubt,' he wrote, 'but they would supply the British Navy with the finest masts in the world.' Byron described a wild world with unbounded forests, white with snow, ripe with potential, enriching his account with tales of 'innumerable parrots and other birds of the most beautiful plumage'.[9] Byron included an extra titillation for the Georgian aristocracy, notorious for their love of blood sports: the land was teeming with potential quarry. 'I shot, every day geese and ducks enough to serve my own table,' he declared, 'and several others, and everybody on board might have done the same.'[10]

Fifty years would pass before the Admiralty had the chance to test Byron's tales, and their enterprise was hurried along in the post-Waterloo world by two audacious voyages. The first was made by William Smith of Bligh in Northumberland who stumbled upon an outcrop of islands in the Southern Ocean, which he christened the South Shetland Islands and claimed for Britain. News of Smith's discovery caused a ripple of excitement in Europe. If navigators as able as Cook and Bligh had overlooked an entire cluster of islands, then what other treasures might exist?[11] This prospect was reinforced by the news that Smith's discovery had sparked a craze in seal hunting. In the two years that followed, 100,000 seals were slaughtered on the South Shetlands for their fur or blubber.[12] Twenty thousand tons of sea-elephant oil were harvested for the London market by two hundred seamen who were employed by the trade. But, irritatingly for the British, it was the American merchants who made the greatest profit. Quicker to the prize, the Americans loaded ships full of sealskins and transported them across the Indian Ocean to the Chinese market where they were sold for five dollars each. Fortunes were made in a single crossing.

One lured to the southern seas was a British sailor, James Weddell, in his brig *Jane*. Accompanied by a fellow seaman Matthew Brisbane in a cutter called *Beaufoy*, Weddell embarked in 1822 on one of the most unlikely voyages of the century. Finding the seal grounds exhausted in the South Shetlands, Weddell continued southwards. He sailed into a dark and inhospitable world of freezing fog and biting winds, dodging through a jigsaw of icebergs. Incredibly, Weddell travelled to within a day or two of the then unknown Antarctic continent, further south than anyone in recorded history. Like Coleridge's ancient mariner Weddell had entered a world of glimmering ice and wondrous light, a place that Ishmael in *Moby-Dick* would dub that 'charmed circle of everlasting December'.[13] On 20 February 1823 he stopped at a latitude of 74°15' S. Surrounded by a thickening, shifting seascape Weddell ordered three cheers, the hoisting of the Union Jack and for the cannon to be fired.[14]

Once home, Weddell published *A Voyage Towards the South Pole, Performed in the Years 1822–1824*. It was filled with descriptions of humpback whales, sea leopards, condors and gigantic albatrosses. On South Georgia Weddell had been charmed by the sight of crowds of penguins looking, at a distance, like 'little children standing up in white aprons'.[15] Weddell's playful descriptions and quixotic ability to extract himself from perilous situations made a winning formula and his account was published in 1825 to enormous success. Having disproved the existence of South Iceland and written up a set of weather guidelines for sailing around Cape Horn, Weddell's final flourish was to dedicate his book to Viscount Melville. His voyage sent a message right to the heart of government.

In 1829 FitzRoy was co-opted on to the South American Survey as the new commander of the *Beagle*. On 27 March, repairs having been completed, he received his orders from King. He was to chart an unexplored stretch of the Strait of Magellan that included a series of bays: Lyell Bay, Cascade Bay, San Pedro Bay and Freshwater Bay. After this he was to navigate the labyrinth of sinuous passages to the west of the Strait. This was a zone poorly known, barely charted and filled with fast tides, hidden currents and concealed rocky outcrops, any of which could send the *Beagle* to the bottom in a minute. She was to roam this world alone during the winter months. For FitzRoy, who had trained half his life for the opportunity, it was to be a baptism of fire. By 19 April 1829 the *Beagle* had ghosted through the narrow entrance to the Strait and taken on provisions at Port Famine. That day she parted from her companion schooner, *Adelaide*, and began her cruise.

FitzRoy was instantly charmed by the landscape around him: the leaden rocks and beech trees on the shore, the silvery hue of the water, the intensity of the light. Glaciers tumbled down the mountainsides, their transparent blue contrasting strikingly with the snow on the mountaintops.[16] 'I cannot help here remarking,' he wrote in his journal, 'that the scenery this day appeared magnificent.' In the distance he saw the outline of Mount Sarmiento, a pyramid of ice and snow, a blend of Ancient Egypt and the Arctic. He noticed the 'continual change occurring in the views of the land, as clouds passed over the sun, with such a variety of tints of every colour, from that of the dazzling snow to the deep darkness of the still water'.[17]

The feeling of exhilaration stayed with FitzRoy. 'The night was one of the most beautiful I have ever seen,' he recalled. Only the gentle lullaby of the water on the timbers, the creaking of the hull, the knock of the anchor chain and the ringing of the ship's bell disturbed the silence. It was 'nearly calm, the sky clear of clouds, excepting a few large white masses, which at times passed over the bright full moon'. Moonlight shone on the rugged snow-covered peaks of the surrounding mountains that 'contrasted strongly with their dark gloomy bases, and gave an effect to the scene which I shall never forget'.[18]

Onshore at Cascade Bay the crew discovered limpets and mussels 'of particularly good quality' to augment their stores of wild celery and cranberries. As April wore on, despite several squalls of snow, FitzRoy found the temperature better than expected, never dropping below 31°F. With members of the *Beagle*'s crew he made daily sallies to shore to botanise in the bays, take measurements and collect specimens. The crew shot wildfowl when they could and, best of all, a black swan, which was kept for a special occasion. FitzRoy revelled in the solitude. 'Though the season was so far advanced,' he wrote in early May, 'some shrubs were in flower, particularly one, which is very like a jessamine and has a sweet smell. Cranberries and berberis-berries were plentiful: I should have liked to pass some days at this place, it was so very pretty; the whole shore was like a shrubbery.'[19]

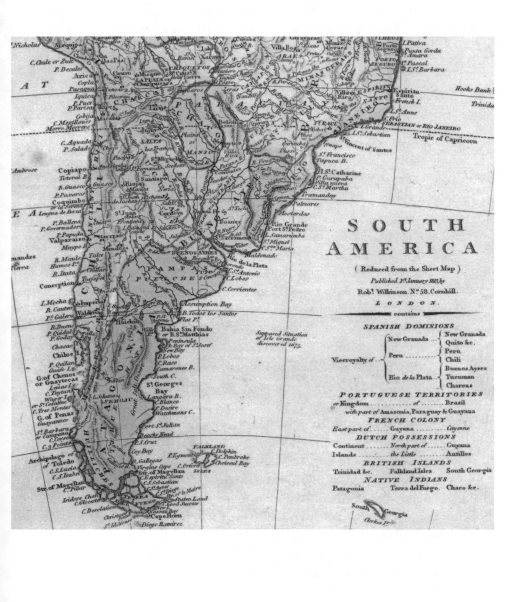

SOUTH AMERICA

(Reduced from the Sheet Map)

Published 1st January 1811 by

Robt. Wilkinson. No. 58. Cornhill.

LONDON.

contains

SPANISH DOMINIONS

Viceroyalty of	New Granada	New Granada
		Quito &c.
	Peru	Peru
		Chili
	Rio de la Plata	Buenos Ayres
		Tucuman
		Charcas

PORTUGUESE TERRITORIES

or Kingdom . . . . of . . . . Brasil

with part of Amazonia, Paraguay & Guayana

FRENCH COLONY

East part of . . . . . . Guyana . . . . Cayenne

DUTCH POSSESSIONS

Continent . . . . North part of . . . . . . Guyana

Islands . . . . the Little . . . . Antilles

BRITISH ISLANDS

Trinidad &c.   Falkland Isles   South Georgia

NATIVE INDIANS

Patagonia   Terra del Fuego.   Chaco &c.

On 7 May, FitzRoy took a group of men in the *Beagle*'s whaleboat and cutter and set out to explore a little-known waterway called the Jerome Channel. They took enough provisions with them for a month's surveying. FitzRoy captained the whaleboat, an open-topped vessel used under oars that was pointed at both ends to allow it to beach, and they glided through the cold, twisting waters 'with all the anxiety that one feels about a place, of which nothing is known, and much is imagined'. In midwinter at the edge of the known world, this was ultimate freedom. FitzRoy bathed in the shallow waters, stopping to measure the temperature (42°F).[20] He and his men scrambled up a nearby hill, called the Sugar Loaf, carrying barometers, theodolites and telescopes. 'It cost a struggle to get to the top with the instruments,' he admitted, 'but the view repaid me.' At the summit a swathe of Tierra del Fuego opened up before them. Far away from Croghan Hill in Ireland, FitzRoy experienced the same thrill of adventure as Beaufort had a quarter of a century before.[21]

It was now mid-May and FitzRoy was working in the depths of a South American winter. With the beauty came danger. The days were shortening. The sun did not rise above the hills till about eleven o'clock and it disappeared again shortly after two. The weather, too, was about to change. On the 17th 'a heavy squall of wind and hail passed over from the south west, so cuttingly cold', FitzRoy wrote, 'that it showed me one reason why these plains, swept by every wind from S.S.W. to N., are destitute of trees.' The next night it was again very cold. Rain fell heavily and a hard gale blew up from the south-west. At night on 19 May the skies were so clear that FitzRoy was able to gaze into the glittering heavens and he woke to a blue dawn. 'Everything was frozen.' The whaleboat was useless until its sails were thawed. Just after midday, FitzRoy ordered it to be rowed out in search of a channel between Otway Water and the Strait of Magellan. It was a fateful decision.

The breeze was already stiff and for two hours they pulled against a rising swell. FitzRoy had no choice but to continue. On either side the shore was unbroken, flat and low, with a great surf breaking against it. 'To have attempted to land, would have been folly.' A further hour passed. The whaleboat was now taking on water: the crew's bags and clothes were saturated. As the sun set at four o'clock the wind was as strong as ever. In the cold darkness, all dangers were multiplied. FitzRoy continued:

Night, and having hung on our oars five hours, made me think of beaching the boat to save the men; for in a sea so short and breaking, it was not likely she would live much longer. At any time in the afternoon, momentary neglect, allowing a wave to take her improperly, would have swamped us; and after dark it was worse. Shortly after bearing up, a heavy sea broke over my back, and half filled the boat: we were bailing away, expecting its successor, and had little thoughts of the boat living, when – quite suddenly – the sea fell, and soon after the wind became moderate. So extraordinary was the change, that the men, by one impulse, lay on their oars, and looked about to see what had happened. Probably we had passed the place where a tide was setting against the wind. I immediately put the boat's head towards the cove we left in the morning, and with thankful gladness the men pulled fast ahead.[22]

FitzRoy was not a man given to melodrama. His account conveys the vulnerability of their situation. That the whaleboat survived this encounter with the weather must be put down to fine seamanship, colossal effort and phenomenal luck. But it came as a warning. For days FitzRoy had been scribbling meteorological notes in his journal, recording his surprise at the 'mildness of the weather'. Now, in the space of a few hours, he had seen how swiftly conditions could change. If the Strait had a beauty, it was a restless, dangerous beauty. In minutes a blue sky could be overrun by powerful winds and squalls, thunder and dark clouds. The dilemma for commanders like FitzRoy was not *if* the weather would turn, it was *when*.

Any navigator who sailed through the narrows at the western opening of the Strait of Magellan was willingly entering this world. To pass them was to cross the Rubicon, and a test of physical and mental ability awaited. The silver river, the blue sky, the gentle breeze from the west, the liquid calm – all these could lull a navigator into complacency. And the speed with which the atmosphere could change was frightening. At midwinter FitzRoy would first encounter a williwaw, a burst of mountain wind that King had experienced a year before. To King these were 'hurricane squalls' that 'rush violently over the precipices, expand, as it were, and descending perpendicularly, destroy everything movable'.[23] Williwaws were a notorious trademark of Tierra del Fuego and, in particular, the Strait of Magellan where winds were tunnelled towards the river by the surrounding mountains. Less energetic than a pampero, a williwaw still had the potential to inflict great damage. King described how 'the surface of the water, when struck by these gusts, is so agitated, as to be covered with foam, which is taken up by them, and flies before their fury before dispersed in vapour'.[24]

Increasingly fascinated by the weather, FitzRoy started to analyse it with closer attention in his journal. On 18 May he wrote, 'For the last four nights I noticed, that soon after sunset the sky was suddenly overcast, a trifling shower fell, and afterwards the heavens became beautifully clear.'[25] Ten days later, he returned to this theme. 'Almost every night I observed that the wind subsided soon after sunset, the clouds passed away, and the first part of the night was very fine; but that towards morning, wind and clouds generally succeeded.' Here is FitzRoy at the start of his interest with weather. He emerges as half scientific observer and half weather-wise sailor, looking to expose telling patterns or templates for future reference.

At the end of May six weeks had passed since FitzRoy embarked on his orders. In his own manner he had established himself as a leader. After the whaleboat incident on 20 May he had noted, 'No men could have behaved better than the boat's crew: not a word was uttered by one of them; nor did an oar flag at any time, although they acknowledged, after landing, that they never expected to see the shore again.' Such an account can only have been left by a man with a natural ability to lead. FitzRoy had his own distinct charisma. He didn't inspire with bravado or buccaneering showmanship; he led by example. The men respected his ability, hard work and even-handedness. As the onshore explorations continued, he abandoned the comforts and privacy of his own tent and slept in the open with 'two or three of the men'. He noted in his journal, with an uplifting flourish, 'My cloak has been frozen hard over me every morning; Yet I have never slept so soundly, nor was in better health.'[26]

In this vast and distant wilderness the *Beagle* and her boats were tiny cells of scientific endeavour. Along with her anchors and cables, hawsers and kedges, she carried a full complement of philosophical instruments that allowed FitzRoy to continue with his endless round of measurements: sextants, quadrants and compasses for navigation, theodolites for triangulation, levels, gauges and balances for soundings and chronometers for calculating longitude. Onboard the *Beagle* were also a variety of meteorological tools: thermometers, hygrometers, a pluviometer for measuring rainfall, several different barometers and a sympiesometer, a lightweight and portable mercury-free barometer that FitzRoy favoured above all.

FitzRoy's education and instinct told him to quantify everything he discovered. The depth of a channel, the incline of a bank, the bearings of a point, the height of a hill, the temperature of the air, its humidity and pressure. Science outdoors had been made fashionable over the previous fifty years by Horace-Bénédict de Saussure, who had taken meteorological readings in the Alps, and the Prussian explorer Alexander von Humboldt, whose scientific exploits in Europe and Latin America, to FitzRoy's generation, were the stuff of legend. Now FitzRoy was taking up the mantle. Bartholomew Sulivan, one of his officers, observed FitzRoy's diligence. In his autobiography he later stated, '[FitzRoy] was one of the best practical seamen in the service, and possessed besides a fondness for every kind of observation useful in navigating a ship.'[27] Traces of his commitment to detail are evident in the *Beagle*'s log. Shortly after the pampero on 30 January FitzRoy made alterations to the standard tables. Unusually for a commander, he had always kept daily readings of the pressure and temperature, but after Friday 20 February, 'a squally day',[28] he started to take barometric readings every three hours.

Weather records were an important part of FitzRoy's daily observations. By the 1820s the meteorological tools – the thermometer, the barometer and hygrometer – were well established. But they still retained a fashionable allure: their strange alchemy distilled nature, stripping it down to its numerical core. Frosty mornings, clear nights and stormy afternoons were not just to be recalled by adjectives and metaphors. Far better was to describe the atmosphere with a series of numbers.

The thermometer was the most trusted instrument. In Britain, Fahrenheit's thermometers had already been in use for about a century. They were renowned for their accuracy and were used at sea to take daily measurements of the water and air. After Franklin's mapping of the Gulf Stream in 1770 thermometers were used to test whether a vessel was labouring to New York against the famous warm current – 'the River in the Ocean' – a small detail that could cost a ship two weeks at sea. The hygrometer was similarly prevalent, a device that enabled measurement of the levels of moisture in the air.

There were innovations too. In 1823 John Frederic Daniell, an experimental philosopher and businessman in London, had grown so fed up with the quality of hygrometers that he had developed a new model. The result was the Daniell Hygrometer. The improved instrument helped to predict frosts, dews and precipitation with accuracy. This hygrometer remained in use for decades and would be celebrated as a 'perfect and elegant instrument'. The *Beagle* and *Adventure* carried some of the first Daniell Hygrometers ever used outside Britain. On the *Adventure* readings were taken at 3 p.m. daily and entered into a meteorological journal.[29]

The barometer, though, was the most intriguing instrument of all. In his *Ample Instructions for the Barometer and Thermometer* (1825), Jeffery Dennis wrote, 'The barometer is probably the most useful, entertaining and interesting of all philosophical instruments.' It allowed him to ascertain the weight of the atmosphere, 'heights of the mountains, and the depths of caverns and mines'.[30] It performed a single, simple task: reading atmospheric pressure. But from this measurement a thousand possibilities sprang. Like no instrument before, it seemed to predict coming weather conditions. A plunging reading, like that FitzRoy noticed off Maldonado Bay in 1829, often presaged wind and rain. Equally, a high, settled measure would imply a spell of fine weather.

In the century and a half since Robert Boyle had introduced the barometer into England, no universal laws of prediction had been concluded from it. As a device it could be fickle and untrustworthy. Centuries of analysis had been ploughed into breaking the barometer's enigmatic code, all to no avail. Driven mad by the complexities John Frederic Daniell lamented in his *Meteorological Essays* (1823) that some had even gone so far as to abandon Newton's laws of physics in their desperation to solve the riddle. Daniell laughed at one of the latest theories: that the powerful force of a horizontal wind disturbed the downward force of air pressure – knocking it off its feet like an old man on a windy pier.

Spotting an opportunity, some barometer manufacturers had begun supplying instruments with lists of hints for the user: code books that helped decode atmospheric signs. Like a forerunner of the twentieth-century instruction manual these books encouraged the user to tab-

ulate a set of signals with their specific geographic position. Dennis'
book included a detailed section on using barometers at sea:

> In winter, spring and autumn, the sudden falling of the mercury,
> say 3-10ths of an inch, always denotes high winds and storms,
> but in summer it presages heavy showers and thunder; it invar-
> iably sinks lowest of all when great winds prevail, though not
> accompanied with rain: it always falls more for wind and rain
> combined than either of them separately. Also, if after strong
> winds and rain together, the winds should change in any part
> of the northern or southern hemisphere, accompanied with a
> clear and dry sky, and the mercury rise at the same time, it is
> a certain indication of fine weather.[31]

Taking accurate measurements was a vital part of being a good
sailor. For all his courage and success Weddell was criticised for his
slapdash approach to record-keeping. Ten days before reaching the
southernmost point of his Antarctic odyssey he broke his thermometer
and, having no replacement, gave up on his measurements, a fact he
later regretted: 'I was well aware that the making of scientific observations in this unfrequented part of the globe was a very desirable
object, and consequently the more lamented my not being well
supplied with the instruments with which ships fitted out for discovery
are generally provided.'[32] It was an important point. Voyages could
cost thousands of pounds and often involved years of work. The loss
of data in a shipwreck was a constant fear. While wintering at the
Falkland Islands in 1823 Weddell had met the French scientific adven-
turer Commodore Freycinet whose ship had been wrecked on a
submerged rock in the bay. Freycinet was returning home after a
'voyage of science almost round the world, and after having spent
nearly three years'. Though Freycinet managed to save his men and
most of his papers and specimens, the disaster was a warning for
others.

There were other signs of nature that were worth studying. In 1827
Thomas Forster published a *Pocket Encyclopaedia* crammed with natural
weather signs. According to Forster, rain was foreshadowed by aches
and pains in the human body, or toothache; ants bustling over their
anthills carrying eggs; asses braying in the fields, cattle gambolling
and candles flaring. Fine weather was coming when larks flew high.

Thunderstorms were presaged by milk suddenly turning sour. An east wind gave the nervous 'headaches and hurrying dreams', while, most vividly of all, Forster pointed out, 'when there is a piece of blue sky seen in a rainy day big enough, as the proverb says, "to make a Dutchman a pair of breeches", we shall probably have a fine afternoon'.[33]

Too late to make Forster's selection was another sprightly fact, reported in 1829 in the *Quarterly Journal of Science, Literature and Art*:

> At Schwitzengen, in the post house, we witnessed for the first time what we have since seen frequently, an amusing application of zoological knowledge, for the purpose of prognosticating the weather. Two frogs, of the species *Rana arborea*, are kept in a glass jar about eighteen inches in height, and six inches in diameter, with the depth of three or four inches of water at the bottom, and a small ladder reaching to the top of the jar. On the approach of dry weather the frogs mount the ladder, but when wet weather is expected, they descend into the water. These animals are of a bright green, and in their wild state, here climb the trees in search of insects, and make a peculiar singing noise before rain.[34]

Amusing to some, to others it was an embarrassment that a frog in a jar or a Herefordshire bull in a farmer's field knew a storm was coming before the urbane men of science in a townhouse, with their instruments at the ready. This was not the only provocative fact. De Saussure wrote that 'it is humiliating to those who have been much occupied in cultivating the Science of Meteorology, to see an agriculturist or a waterman, who has neither instruments nor theory, foretell the future changes of the weather many days before they happen, with a precision, which the Philosopher, aided by all the resources of Science, would be unable to attain.'[35]

As if to confirm the enduring attraction of weather-wisdom, in 1827 Hurst and Chance, a London publisher, had released a new edition of an old book with a tumbling title, John Claridge's *The Shepherd of Banbury's Rules – to judge the Changes of the Weather, Grounded on Forty Years' Experience; By which you may know The Weather for several Days to come, and in some Cases for Months*. First published in 1670 for two centuries the book had been a popular reference work for weather prognostication. In the 1740s an artful introduction had been added, supposedly by a man called John Campbell, that sought to justify its place alongside scientific treatises.

> The Shepherd whose sole Business it is to observe what has a Reference to the Flock under his Care, who spends all his Days and many of his Nights in the open Air, and under the wide spread Canopy of Heaven, is in a Manner obliged to take particular Notice of the Alterations of the Weather, and when once he comes to take a Pleasure in making such Observations, it is amazing how great a Progress he makes in them, and to how great a Certainty at last he arrives by mere dint of comparing Signs and Events, and correcting one Remark by another. Every thing in Time becomes to him a Sort of Weather-Gage. The Sun, the Moon, the Stars, the Clouds, the Winds, the Mists, the Trees, the Flowers, the Herbs, and almost every Animal with which he is acquainted. All these I say become to such a person Instruments of real Knowledge.[36]

This knowledge was presented in about thirty weather maxims of different layers of complexity, all derived from years of experience out in the Oxfordshire pastures.

> If the Sun rise red and fiery – Wind and Rain
> Clouds like Rocks and Towers – Great Showers
> Clouds Small and round, like a Dapple-grey, with a North-Wind
>   – Fair Weather for 2 or 3 Days
> Mists. If they rise in low Ground and soon vanish – Fair Weather

Campbell argued that it was wrong to create a division between scientific research and the wisdom of shepherds like Claridge. 'Men who derive their knowledge entirely from Experience are apt to despise what they call Book Learning, and Men of great Reading are apt to fall into a less excusable mistake, that of taking the Knowledge of Words for the Knowledge of Things.' It was a fair point and one that crystallises the contrasting figures of the philosopher in his study and the watcher – the sailor, the shepherd, the ploughman – observing the world at first hand. On HMS *Beagle*, FitzRoy was a mixture of these archetypes. He had learnt at Rottingdean, Harrow and Portsmouth and he had studied on the world's wildest oceans. He was a philosopher but at the same time he was weather wise, with the skill to draw conclusions from the world around him.

By July FitzRoy had completed his survey of the Strait. He sailed west into the Pacific and then to the north up the craggy South American coastline to rendezvous with Captain King at San Carlos on the island of Chiloé. At Chiloé the *Beagle* was refitted and replenished and a replacement boat was built.

FitzRoy received instructions from King on 18 November 1829. He was told to chart the southern coastline of Tierra del Fuego, starting at the western opening of the Strait of Magellan, skirting the bottom of South America, rounding Cape Horn, and returning up the eastern coast to Montevideo. King told FitzRoy he planned to be in Rio on 1 June the following year. Until then, for almost seven months, FitzRoy was on his own. The next day the *Beagle* sailed.

It was a challenging task. The wind-ravaged western coast of Tierra del Fuego brought dangers even more pronounced than those in the Strait of Magellan. In the Strait, at least, he had remained somewhat sheltered from the ocean winds, but on this jagged shoreline gusts of cold Pacific air blasted ships day and night. It was the weather of this part of South America that had driven FitzRoy's predecessor as captain of the *Beagle*, Pringle Stokes, to despair.

Stokes, a good man and talented navigator, had been tormented by this wild world in 1828. Little by little he had been worn down, physically and emotionally. Stokes had charted his downfall in his journal, where he described the Strait of Magellan and the west coast of Tierra del Fuego with Gothic intensity. He saw beaches littered with whale skeletons, albatrosses wheeling in the air, shorelines 'lashed by the awful surf of a boundless ocean, impelled by almost unceasing western winds'. For weeks and months Stokes had carried stoically on, drawing on all his mental reserves. But there was no escape from the weather. Inside him, the shadows lengthened. He came to dread the morning blizzards, the sleet and hail that accumulated in dense layers, covering the *Beagle* in a crust of ice 'about the thickness of a dollar'.[37]

In early June 1828 Stokes was surveying the Gulf of Peñas – the Gulf of Distress. 'Nothing could be more dreary than the scene around us,' he wrote. 'The atmosphere seemed to have assumed a horrid, mocking tangibility. 'The lofty, bleak and barren heights, that surround the inhospitable shores of this inlet, were covered, even low down their sides, with dense clouds, upon which the fierce squalls that assailed us beat . . . They seemed as immovable as the mountains where they rested.'[38]

It was a place, Stokes concluded, where 'the soul of man dies in him'. Sinking under the strain, he despaired, for his crew were beset by pulmonary complaints, hacking coughs and wheezing chests. By midwinter Stokes had locked himself in his cabin and was refusing to come out. Six weeks later, back at Port Famine, he pulled out a pocket pistol and shot himself in the head. He died eleven days later.

Captain King had written:

Thus shockingly and prematurely perished an active, intelligent and most energetic officer, in the prime of life. The severe hardships of the cruize, the dreadful weather experienced, and the dangerous situations in which they were so constantly exposed – caused, as I was afterwards informed, such intense anxiety in his excitable mind . . .[39]

As Stokes' replacement, FitzRoy was to face the same challenges, the same hostile landscape and weather. It was not long before he encountered them. At Cape Pillar – the western entrance of the Strait of Magellan – FitzRoy experienced 'gloomy days, with much wind and rain' and gusts that blew violently from the mountains.[40] The coast, he decided, had 'a dangerous character'. Christmas passed, with the ship sheltered from the north-westerly winds in a weather-beaten cove he called Latitude Bay. These early experiences set a template for the months that followed, with FitzRoy using all his skill to navigate gales that constantly drove them towards the shoreline. Only on clear days could he take measurements or organise boat trips to the land. For the most time the crew were marooned on the *Beagle*, embroiled in skirmishes with thieving native Fuegian tribes and left to gaze at the 'multitudes of penguins' and swallow-like birds that skimmed and twisted over the ocean.

The bad weather continued into March – 'rainy and blowing'. By now the *Beagle* had reached the southern tip of the continent. They had seen the striking outline of York Minster – 'a black irregular shaped rocky cliff, eight hundred feet in height', rising like an incisor almost perpendicularly from the sea – named by Cook, on his circumnavigation fifty years earlier, after the cathedral in his home county. At the end of the month they were passing the islands at the far south of the continent, winds still blowing cold and fresh. 'From the season, the appearance of the sympiesometer, and the appearance of the weather,' FitzRoy noted on 25 March, 'I do not expect any favourable change until about the end of the month.'[41]

Perhaps FitzRoy had gained a feel for the atmosphere or had noticed a trend in his meteorological journal, because his hunch turned out to be right. At the end of March the weather did improve as FitzRoy steered the *Beagle* into Orange Bay – 'a large, roomy place, with an even bottom'. After almost five months at sea many of the men had colds and rheumatic pains and FitzRoy decided the time was right to rest. A quiet spell, he reasoned, would be not just timely but prudent, as his barometer and sympiesometer were giving unusually low readings. With his experiences of the Rio de la Plata pampero still fresh in mind, the possibility of navigating into another storm was too much to risk.

Yet, oddly, no storm came. Despite the readings the weather was now as settled as at any time in the last few months. The pleasant spell continued the next day, and the next, in complete variance with his pressure readings. By 5 April FitzRoy was mulling the contradiction openly: 'Two more days with a very low glass, shook my faith in the certainty of the barometer and sympiesometer.' They were now giving unusually low readings of 28.94" and 28.54".[42]

This peculiar weather endured until the middle of the month. Attempts to sail were frustrated by a lack of wind, as laughable a problem at Cape Horn as a want of sand in the Sahara. Finally on 17 April FitzRoy managed to guide the *Beagle* into open sea, passing the little outcrop of rocks that marked the extreme end of the South American continent, rounding Cape Spencer in so much fog that he mistook it for Cape Horn itself, the lower part of the rock looking, he noted, like 'the head of a double horned rhinoceros'.

Anchored at the foot of the continent and the weather staying fair, FitzRoy decided to organise a bold boat expedition to Horn Island. Sir Francis Drake had claimed to have landed there on his circumnavigation in 1577, though this may have been a hollow boast. Certainly few had set foot on the island in the years since. On 18 April FitzRoy carried out a scouting trip. 'Many places were found where a boat might be hauled ashore,' he discovered. It seemed possible that instruments might be carried to the summit. By the morning of the 19th FitzRoy had organised a party to visit the island to complete a set of readings for the survey. At noon the next day they set off. They carried five days of provisions, a chronometer for measuring longitude and a range of meteorological instruments. They landed before dark, hauled their boat to safety on the north-east side of the island and, as FitzRoy added with a note of triumph, 'established ourselves for the night on Horn Island'.[43]

In the quiet of that night FitzRoy and his men looked out across a stretch of water that was infamous among sailors. It was here, at 56 degrees south, that the Pacific and Atlantic Oceans crashed into one another with tremendous force. At Cape Horn powerful westerly winds raged uninterrupted by land around the foot of the globe, combining with gales that blew south off the Andes. To aggravate matters the wind agitated the shallow waters of the Cape, whipping up a short sea that was particularly venomous to shipping. In a storm the Cape was transformed into a monochromatic nightmare of spray and motion. To navigate it from the east was to confront a hundred dangers: waves, rocks and winds that could rip sails in two. On his voyage to the Antarctic James Weddell had met an American commander who had attempted the route in 1814. He cautioned Weddell, 'Indeed our sufferings, short as has been our passage, have been so great, that I would advise those, bound into the Pacific, never to attempt the passage of Cape Horn, if they can get there by any other route.'[44]

Weddell had declared the middle of February to the end of May to be the worst time to attempt the Cape. Yet now, on 19 April, FitzRoy was camped out on Horn Island in comparative serenity. It was eerily still. At daybreak his party set off for the peak in fine, bright weather. As the sun reached its meridian he stopped to take a series of angles. Then they pushed on, reaching the summit shortly after. Here FitzRoy gazed across the infamous water that lay quiet like a becalmed beast. In the distance he could see as far as the Diego Ramírez Islands, sixty-five miles away. At noon he took a further set of measurements: 'A round of angles, compass bearings for the variation, and good afternoon sights for the time completed our success.' FitzRoy then turned his attention to a commemoration of their visit, a tower of stones eight feet high. Once it was built the men crowded around a Union

Jack, drank the health of King George IV and gave three hearty cheers.

For FitzRoy it was a triumph. With courage and seamanship he had transported the *Beagle* and his men to the tip of South America. Here, on the wind-blasted surface of Cape Horn, he had been able to conduct observations with the ease and accuracy of a Cambridge don in his study.

FitzRoy spent the following days back on the mainland, attempting to climb a nearby hill called Cater Peak. On 25 April he scrambled to the summit but 'found so thick a haze, that no distant object could be seen'. With his instruments at his side FitzRoy appears in the imagination as the real-life embodiment of Caspar David Friedrich's *Wanderer Above the Sea of Fog*.

Friedrich's 1818 oil composition shows an enlightened man on the summit of a rocky precipice, gazing into a landscape wreathed in fog. It shows the ascent of man. The figure is not fatigued by his climb, dwarfed or awed by the scene before him. He is in control. Transport Friedrich's wanderer from the Alps to the southern coast of Tierra del Fuego. Turn the triangular snow-capped peak in the distance into Mount Sarmiento. Transform the walking pole, which the wanderer grips, into a sympiesometer. And then here is no anonymous *wanderer*, but Robert FitzRoy on 25 April 1830, standing defiantly at the end of the world.

Eighteen months later Captain Francis Beaufort was studying FitzRoy's charts of Tierra del Fuego at the Admiralty in Whitehall. The Portland stone, clattering horses and coaches, politicians, bustling clerks and newspaper boys of London were completely at odds with the distant world Beaufort scrutinised in his office, drawn neatly out on an Admiralty chart.

By now, Beaufort was a respected figure at the apex of government and two years into his role as Hydrographer to the Royal Navy. He was respected and well-connected. He counted as friends the Arctic explorers John Franklin, George Francis Lyon and James Clark Ross, oceanographer James Rennell, mathematician Charles Babbage, and engineer Davies Gilbert. Over the last few months the Royal Society had invited him to sit on a steering committee to renew their charter. He had also served on the council of the Royal Astronomical Society for the last seven years. He was a founding member, alongside such talents as Humphry Davy and

J. M. W. Turner, of the ultra-fashionable Athenaeum Club in Pall Mall, a haven for the scientific, literary and artistic elite. At last he had become the man he hoped to be.

Beaufort's rise through the social and professional ranks had begun with the publication of his Asia Minor charts. Drawn with a fastidious commitment to accuracy, all twelve of them, supplemented by twenty-four plans and twenty-six views, were lauded as gems of draughtsmanship. To this success in 1817 Beaufort added a descriptive narrative account of the voyage entitled *Karamania*. A richly imagined blend of travelogue, geological and archaeological detail, it appealed to a Regency elite with a taste for antiquities. On its publication he sent a copy to Edgeworthstown for the verdict of his old friend. On 17 May 1817 Edgeworth replied.

> My dear Francis,
> How great would have been my mortification, had I been disappointed with *Karamania*. Had I been obliged to blame, or have been silent – On the contrary I have read it through, with pleasure & with much care . . . I think the book is written in a good & appropriate style, free from exaggeration, free from false ornaments & from pretension of any sort.[45]

From Edgeworth, who had mentored his daughter Maria's literary works for many years and critiqued the verse of Erasmus Darwin, this was true praise. It was also a parting note. Less than a month later, on 13 June, Edgeworth died at his family home. The news soon reached Beaufort, who was left to mourn 'my warmest and most anxious friend'. To his sister Fanny, now Edgeworth's widow, he confessed, 'Whatever improvement I have made in my mind must be ascribed to the impact he gave it. It was he who taught me that true education begins but with the resolution to improve, and it was he alone of all my friends who tried to wind me up to that resolution and sustain it.'[46]

Karamania had made Beaufort. Sir John Barrow would later describe it as 'a book superior to any of its kind in whatever language, and one which passed triumphantly through the ordeal of criticism in every nation of Europe'.[47] Thereafter Beaufort had advanced into circles of ever-expanding influence.

Full recognition, though, had come slowly. He had to wait until

May 1829 before Lord Melville offered him the job of hydrographer. He was awarded £500 a year and an official address at Somerset Place on the banks of the Thames. On 12 May he had written, 'Took possession of my new Hydrographer's room. May it be a new era of industrious and zealous efforts to do my duty with sincerity, impartiality and suavity . . . not for worldly motives but from a sense of the far higher duty I owe to that Providence who placed me in this vocation.'[48]

Beaufort's Hydrographic Department would become the nineteenth-century equivalent of NASA. On behalf of the wealthiest and most powerful nation on earth he was conducting explorations at the very edge of human knowledge. But instead of looking out to space, Beaufort organised voyages to vibrant, tangible worlds teeming with life. With great personal authority, from his suite of offices at Charing Cross Beaufort was conducting an exploration of the world.

From his first day in post he injected a sense of purpose into the Hydrographic Department. Harriet Martineau, the writer and journalist, later described how he transformed the office from a forgotten map depot, a 'small, cheerless, out of the way' place, into a hotbed of ideas and enterprise. Martineau noted Beaufort's 'miraculous' stamina. 'Day by day for a quarter of a century he might be seen entering the Admiralty as the clock struck: and for eight hours he worked in a way which few men ever understand.'[49] Out of principle he brought his own writing-paper and pens for private correspondence. In his spare time he was equally productive. Rising at five o'clock each morning before his official work began, Beaufort would spend an hour or more working without payment for the Society for the Diffusion of Useful Knowledge. His vision was to create a series of affordable and high-quality maps for the general public and day after day, year after year, he strove to achieve this ambition. In time, Beaufort's SDUK charts would become among the most widely circulated of their day with more than one hundred available for sixpence each.

Beaufort met FitzRoy on his return from Tierra del Fuego in the autumn of 1830. There was an instant bond between the men. To FitzRoy, the hydrographer was able and proven, a contemporary of Nelson and one of the dwindling few to have served at the Glorious First of June. He also had a glamorous edge. His picaresque career seemed like something out of a Smollett novel to FitzRoy. Beaufort himself admitted that the memory of his naval career was like a 'sort of favourite romance'.[50]

In FitzRoy Beaufort saw ability and potential. He was impressed by FitzRoy's surveying work. The South American survey 'will be acknowledged to be one of those which has eminently contributed to the credit of the country and to that of the officers employed in it', he wrote to his superiors at the Admiralty. Soon there was talk of a new scheme. Beaufort knew FitzRoy was eager to sail to Tierra del Fuego to return three native Fuegians he had captured in skirmishes with thieving tribes. He had envisaged this as a private matter, but after conversations with Beaufort the project had evolved. Over the summer, plans for a new South American survey emerged.

Suddenly a voyage of six months had become a circumnavigation of the globe, an enterprise of several years. Not wanting to be intellectually isolated for such a long time he had asked Beaufort to find him a gentleman companion. The obvious candidate to accompany FitzRoy was a young naturalist who could assume a role similar to that of Joseph Banks on Cook's *Endeavour* voyage. Beaufort wrote to a friend, Professor Henslow at Cambridge University, offering him the opportunity. Too busy, Henslow had passed the letter on to George Peacock, a fellow don. From Peacock news of Beaufort's search extended to a recently graduated theology student and talented botanist called Charles Darwin. At just twenty-two Darwin was full of potential. Grandson of the famous Erasmus, Edgeworth's good friend, Charles had a reputation for his 'rejoicing enthusiasm' and his love of collecting beetles. 'Entomology, riding, shooting in the Fens, suppers and card-playing, music at Kings' had been the core of Darwin's life for the past three years. Though a very different character to the sailor, he seemed a potential match for Robert FitzRoy.[51]

News of Beaufort's offer reached Darwin in late August and after a period of indecision he accepted. On 1 September 1831, Darwin wrote to Beaufort in London: 'If the appointment is not already filled up, – I shall be very happy to have the honor of accepting it.' The news was relayed from Beaufort to FitzRoy:

I believe my friend Mr Peacock of Trin^y College Camb^e has succeeded in getting a 'Savant' for you – A Mr Darwin grandson of the well known philosopher and poet – full of zeal and enterprize and having contemplated a voyage on his own account to S. America. Let me know how you like the idea that I may go or recede in time.[52]

The spate of intellectual matchmaking concluded with a dinner between FitzRoy and Darwin in London. It went well. 'All was soon arranged,' Darwin later wrote. Although, 'afterwards, on becoming very intimate with Fitz-Roy, I heard that I had run a very narrow risk of being rejected, on account of the shape of my nose! [FitzRoy] was an ardent disciple of Lavater, and was convinced that he could judge of a man's character by the outline of his features; and doubted whether any one with my nose could possess sufficient energy and determination for the voyage. But I think he was afterwards well satisfied that my nose had spoken falsely.'

Darwin would later laugh at how the course of his life had hung, for a moment, on 'such a trifle as the shape of my nose'. He would also recall his first impressions of FitzRoy:

Fitz-Roy's character was a singular one, with his many noble features: he was devoted to his duty, generous to a fault, bold, determined, and indomitably energetic, and an ardent friend to all under his sway. He would undertake any sort of trouble to assist those whom he thought deserved assistance.[53]

By 24 October Darwin had joined FitzRoy at Plymouth for final preparations as Beaufort wrote up the hydrographic orders in London. Beaufort's instructions were notoriously detailed. They reflected his microscopic knowledge of distant shorelines, waterways and currents. He wrote his orders like a benevolent headmaster encouraging his surveyors on to greater feats: to sketch views of the coasts, bays and anchorages, to take soundings of the shallows and the deep, list prevailing winds and fix longitudes in disputed areas. On 15 November his orders arrived in Plymouth. FitzRoy was to establish the longitude of Rio de Janeiro beyond doubt, fill in geographic gaps south of Rio de la Plata and especially around Tierra del Fuego and the Falkland Islands.

Something else had caught Beaufort's attention as well. King, FitzRoy and Stokes' journals had been filled with descriptions of wind: 'half a gale', 'a furious gale'. What did these mean? How fast did the winds blow during a pampero or a williwaw? From his own visits to

South America Beaufort was familiar with pamperos, and now he felt the time was ripe to test his long-treasured weather system on a grander scale. For years he had used it in his own pocketbooks, but he had never advocated its wider use until now.

Beaufort asked FitzRoy to keep a careful meteorological register, with twice-daily barometric and temperature readings:

> In this register the state of the wind and weather will, of course, be inserted; but some intelligible scale should be assumed, to indicate the force of the former, instead of the ambiguous terms 'fresh', 'moderate' &c, in using which no two people agree; and some concise method should also be employed for expressing the state of the weather.[54]

To the end of his instructions Beaufort attached his wind scale, barely altered from the one that he had jotted down in his journal twenty-four years before. It involved four escalating strengths of wind from 0 – *calm* to 12 – *a hurricane*. To help FitzRoy distinguish between the levels Beaufort added a quantifiable guide. Level 2 was equal to 1–2 knots. Level 6 was reached when single-reefed topsails and topgallant sails were flown. Level 12 (a hurricane) was a force 'which no canvas could withstand'.

In December the *Beagle* was ready to sail. 'Every thing is on board & we only wait for the present wind to cease & we shall then sail. —This morning it blew a very heavy gale from that unlucky point SW,' wrote Darwin on 7 December. The storms continued for five days more. Then on 10 December the skies brightened. 'Accordingly at 9 oclock we weighed our anchors, & a little after 10 sailed.' All went well until they reached the breakwater, where his misery began.

> I was soon made rather sick, & remained in that state till evening, when, after having received notice from the Barometer, a heavy gale came on from SW. The sea run very high & the vessel pitched bows under. —I suffered most dreadfully; such a night I never passed, on every side nothing but misery; such a whistling of the wind & roar of the sea, the hoarse screams of the officers & shouts of the men, made a concert that I shall not soon forget.[55]

In the face of the winds FitzRoy turned the *Beagle* back to Plymouth to wait for better luck. The boisterous weather continued for the next weeks, gale upon gale, storm upon storm. Pent up in Plymouth FitzRoy could do nothing but wait, since in 1831 there was still no understanding of how or why storms came. For so long a mystery to the scientific community, storms were about to be analysed like never before.

Morning

A blue dawn has progressed into a bright morning. A few hours into the day and the night's dew has evaporated into the atmosphere, making the air damp. As the sun climbs the ground begins to warm. Energy is radiated into the air above it and soon a thermal column is rising.

Although it is not immediately obvious, the laws of physics dictate that warm damp air like this will always rise upwards towards cooler, drier altitudes in the atmosphere. This is because atmospheric air – which is mostly formed of nitrogen and oxygen atoms – becomes lighter when water vapour molecules are mixed with it. As water vapour (H_2O) is chiefly comprised of hydrogen, the lightest of all elements, it means that the damper a parcel of air, the lighter it will be and the faster it will rise. This is a fundamental principle of atmospheric science. Another is that warm air is less dense than cold air and will also rise. Anyone can observe this in their own home, opening the bathroom door after a shower and feeling the warm air escape upwards while cold air rushes in below.

On days like today strong thermals can form. From the meadow warm damp air travels upwards at surprising speeds of up to two metres a second. Invisible to us, the thermal is located by a lone hawk, who opens his wings and, without effort, soars upwards.

The thermal rises over the meadow towards cooler air. The temperature drops at around 3.0°C per 1,000 feet, what is known as the dry adiabatic lapse rate. Soon the air temperature reaches a crucial juncture: the dew point. When it does water vapour begins to condense.

The process is taking place in a microscopic world. Specks of condensed water settle on minute condensation nuclei that are often no more than one ten-thousandth of a millimetre across: tiny flecks of salt, dust particles, compounds of ammonium nitrate. All have the potential to become the core of a cloud droplet. It can take billions of collisions for water vapour molecules to stick to a condensation nucleus; often they bounce off. But little by little the particles accumulate. Soon a droplet, a hundredth of a millimetre in diameter, has formed. More join. As much as 100 million of these droplets cram into a cubic metre of air. It is the beginning of a cumulus cloud. On a fine day like this, with a steady breeze, the thermals generate a production line of cumuli. They form at a fixed altitude, their flat bases tracing a dew- point line across the sky.

The cloud has a short life ahead of it. Perhaps five minutes, perhaps half an hour. The droplets of water are heavy enough to fall – the average cumulus cloud is one kilometre cubed in size and, if gathered together, would weigh more than 500 tonnes, as much as 100 African elephants – but they are kept in a state of delicate stasis by wind resistance and updraughts of rising air. Instead the cloud glides horizontally through the atmosphere on the breeze, throwing shadows over the landscape below.

PART TWO

Contesting

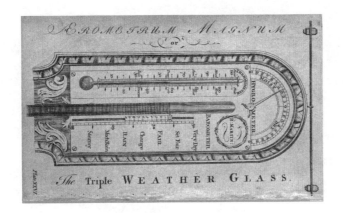

CHAPTER 4

Detectives

As the *Beagle* turned back towards Plymouth on 10 December 1831 another ship, carrying Captain William Reid of the Royal Engineers, was battling west through the sort of weather that had earned the English Channel its nickname as the sea of sore heads and sore hearts. Like FitzRoy and Darwin, Reid was sailing on government orders. He had left London two days earlier, bound for the Caribbean. His task was more urgent than that of the *Beagle*. He was travelling in response to news of an enormous West Indian hurricane, reports of which had been filling the papers for weeks.

It was four months since the hurricane had struck on 10 August but it had taken that long for the enormity of the crisis to be absorbed in Westminster. News had arrived piecemeal. A French vessel, the *Martial*, had docked in Le Havre on 15 September carrying initial reports and more details had arrived with the British army schooner *Duke of York* a day later. Both captains reported that a fierce six-hour gale had destroyed many plantations in Barbados, a strategically significant colony at the eastern tip of the Caribbean. Other accounts followed. One from the US Consul in Martinique stated bluntly, 'This island, I much fear is ruined, and it will be impossible for it to recover.'[1]

The worst fears were confirmed shortly after, with the syndication of an emotional article by the editor of the *Barbados Globe*. Still in shock, he had nonetheless managed to set down, in jittery prose, his experience of what he called a 'dreadful calamity'. On the evening of 10 August the wind had been blustery, he wrote, banks of clouds had amassed on the horizon suspending Barbados in gloom. At midnight a squall had blown up. A smart breeze and heavy showers had swept over the bays. Then for two hours a tremendous gale had blown, which by three in the morning had increased to 'a perfect hurricane'.

Now the work of destruction commenced. From this hour until five, it raged with unparalleled violence, whilst the lightning would at intervals cast a momentary but horrid glance on the mangled objects around. The houses were either levelled to the earth or uproofed – the largest trees torn from their roots, or broken as reeds. Numerous individuals were buried under their ruins, or exposed without a shelter to the peltings of the storm, and threatened with instant death at each successive blast that hurled the shattered fragments in every direction.[2]

When morning came, it revealed a scene of total destruction. The capital, Bridgetown, had barely a house left standing. Every vessel in Carlisle Bay, south of the city, had been wrenched from its anchorage and driven up the beach. St Anne's Barracks, the seat of government and the boys' and the girls' schools were all 'one mass of ruin', while out in the grasslands, banana and breadfruit trees, fields of sugar cane and corn were flattened. 'In the country the whole face of nature has been completely changed,' wrote one witness. Another estimated that four or five thousand merchants, soldiers, mariners, masters and slaves were dead. Bodies were piled in the streets.

News had reached Whitehall in a letter to Viscount Goderich, Secretary of State for War and the Colonies, from Sir James Lyon, Governor of Barbados. 'On the evening of the 10th the sun set on a landscape of the greatest beauty and fertility and rose on the following morning over an utter desolation and waste.'[3] Lyon and his family had only survived by sheltering in his cellar as their home crumbled about them.

For Lord Grey's Whig government this was a disturbing development. Barbados was a valued component of the empire. It had been built up over the centuries into a hugely profitable source of sugar. The hurricane presented the government with an economic and humanitarian crisis. In months to come a bill would be raised in the Commons by Viscount Althorp, Chancellor of the Exchequer, allocating £100,000 for rebuilding and compensation costs. Althorp argued successfully that it was 'incumbent on the mother country to give assistance to the distressed colonists'.[4] But it would take months if not years for this financial aid to materialise and as an immediate measure the army were asked to allocate a brigade to oversee the reconstruction of governmental buildings. They had turned to William Reid of the Royal Engineers.

Reid was a hardened, intelligent Scotsman. Born at Fife in 1791 the eldest son of a church minister and grandson of Thomas Fryers, the chief engineer of Scotland, he had a quarter of a century's military service behind him as a surveyor for the Ordnance Survey and Royal Engineers in the Napoleonic Wars. When reports of the hurricane arrived in the autumn of 1831 Reid was summoned to Whitehall and ordered to oversee the rebuilding process. The government could not have chosen a better man.

The Caribbean hurricane was not without precedent. Violent weather was a characteristic of the western Atlantic during the late summer and autumn months. This 'Hurricane Season' was recognised by ships' captains, who usually stayed in port between 1 August and 22 October, and underwriters, who raised their premiums during that period. Bred in the hot, moist air of what Georgian geographers called the 'torrid zone' – the belt that stretched from the equator into the tropics – hurricanes were known to have a power far beyond that of a European storm. Like the venom of a snake or the terror of yellow fever, a hurricane was an example of the additional potency of nature in these sweltering lands.

For those Europeans who had never felt the power of a West Indian storm *The Seaman's Practical Guide for Barbados* (1832) explained:

> In Europe we sometimes use the word hurricane, to denote a storm of uncommon violence, but we must not imagine, that a hurricane in Europe resembles a hurricane in the West Indies. The most furious tempest experienced here is a calm, when compared to those terrific elementary conflicts. He who has not felt them, can scarcely conceive the awful scene, much less describe it.[5]

To sail in the Caribbean in Hurricane Season was a perilous practice. Christopher Columbus was the first European to taste a true West Indian storm in 1494, and in the years since, many a ship had been mauled by their might. In his patriotic *History of Barbados* (1808) John Poyer singled out the two most devastating hurricanes. One in August 1675 ploughed over the plantations sparing 'neither the palace, nor the cot'. Poyer included the memorable tale of a newly married officer, 'Major Streate'. 'The pitiless storm, regardless of the sanctity of the marriage bed, blew them from their bridal chamber,' Poyer explained, 'and with relentless fury lodged them in a pimploe hedge. In this bed of thorns they were found next morning, incapable of manifesting those tender affections which their new-formed relation demanded, or affording each other the assistance which their comfortless position required.'[6]

But there was no leeway for light-hearted asides when Poyer recalled the Great Hurricane of 1780. It had descended on the island on Tuesday 10 October. Poyer described it in the grandest terms as a 'violence unparalleled in the history of the world'.[7] He had fled his house with his baby daughter in his arms. He saw coconut trees snapped in two and cattle blown off their feet. Later he heard that at Bridgetown a cannon 'of twelve pound ball' was blown cartwheeling from the pier-head to the wharf at the far side.

The Barbados Hurricane of 1780 left more than 4,000 dead. Now, fifty years on, the editor of the *Barbados Globe* suggested that this disaster was worse.[8] As Reid sailed through the Channel he must have steeled himself. The flattened houses, wrecked boats or tales of human suffering he might have expected, but little did he know that once in Barbados his attention would shift. Reid would start to study the hurricane: How did it form? Where did it come from? How did it move? How fast did it travel? Where did it go? Was there a formula that might be repeated?

Only a few had tackled such questions scientifically before. Long into the nineteenth century the vast majority believed that storms were the work of God. Sir James Lyon had already taken the customary step in Barbados, setting 7 October aside as a day of solemn humiliation and thanksgiving to God, who 'in the midst of Judgment was pleased to remember Mercy, and to stay the fury of the Hurricane'. It was a natural response. For Lyon, as for so many others, hurricanes lay beyond the bounds of science. At a public meeting in 1838, William Reid himself would affirm his personal conviction about 'the operations of the Deity' in the 'fixed laws' of nature: 'designed by incomprehensible wisdom, arranged by supreme power, and tending to the most benevolent ends'.[9]

Religious dogma had stifled meteorology for centuries. The weather was a potent symbol of divine vengeance and mercy, and central to the Creation story: the antediluvian paradise of Eden, the Flood, the covenant of the first rainbow. Storms were the ultimate example of God's might. As Psalm 29 asserted, God did not just direct the weather, he *was* the weather.

> The voice of the Lord is upon the waters: the God of glory thundereth: the Lord is upon many waters.
> The voice of the Lord is powerful; the voice of the Lord is full of majesty.
> The voice of the Lord breaketh the cedars; yea, the Lord breaketh the cedars of Lebanon.[10]

For more than a millennium after Christianity's spread across Europe in the fourth and fifth centuries this conception of divine weather held. Curiosity was a vice and any rational investigation was eschewed and suppressed. It was not until the rediscovery of Aristotle and the translation of his major works, including *Meteorologica*, into Latin in the twelfth and thirteenth centuries that his theories about exhalations would be reconsidered. Even then there was friction. The Bishop of Paris, Etienne Tempier, issued condemnations in 1270 and 1277, intended to halt the dissemination of his ideas. In particular,

Tempier was unnerved by the notion that the so-called 'secondary causes' like weather would continue to function properly if the first cause 'God' was removed from the scene. It was the beginning of a schism between religion and rationalism that would rumble on and on.

Over the next two hundred years Aristotle's ideas were assimilated into medieval Christendom in little tottering steps, the two great forces settling into an uneasy alliance. While God remained paramount, Aristotle's reputation steadily grew. Many referred to him as 'the Philosopher'. Dante declared him 'the master who knew all'.[11]

With his ideas cloistered away in universities Aristotle did not find a wide audience until the invention of movable-type printing in the fifteenth century. And it was not for another hundred years, after at least twenty-eight printings of *Meteorologica*, that his weather ideas would finally be produced for the average reader. Published in 1563, William Faulke's *A Goodly Gallerye* was written for a lay readership and featured instructive sections:

Of Wyndes
Of Storme Wyndes
Of Thonder
Of the Fryst
Of the Raynbowe

Although Faulke's ornate Tudor language gave a modern gloss, the theories were lifted straight from Aristotle. *A Goodly Gallerye*, though, did present a rational explanation of the weather to a wide audience for the first time in a millennia. Faulke was careful not to overplay his hand. God, he cautioned, was still at the helm. 'The first and efficient cause is god, the worker of all wonders, according to that testimony of the Psalmist, whiche sayeth. Fier, haile, snowe, yse, wynde and stormes, do his will and commanaundement.'[12]

The clearest proof of this was just around the corner. Twenty-five years after *A Goodly Gallerye* was published Philip I of Spain ordered his *Grande y Felicísima Armada* to set sail. Comprising 130 ships and 30,000 men, the fleet was one of the greatest military forces ever assembled and it was charged with invading and conquering Protestant England. At sea by the end of May, the Duke of Medina Sidonia, who commanded the fleet, might have expected a smooth passage. But wild weather shadowed them from the start. They were delayed in the Western Approaches by gales, then mauled by English battleships off Plymouth in July. August passed with Medina Sidonia caught in skirmishes and clandestine attacks. By September hopes of an invasion had been abandoned and the Armada was fleeing along the eastern English coast. The plan was to loop around Scotland and sail south towards Spain. But yet again, matters went dreadfully awry. Miscalculating their longitude, the fleet turned south too soon. Rather than the open sea the commanders were faced with the jagged outline of the northern and western Irish coast. The splintered rocks and beating winds were not unlike what FitzRoy would experience in Tierra del Fuego two and a half centuries later. More than twenty ships were lost, thrown on to the shore by Atlantic gales. Only half of Philip's invincible fleet saw home again.

For the English population, the story was rich in religious resonance. It was clear sign that God favoured the reformed Church of England over the old, corrupt Catholic faith. Much was made of the 'Protestant winds' that had come to England's aid. A series of commemorative medals and coins were minted in celebration, carrying the inscription *Flavit Jehovah et Dissipati Sunt*' – 'God blew, and they were scattered'.

The defeat of Philip's Armada became important to England's self-image, and it underscored the role of God in everyday life. Shakespeare's plays, written in the wake of the victory, reflect this. They often feature weather as a dramatic device. The fogs of *Macbeth*, King Lear on the Heath in the storm, the shipwreck in *The Tempest*, in all these weather is a brooding force, just as it was in everyday life.

In March 1599 Robert Devereux, the Earl of Essex, set out from London to crush an Irish uprising. Essex's theatrical departure through the capital's streets was undermined by a fierce thunder storm that many interpreted as an ominous augury. Essex's Irish campaign turned out a disaster and within two years he was dead, beheaded at Tower Green, convicted of treason. People remembered the storm which had preceded his change in fortunes. In his *Dictionarie of the Italian and English Tounges* (1611), John Florio used the word Ecnéphia to explain what had happened that day. Florio defined Ecnéphia as 'a kind of prodigious storme comming in Sommer with furious flashings, the firmament seeming to open and burne, as hapned when the Earle of Essex parted from London to goe for Ireland'.[13]

Amid such superstition, science could never flourish. It was not until the second half of the seventeenth century during the Scientific Revolution that rationalism found a foothold with the establishment of Deism and Natural Religion – movements which acknowledged God as Creator, but argued that he had left the universe thereafter to progress according to the laws of science. This reconfiguring of the role of God in daily life was a profound shift. It prepared the ground for the 'ingenious pursuers' of the seventeenth century like Newton, Boyle, Harvey, Galileo and Wren. 'I do not feel obliged to believe that the same God who has endowed us with sense, reason, and intellect has intended us to forgo their use,' Galileo famously declared.[14]

In the vanguard of the rise of the empirical sciences was the newly established Royal Society of London with its motto *Nullius in verba* (take nobody's word for it) and its bi-monthly journal, *Philosophical Transactions*. Soon descriptions of strange meteors were appearing in its early editions, of brilliant rainbows, dreadful thunderclaps, vivid lightning and red snow. When the Great Storm hit England and Wales in November 1703 it presented a perfect subject for rational analysis. The *Philosophical Transactions* promptly published a bumper issue replete with accounts of the damage. A Mr John Fuller of Sussex wrote to explain: 'We live ten miles off the sea in a direct line, and yet can scarce persuade the Country People, but that the salt water was blown thus far, or that during the Tempest of Rain was salt, for all of the Twiggs of the Trees the day after were white, and tasted very salt.'[15]

The Great Storm blew from the Welsh coast near Aberystwyth and cut a line through the Midland counties with the most powerful winds on its southern flank. It stripped lead from church roofs, set windmills on fire and swept farm animals away. Even London, newly rebuilt after the Great Fire of 1666 had destroyed three-fifths of the ancient city, was devastated. Haymarket, Leicester Fields, Soho, the Seven Dials, Red Lion Square, embodiments of civilisation, charm and beauty and built stoutly in pink or red brick, were decimated. Such destruction plunged England into a fit of introspection. Scandal, greed, profanity, the War of the Spanish Succession and clerical corruption were cited as possible reasons for the disaster. One anonymous sermon, *The Terrible Stormy Wind*, attacked the spread of the dangerous pursuit of science as proof of Man's ambition and vanity. It scorned the 'philosophical' claim that the storm 'was nothing but an Eruption of Epicurus's Atoms; a Spring-Tide of Matter and Motion: a Blind Sally of Chance so throwing *Providence* out of the Scheme'.[16] As penance, an official day of fasting was declared.

The most famous popular work to emerge in the aftermath was Daniel Defoe's masterpiece of early reportage, *The Storm* (1704). Always a daring writer, Defoe tried to leaven his traditional interpretation of events with enlightened analysis. He opened his introduction like a storm preacher – '*I cannot doubt but the Atheist's hard'ned Soul trembl'd a little as well as his house, and he felt some Nature asking him some little Questions; as these – Am I not mistaken? Certainly there is some such thing as a God – What can all this be? What is the Matter in the World?*'[17] – yet soon proved himself willing to test new ways of thinking. His first chapter was devoted to an assessment 'Of the Natural Causes of the Winds'. Such investigations, he argued, were both justified and necessary because:

> To search after what God has in his Sovereignty thought fit to conceal, may be criminal, and doubtless is so; and the fruitlessness of the Enquiry is generally Part of the Punishment to a vain Curiosity: but to search after what our Maker has not hid, only cover'd with a thin Veil of Natural Obscurity, and which upon our Search is plain to be read, seems to be justified by the very Nature of the thing, and the Possibility of the Demonstration is an Argument to prove the Lawfulness of the Enquiry.[18]

But Defoe did not arrive at any conclusion. Wind was beyond his comprehension. He lifted ideas from Ralph Bohun's *Discourse Concerning the Origine and Properties of Wind* (1671), already a generation old and fantastically vague, and mentioned Aristotle's ancient theories before giving up. '*We hear the Sounds, but know not from whence they come.*' He rounded off:

> From this I draw only this Conclusion, That the Winds are a Part of the Works of God by Nature, in which he has been pleased to communicate less of Demonstration to us than in other Cases; that the Particulars more directly lead us to Speculations and refer us to Infinite Power more than the other Parts of Nature does.[19]

That Defoe was free to set out a rational and scientific response was nevertheless a sign of sure and steady progress. In the years and decades that followed, philosophers found themselves with an ever-increasing latitude to speculate about how storms formed as Deism became an accepted force in British life. Consolidating the Deist position in 1730 in his *Christianity as Old as the Creation*, Matthew Tindal would argue that, 'If God will judge Mankind as they are accountable, that is, as they are rational; the Judgment must hold an exact Proportion to the Use they make of their Reason' and ''Tis Reason alone which must judge; as the Eye is the sole Judge of what is visible; the Ear of what is audible; so Reason of what is reasonable.'[20]

But what was reasonable remained contentious. Various theories had their moments during the 1700s: that sulphurous minerals in the soil engendered a volatile atmosphere like that found by Mount Vesuvius; or that storms were aerial battles between noxious gases. The most enduring argument was that weather was dictated by the orbit of the sun, moon and five known planets: Mars, Venus, Mercury, Saturn and Jupiter. This astrological meteorology could be traced back to ancient history and the writings of the Graeco-Roman philosopher Ptolemy. In the 1680s it was rejuvenated by Dr J. Goad in his *Astro-Meteorologica* (1686). The title was presumably chosen to underscore lineage with Aristotle, and Goad emphasised his credentials, claiming his calculations were based on thirty years of observation. Goad's ideas were extremely influential. He became something of a prototype weather forecaster to King Charles II, providing him 'and several persons of Quality of this Nation'[21] with monthly outlooks – with which they could presumably plan their hunting trips.

Presented as an empirical science *Astro-Meteorologica* set out to unravel the great complexity of God's celestial design. With Newton having proved that tides operated logically, pushed and pulled by the moon's gravity, Goad set out to show that the planets had a similar effect on earth's atmosphere. For Goad the moon in particular had a beguiling, little-understood power. It was well known, he claimed, that its rays caused flesh to putrefy with increased speed; lobsters, oysters and crabs to swell and taste sweet; and epileptics to suffer fits.[22] All this, Goad claimed, was evidence of the moon's power, which extended beyond simply lighting up the night sky. 'If the moon were made for illumination, she would never appear by day, when there is no need for her Light, nor disappear at night when there is need,' he reasoned.[23]

But the sun and moon did not act alone, Goad argued. Instead they worked in congress with the five other planets, generating an infinite number of atmospheric combinations. Mars brought heat, drought, thunder and stormy winds, while Jupiter ushered in 'healthful and temperate Air, yet with Wind and competent Moisture'.[24] Goad produced pages of observations to support his arguments. At times he soared to heights of pomposity: 'Planetary Aspects are no vain Terms of a Bawbling Art, but are Mysterious Schematisms of a Secret Force and Power toward the Alteration of the Sublunar World, especially the Air, and those Great issues that depend thereon.'[25]

Astrological meteorology won some important supporters in the seventeenth century, among them Francis Bacon, the Astronomer Royal John Flamsteed and Robert Boyle, who considered it likely 'that all of the planets had some influence on atmospheric composition by drawing forth effluvia from the earth'.[26] Throughout the eighteenth century and on into the 1800s the idea persisted that the moon, 'that moist star', was the key to the atmosphere. A new moon on the fourth day was thought to bring storms at sea and many sailors knew the ancient words of the Venerable Bede: 'If she looks like gold in her last quarter there will be wind, if on top of the crescent black spots appear, it will be a rainy month, if in the middle, her full moon will be serene.' Luke Howard was still tinkering with the idea in the early 1800s, as was James Weddell.[27]

Captain King of the *Adventure* was another believer. King thought the moon's influence particularly strong in the waters around Cape Horn, something he wrote up in his meteorological notes:

Being to the north of Staten Island for three days preceding full moon, which occurred on the 3d April (1829), we had very foggy weather, with light winds from the eastward and northward, causing a fall of the mercury from 29.90 to 29.56. On the day of full moon the column rose, and we had a beautiful morning, during which the high mountains of Staten Island were quite unclouded, as were also those of Tierra del Fuego.[28]

Whilst planetary meteorology endured, it was never completely embraced by the scientific establishment who thought it too tainted by supernatural association. Instead philosophers preferred the firmer grounds of chemical or electrical studies, the latter of which went through a period of tremendous popularity after Franklin's kite experiment and invention of the thunder rod. Franklin's thunder rod, or electrical conductor, was heralded as a dazzling achievement and it caught the imagination of his generation. It was 'an operation which seems to impart to the feeble hand of Man, a portion of the power of the Supreme', wrote one admirer.[29]

Franklin's electrical researches not only provided people with a way of deflecting lightning safely to earth – the thunder rod, or conductor – it also explained scientifically why elevated objects were most susceptible to a strike. For years this had been perplexing. Why should God strike out at church spires and not, for example, at the squat little alehouses or any other den of vice? It is an irony of science that the higher church spires rose to proclaim the glory of God, the more likely they were to be hit as a result.

Franklin's explanation of lightning would finally bring logic to the puzzle in the 1750s. But years passed before his ideas were fully accepted. Many towns continued with the old tradition of stockpiling weapons in church towers, convinced God would not rain his fire on such a place. In Brescia in Italy, two million pounds of gunpowder had been locked up in the tower of St Nazaire. On 18 August 1769 the church was struck by lightning, 'the tower of Saint Nazaire was projected entire into the air, and fell down in a shower of stones', wrote one report. In a flash a sixth of the town was levelled and 3,000 were dead.[30]

Interpreting such disasters had long been part of the Christian's lot. A God that sent Protestant winds to vanquish an invading force was one thing, but a God that killed innocents, levelled cities and destroyed churches was another. Some reasoned that God sent bad weather as a test for humanity, to make people strive to greater piety. The trick was to put your trust in God's grace. As a sermon reminded worshippers in 1827, King David had been exposed to the raging of a 'terrible tempest' but he had retained his 'strong confidence and was graciously preserved'. It urged Christians to follow David's lead. 'Confidence is felt by good men, when winds blow, when thunders roar, when lightnings blaze, and when earthquakes shake the ground.'[31]

For many others the poet William Cowper had summed it up best.

God moves in mysterious ways
His wonders to perform;
He plants His footsteps in the sea
And rides upon the storm.[32]

William Reid arrived at Barbados in January 1832. He found the
islanders mentally recuperating and beginning a long period of
rebuilding. The damage was severe. The harbour was wrecked, split
open by a storm surge. Just two vessels had survived and the wharf
had crumbled into the sea. Behind the seafront the destruction
cascaded into the distance. Entire streets had disappeared, crumpled
in a mass of ruin or covered by shingle blown up from the beach.
The Commissariat, slave prison, St Michael's Cathedral, King's House
and several churches were badly damaged and the House of Industry,
the Boys' and Girls' Schools, Governor's House, the Theatre Royal,
Lunatic Asylum and alms houses were destroyed beyond repair.
Perhaps most disturbing of all was the fact that the Royal Gaol had
been flattened and one of its 'massive gates torn from its hinges and
cast, broken to pieces, into the road'.[33] All but three of the prisoners
had fled into the ruins, slipping invisibly into the chaos.

A journalistic account of the hurricane was already circulating. *An
Account of the Fatal Hurricane by which Barbados Suffered in August 1831*
had been published in December as Reid crossed the Atlantic. Reid
bought a copy and read it closely. Its anonymous author had collected
a series of eyewitness accounts. The opening description explained
how the weather had been wet in the weeks beforehand and wrote
that 'electric clouds' had blown over the island. At the start of August
a flash of lightning had hit a house in the countryside, killing an
infant, wounding the mother and destroying some poultry. But on
10 August, the date of the hurricane, the morning had begun as
serene as the finest summer's day. 'The sun arose without a cloud,
and shone resplendently through an atmosphere of the most trans-
lucent brightness.'[34]

Thereafter the hurricane had approached in familiar fashion. Black
clouds had gathered on the northern horizon. There had been a
fleeting shower of rain, sudden stillness, darkness, and scud clouds
– the messengers Constable had described – had flown overhead. At
eleven the breeze freshened to a gale. At midnight the hurricane was
directly overhead, with lightning strobing out of the black and the

wind blowing 'prodigious beyond conception, hurling before it thousands of missiles'. The commotion was so loud that 'No thunder was at any time distinctly heard; had the cannon of a hundred contending armies been discharged, or the fulmination of the most tremendous thunderclaps rattled through the air, the sounds could not have been more distinguished'.[35]

The accounts were dramatic, but what really caught Reid's interest were the factual details. In particular he noted down the shifting directions of wind. Everyone agreed that it had come from different points of the compass at different stages of the night. At one moment it had seemed to blow from the north-east, then it shifted to north-west. An hour later it had died away momentarily, then blew with renewed force from the south-west, then the west and the west-north-west.[36] These veering winds had long been a recognised feature of a hurricane. As far as Reid knew, no one before had ever explained this behaviour.

Over the coming months Reid began to fill notebooks with memoranda. A decade before Edgar Allan Poe unleashed Dupin, the archetypal detective, in *The Murders in the Rue Morgue* (1841), Reid was embroiling himself in the minute study or ratiocination that would become a hallmark of a coming age. But, unlike Dupin, Reid was seeking not a human offender but an atmospheric one. He later wrote of how he started to 'search everywhere for accounts of previous storms in the hope of learning something of their causes and mode of action'.[37]

The art of tracking storms stretched back about a century – once again to Benjamin Franklin. On 21 October 1743 Franklin had planned to watch a lunar eclipse from his Philadelphia home. Franklin's hopes had been dashed by a thick north-westerly gale that obscured the skies – 'Neither the Moon nor the stars could be seen.' Shortly afterwards he had learnt from the Boston newspapers that the eclipse had been perfectly visible in New England and that the bad weather had not arrived until half an hour later. The accepted theory was either that storms rose and fell in a single geographic location or that they were carried along by the prevailing wind. Yet here the evidence suggested the opposite, as the storm had travelled north-east from Philadelphia to Boston. Franklin wrote up his observation and it became known as yet another snippet of inspired Franklinia. Since Franklin's death in 1790 a few notes had been published on storms, a Colonel James Capper of the East India Company and a Harvard professor called John Farrer had written personal experiences, but no solid theory had emerged. At first, Reid's ambitions did not stretch to theory; instead he simply wanted to create a complete biography of the Barbados hurricane, documenting its growth from a boisterous infant to a roaring adult to an old and fading creature.

Realising that good data was the key, Reid started to assemble notes from the logs of vessels in the area, records of wind direction against time so that he could plot real-time wind diagrams of the hurricane. He studied meteorological theory and travelled eighty miles from Barbados to St Vincent to compare how that island had coped. In St Vincent he interviewed a Mr Simmons who had watched the hurricane approach at dawn on 11 August. Simmons had seen 'a cloud to the north of him so threatening in appearance that he had never seen any so alarming during his long residence in the tropics and he describes it as appearing an olive-green colour'. Simmons had nailed his doors and windows shut and sheltered as the storm passed over at seven o'clock in the morning, coming from the north.[38]

From this Reid established that the hurricane had travelled 'the nearly eighty miles' from Barbados to St Vincent in about seven hours, and he estimated that its speed of progression was therefore a little over ten miles an hour.[39] Now that his pursuit of facts was becoming well known more information was forwarded to him. One story came from two slaves who claimed to have been terrified by sparks flying off each other in early August. Retrospectively they thought these sparks might be a sign of a highly charged atmosphere. Reid discounted this as having anything to do with the hurricane, as well as a separate account of an earthquake. Instead Reid concentrated on the facts. Why had the winds spun around? Why in the middle of the hurricane had there been a pause of relative quiet? What did the fall in barometric pressure mean?

The answers to these problems eluded Reid until in 1832 he received a copy of the previous year's *American Journal of Science*. The journal included an ingenious article on storms written by an American called William C. Redfield.

Redfield's article, 'Remarks on the Prevailing Storms of the Atlantic Coast', was just what Reid had been seeking. For the past ten years Redfield had been working on the same problem from his base in

New York City. Already years ahead of Reid in his conclusions, Redfield's arguments were compelling. Hurricanes, he wrote, were not entirely chaotic. Instead they operated to strict atmospheric rules regarding the motions of the wind and the direction of progress. In the course of his researches, Redfield announced, he had come to the conclusion that hurricanes were giant whirlwinds that hurtled across the sky like flying discs. It was a bold and original claim. And as a theory, for William Reid at least, it fitted the facts perfectly.

Very few in scientific circles would have heard of William C. Redfield's name before the publication of his storm paper in 1831. A New York businessman, he had made his name with his Steam Navigation Company. Redfield's steamers plied up and down the Hudson, from New York to Albany, carrying passengers and freight. Redfield's success had come through his natural instinct for innovation. In the 1820s, the early years of steam, passengers had been wary of travelling too close to the engines, worried that they might explode – as they often did. Redfield's solution to the problem had been simple but effective. He had designed 'safety barges' for the passengers to travel in, precursors of the railway carriages of the future, drawn in strings behind the steamer. Over time, as safety standards had improved and passengers had become more confident, he had switched his tactics: moving the passengers back into the steamer and filling the barges with cargo.

But Redfield was more than a wily businessman. He had worked as a mechanic in his youth in small-town Connecticut, and he had retained his interest in engineering. He relished the challenge of invention and was often involved in the innovation of his steamers, striving to produce his own 'simpler, cheaper and safer forms of apparatus'. In time, others had come to appreciate the quality of Redfield's work. He was that rare blend: a businessman with the knack of getting things done, and an inventor with an independent mind and a Yankee love of detail.

Travelling on a steamer from New York to New Haven one day in 1831 Redfield chanced to meet Denison Olmstead, Professor of Mathematics and Physics at Yale. Spotting Olmstead on deck he had approached and 'modestly asked leave to make a few inquiries' about a paper Olmstead had recently published on hailstorms in the *American Journal of Science*. Soon Olmstead and Redfield were talking about storms and it was then, for the first time, that Redfield unveiled his theory of whirling winds. It was a pivotal moment in the history of meteorology.[40]

Redfield had devised this idea a decade before, after the 'Great September Gale of 1821'. The gale – as the hurricane was then called – had sparked panic right along the north-eastern coast, causing a storm surge that had flooded the New Jersey coastline as well as several streets on Manhattan Island. In the days afterwards Redfield had been out walking with his son in rural Connecticut, which had been equally affected. He noticed that near Middletown, in the centre of the state, trees had been blown over towards the north-west. But in neighbouring Massachusetts the trees had fallen in the opposite direction, pointing towards the south-east. Redfield took this as proof that in just seventy miles the winds had reversed direction. To check his facts, he had collected newspaper reports and soon had sketched the path of the storm. It was then that 'the idea flashed upon his mind that the storm was a *progressive whirlwind*'.[41]

Not having any connection to the scientific establishment, Redfield kept his ideas to himself until his chance meeting with Olmstead in 1831. Intrigued, Olmstead persuaded him to write an article on the subject for the *American Journal of Science*. Redfield agreed, on the condition that Olmstead revise the manuscript and oversee its production. Several months later, Redfield's article was published.

'Remarks on the Prevailing Storms of the Atlantic Coast' appeared in the July 1831 issue just a month before the Barbados hurricane. Redfield, hitherto unknown, set out his case confidently. He progressed in logical steps: defining simple terms, distinguishing between winds, calms, storms and hurricanes, *'a wind or tempest of the most extraordinary violence*. It has been stated as a distinguishing characteristic of hurricanes, that *the wind blows from different points of the compass* during the same storm.'[42]

It was this confusion, Redfield announced, that he set out to explain. Using the September Gale of 1821, he presented his case like a prosecution barrister: noting where the storm had been at a specific time, revealing details of where the winds were blowing:

In reviewing these facts, we are led to inquire how, or in what manner it could happen, that the mass of atmosphere should be found passing over Middleton for some hours, with such exceeding swiftness, towards a point apparently within thirty minutes distance, and yet never reach it; but a portion of the same or similar mass of air, be found returning from that point with equal velocity? And how were all of the most violent portions of these atmospheric movements which occurred at the same point of time, confined within a circuit whose diameter does not appear to have greatly exceeded one hundred miles? To the writer there appears but one satisfactory explication of these phenomena. *This storm was exhibited in the form of a great whirlwind.*[43]

Redfield supported his claims throughout. 'If our position be conceded then it is no longer difficult to explain the paradox, or mystery, which otherwise pertains to the phenomena exhibited by this storm . . . We can discern the reason why, in seamen's phrase, "a north-wester will never remain long in debt to a south-easter."' He showed that the gale had begun in the West Indies on 1 September, then had curled up along the coast, over Charleston in South Carolina and then Norfolk in Virginia and Delaware before thundering across New York. All the while the wind blew from different points of the compass, a fact he illustrated with an accompanying map.

This wasn't all. Redfield claimed his whirlwind theory solved the century-old barometric conundrum. He asked the readers to perform a simple experiment:

Let a cylindrical vessel of any considerable magnitude, be partially filled with water, and let the rotative motion be communicated to the fluid by passing a rod repeatedly through its mass, in a circular course. In conducting this experiment we shall find that the surface of the fluid immediately becomes depressed by the centrifugal action, except on its exterior portions, where owing merely to the resistance which is opposed by the side of the vessel, it will rise above its natural level, the fluid exhibiting the character of a miniature whirlpool.[44]

If true, Redfield had not only explained away the lowering reading that accompanied the centre of a storm, he had also demonstrated why the barometer should rise in the moments before it arrived. The whole mechanics of a storm could be reproduced in any house in America at any time, with apparatus no more sophisticated than a cup of water and a spoon.

For the first time, too, Redfield had demystified the eye of the storm – that moment of comparative and eerie quiet when the wind died away. According to his theory this was the vacuum at the core of his whirlwind:

> Every experienced navigator will shrink with instinctive appre-
> hension from the very idea of moments of awful and treacherous
> stillness which place him in the central vortex of the hurricane,
> ready to be overwhelmed by the rapidly advancing and seemingly
> impenetrable line of spray which envelops the onset of the last
> and most dreaded portion of the receding storm.[45]

Though no more than a theory, Redfield's article nonetheless marked the greatest contribution to storm mechanics in decades. The following year, surrounded by the chaos in Barbados, Reid would come by a copy. Redfield's map was the first detail to catch Reid's attention – showing the Great September Gale curving through the Caribbean and up along the American east coast. 'Strongly impressed with the belief that Mr Redfield's views were correct, I determined to verify them by making charts on a large scale,' Reid later wrote, 'and on these laying down the different reports of the wind at points given in the *American Journal of Science*.'[46]

Reid's project was a grand one. He wrote to residents for personal accounts, diary-keepers for written reports, ships' captains for logs and examined all the meteorological records he could lay his hands on. 'The more exactly this was done,' Reid later explained, 'the nearer appeared to be the approximation to the tracks of a progressive whirl-wind.' Soon Reid's project had expanded to other storms. He examined the Savannah la Mar Hurricane of October 1780, Poyer's Great Barbados Hurricane of the same year and the Great September Gale of 1821. At length he managed to define the track of the Barbados Hurricane of 1830, demonstrating that after it had left Barbados it had progressed through the Bahamas and Florida and along the western American coast before dying out just south of St Pierre 'in the longitude 57° West, and latitude 43° north'.

He continued:

It performed this long journey in about six days, at the average rate of about seventeen geographical mph. The general width of the tract, which was more or less influenced by the Hurricane, was from 500 to 600 miles; but the width of the tract where the hurricane was severe was only from 150 to 250 miles. The duration of the most violent portion of the storm at the several points over which it passed, is from seven to twelve hours, and the rate of its progress from the island of St Thomas to its termination between this coast of Nova Scotia, varied from fifteen to twenty miles per hour.[47]

For two years Reid's investigations continued as he oversaw the rebuilding of Barbados. His idea was not to establish a theory to explain why the winds should whirl about. He simply hoped to follow Francis Bacon's dictum to collect good facts from observation and wait for patterns or laws to emerge. The project consumed Reid for the rest of his spell in Barbados. By the time he left in May 1834 he had gathered enough material to begin a book. On his return he was granted a period of leave, which he ploughed into more storm and hurricane research, a continual and industrious sourcing and analysis of data that grew as the months passed.

By 1836 Reid's sabbatical had come to an end and he had been redeployed by the Engineers to fight in the latest phase of the Spanish Carlist War. So he must have missed a curious report syndicated by a handful of British publications from the *Boston Paper*. Under the title 'Important Discovery', it read:

A gentleman of the city has made one of the greatest discoveries, which has been made since the days of Franklin – he has discovered the laws which govern the weather! This great secret, which mankind have been for thousands of years endeavouring to penetrate, has been found out by James P. Espy, Esq.

Like most other operations of nature, he has ascertained that the weather is regulated by fixed unalterable rules – simple, easily understood. He can inform a captain of a vessel how to tell whether there is a storm raging anywhere within 500 miles of him, and how he must steer to get just as far into the storm as he pleases, and of course have just as much or as little wind as he pleases.[48]

Beneath one syndicated copy of this piece, in the *Manchester Times and Gazette*, the editor had included his incredulous analysis. 'It is not because we attach any belief to the reported meteorological augury of the American philosopher that we have copied the following paragraph. We believe that the story is as true as that related of the vendors of the wind in Lapland who sell the article pent up in bags, for the use of mariners.'

It was not unusual for the British papers to treat American news in this way. Still only half a century old, America's spirit of optimism and devotion to progress made it appear especially prone to the chimerical, the nonsensical or ridiculous. But in this case the scepticism would prove ill-placed. James Espy had succeeded in explaining a mysterious process: one that revealed how clouds formed and rain fell. A bright new star had appeared in the scientific sky.

CHAPTER 5

Trembling Air, Whirling Winds

As reports of Espy's weather discovery appeared in the British newspapers in July 1836 Albert Barnes, Pastor of the First Presbyterian Church of Philadelphia, was preparing to address the alumni of Hamilton College in New York State. He delivered his speech on 27 July, a rousing assessment of America's intellectual health. 'It is no discredit to us to admit that our literature and science may fall short, in many respects, of the attainments in the old world,' he began. 'No American need be reluctant to confess that in philology and criticism we may be behind the German; in chemistry and medicine, we may be inferior to France; in classical learning and the exact sciences, inferior to England and Scotland.' But what should be expected, he wondered, 'for an infant people'? 'No nation has had a task to perform so arduous as we have, or has done it so well. We had a vast, an almost illimitable territory to occupy, to subdue, to cultivate.'

This task was now under way, Barnes said. Now was the time for Americans to turn their minds to intellectual improvement. Did not, he pointed out, the same sun shine over American heads as shone on Galileo and Herschel? Could not the American air or water be studied with the same facility as it was by Davy? Indeed, he continued, the American continent seemed especially prepared for scientific investigation:

Nature has here exhibited herself in some respects on a broader scale, and in a more magnificent manner, than in the old world. There is a freshness and vastness in her works here which is fitted to expand the mind, and elevate the soul and fill it with grand conceptions and to invite to successful investigation. It seems almost as if God, in favor to science and the enlargement of the human mind, had reserved the knowledge of the western world, until almost the last felicitous investigations that could be made had been made in the old world.[1]

Barnes' words were echoed a year later by Ralph Waldo Emerson in his *American Scholar* address to the Phi Beta Kappa Society at Harvard University. In his famous speech Emerson would urge the scholars to forge their own path in an act of intellectual independence from the old savants. 'We have listened too long to the courtly muses of Europe,' he said. The danger was that Americans were becoming 'timid, imitative, tame'. This was a prospect that, if America was to be great, had to be guarded against.[2]

Barnes and Emerson's speeches came as America was entering a new phase in its development. It was no longer just a brave social experiment. It was ready to make its voice heard on the international stage. And few of its citizens were more eager to take this leap than James P. Espy, a celebrated teacher, classicist and mathematician-cum-meteorologist from Philadelphia.

In July 1836 Espy was fifty-one years old and in the prime of his career. A portrait from this time shows him in formal dress, his unkempt, greying hair swept to one side. But it is Espy's eyes that command attention. They are full of vigour and tilted a fraction downward. It shows Espy in his familiar role as master surveying his class or orator gazing at his audience.[3]

James Espy was born on 9 May 1785 in Westmoreland County near Pittsburgh in western Pennsylvania, just east of the Ohio River. This was frontier land, an uncertain, dangerous territory. While Espy was in the crib US troops were still marching west to Fort Washington to quell uprisings by Shawnee and Miami Indians. News from the wars filtered back in terrifying dispatches. Murderous ambushes, scalped soldiers, raids – this was Espy's childhood world, far away from the settled east coast where the majority of the population lived in tight, secure pockets. It was a nomadic childhood, too. As the troops advanced west, so the Espy family followed – first to the Miami Valley in newly admitted Ohio and then on to the vast savannahs of the Bluegrass region of Kentucky. It was there that Espy displayed early signs of the conviction that was to characterise his life. Antagonised by slavery on the tobacco plantations after he finished a law degree at Transylvania University in Lexington, he resolved to leave the frontier for good.

Espy was ambitious and fiercely determined. Having progressed quickly through academic ranks, by the 1830s he had become head of the Classics Department at the Franklin Institute in Philadelphia. There Espy worked alongside Professor Alexander Bache, the great-grandson of Benjamin Franklin and future Professor of Natural Philosophy at the University of Pennsylvania. To Bache's mind he was 'one of the best classical and mathematical instructors in Philadelphia'. His talents did not end there. In the late 1820s he had discovered his great passion: meteorology.

Initially Espy's meteorological speculations were confined to short, niche papers discussing the daily fluctuations of the barometer. But using his position at the Franklin Institute, he soon established himself as an authority. By 1834 he had risen to become the chairman of a joint committee of the American Philosophical Society and the Franklin Institute, formed to analyse storms. One of the key ambitions of the committee was to collect atmospheric data, and Espy eagerly set to work, searching for potential correspondents. He wrote to government officials, college professors, lawyers and journalists in towns and cities across the states, petitioning them to keep weather journals. Like Beaufort with his hydrographic instructions for FitzRoy, Espy cajoled his correspondents to record the wind direction and the 'character of clouds' three times a day. No one had done anything like this in America before. Soon Espy had fourteen regular weather diarists sending him data from Maine in the north to Tennessee in the south. In recognition he was elected a member of the American Philosophical Society, a sign of a man on the rise.

Above all, Espy had one specific instruction for his correspondents. 'We also particularly request,' he wrote, 'that if you hear of any storms occurring in your neighbourhood, you will collect all the information concerning them in your power.'[4] Espy had a reason for making such a request. He had read Redfield's essays in the *American Journal of Science* with increasing scepticism. Could storms really be great whirlwinds? This seemed to run against everything he had deduced in his own calculations. With the data from his network as proof, by the early 1830s Espy was already plotting to catch Redfield out.

By 1836, it was seven years since Espy had begun to formulate his own theories. Later he would describe his meteorological awakening in grandiose terms. It had come in two distinct stages. First when he was reading an essay by the British meteorologist and chemist John Dalton:

I was much struck with one of his results; namely, that the quantity of vapour in weight, existing at any time in a given space, could be determined with great accuracy in a few minutes, by means of a thermometer and a tumbler of water cold enough to condense on its outside a portion of the vapour in the air. It occurred to me at once, that this was a level with which the meteorologist was to move the world.[5]

Excited, Espy had begun a forensic analysis of vapour. It turned out to be a fruitless enquiry. He later wrote of the labyrinth of contra dictions he encountered as he battled to explain how water remained suspended in air. Here he floundered until his second burst of inspiration – a moment on which his life shifted. It came while he was studying the precise moment when vapour condensed into the drop-lets that formed a cloud. Espy focused on what he termed the 'latent caloric': hidden energy trapped inside vapour as it underwent the change in state from a solid to a liquid and from a liquid to a gas. What Espy discovered startled him:

The result was an instantaneous transition from darkness to light. The moment I saw that a rapidly forming cloud is specific-ally lighter in proportion as it becomes darker, a thousand contradictions vanished, and the numerous facts, 'a rude and undigested mass', which had been stowed away in the recesses of my memory presented themselves spontaneously to my delighted mind, as a harmonious system of fair proportion.[6]

What Espy had realised was of crucial importance. The sudden liberation of the vapour's 'latent caloric'* explained how the clouds expanded outwards and formed distinct shapes. He conceived an atmospheric vision of invisible columns of air rising upwards like modern-day fuel pumps to feed the clouds with condensing water vapour. This process completed a vital link in what we today call the

---

* Latent caloric is an antiquated term best thought of as meaning hidden energy. Today we call it latent heat: energy released or absorbed by a body during a constant temperature process, like boiling water or melting ice. Espy understood that when snow (a solid) is heated into water (a liquid) and afterwards evaporated into vapour (a gas), it gains latent heat. Conversely when vapour condenses into water drops, as it does in the formation of a cloud, this latent heat is released.

hydrological cycle. He had shown that the atmosphere was not bound together by an electrical field or agitated into action by fumes. Instead he had discovered that it functioned according to strict mathematical principles, with water vapour the essential component in a circulatory system – like blood being pumped around the human body.

Studying water vapour was nothing new. Ever since René Descartes had challenged the orthodoxy that all invisible gas was air in the 1630s, philosophers had considered the study of vapour an important enquiry. Descartes had argued that vapour existed independently of air, and that it somehow had the ability to rise upwards. But it was difficult to explain where it got its buoyancy from. And, from there, more troubles emerged. Once vapour had transformed into cloud droplets, how did it stay afloat? If clouds contained thousands of tonnes of water, how could they possibly be suspended in the sky when water is a hundred times denser than air? For centuries this had been one of nature's great paradoxes.

The most widely held theory was that clouds comprised minute particles called 'vesicles', fragments with their own distinct gravity that kept them buoyant. Vesicles were 'innumerable small globes filled with damp air, analogous in some way to soap bubbles', or better imagined as minute aerial zeppelins that roamed the skies. De Saussure claimed to have seen vesicles on an Alpine hike, the drops floating 'slowly before him, having greater diameters than peas, and whose coating seemed inconceivably thin'.[7] It was a vivid image, but the theory was dogged with problems. How did vapour transform into vesicles? Where did this gravity-defying power come from? What were vesicles? And how did they transform into rain, snow or hail?

The accepted theory of rain was proposed by James Hutton to the Royal Society of Edinburgh in 1784. Hutton believed rain was caused by the 'mingling together of great beds of air of unequal temperatures, differently stored with moisture'.[8] The mixture of these beds of air created what Hutton called an atmospheric imbalance that resulted in rain. Espy's idea, though, was different from the start. He concentrated on columns of rising air. This notion of an ascending current had appealed to several scholars before, particularly the French physicist Du Carla and more recently another Frenchman, Joseph Louis Gay-Lussac. Both Du Carla and Gay-Lussac had published papers on the subject, catching the attention of the German geologist Leopold von Buch. In turn von Buch had told the Berlin Academy of Science that 'the principle of ascending currents of air should really be called the key to the whole science of meteorology'.[9]

That irregular columns of rising air existed was not in doubt. A beautiful example had been written up by a naturalist called John Blackwell. In a piece for the *Transactions of the Linnean Society* in 1828 Blackwell had detailed his observation of a spider in a field of stubble.

> All the spiders were intent upon traversing the regions of the air. Accordingly, after gaining the summits of various objects, as blades of grass, stubble, rails, gates, &etc, by the slow and laborious process of climbing, they raised themselves still higher by straightening their limbs, and elevating the abdomen, by bringing it from the usual horizontal position into one almost perpendicular, they emitted from their spinning apparatus a small quantity of the glutinous secretion with which they construct their webs. This viscous substance being drawn out by the ascending current of rarefied air into fine lines several feet in length, was carried upward, until the spiders feeling themselves acted upon with sufficient force in that direction, quitted their hold of the objects on which they stood, and commenced their journey by mounting aloft.[10]

Espy had witnessed these currents himself. A member of the Kite Club of Pennsylvania – an organisation with impeccable meteorological credentials – he had often noticed how his kite was tugged up by rising air. These columns (today known as thermals) were most prevalent on days when the cumulus clouds formed 'briskly and numerously'. The updraughts become 'so familiar', he wrote, 'during the course of their [the kite-flyers'] experiments that, on the approach of a columnar cloud just forming, they could predict whether it would come near enough to affect their kites'.[11]

Certain he had unlocked the process behind the formation of clouds, Espy was confident enough to dispel the old belief that they floated at all:

> It is not necessary to inquire, as is frequently done, by what power are the clouds suspended in the air, unless it can be shown that they are suspended, of which I think there is no probability . . . We have every reason to believe . . . that the particles of

cloud begin to fall through the air . . . as soon as they are separated from the up-moving column of air, by means of which the cloud was formed.[12]

By now Espy was envisaging an entire dynamic weather system: ascending currents of warm air drawing up vapour, the vapour condensing at a specific height, expanding and forming clouds, and then the water droplets falling back to earth. Under different atmospheric conditions, Espy realised, this system of vaporous circulation was capable of producing every type of precipitation: rain, snow or hail. His task then became to calculate specific circumstances. What temperature was needed to produce snow? What expansion of water vapour would be required to generate a twenty-mile-wide hailstorm?

He worked at these calculations in the backyard of his Philadelphia home in Chestnut Street, and as time passed meteorological equipment began to accumulate. To help him model his calculations he invented an instrument he called a 'nephelescope'. It was little more than an air pump attached to a barometer and a tubular vessel – something of an early cloud chamber – but at least it gave tangible form to his studies. A niece who visited him during these years saw an outdoor space transformed into an atmospheric laboratory, filled with vessels of water, numerous thermometers and hygrometers for determining the dew point. To work with maximum speed he had painted his fence white, so he could use it like an enormous notebook. 'It was so covered with figures and calculations that not a spot remained for another sum or column,' she remembered.[13]

Espy's background as a mathematician made him unusually adapted to this work. Everything became a problem to be solved through minute study. His elevation to chairman of the Meteorological Joint Committee gave him the opportunity to model his theories using real-life data and he sought to augment this data store whenever he could. In July 1835 he raced to the scene of a whirlwind in New Brunswick, determined to write up the case. William Redfield, too, visited the site in the hope of gleaning evidence for his theory of whirling winds. Although Espy did not meet Redfield he was well aware of his ideas. He had studied them in the scientific journals. Espy bided his time, saying nothing.

Redfield's idea of circular winds was clearly perplexing. Espy could find no reason why winds should dart about the central axis of a storm. Eventually he concluded that Redfield was wrong. A more logical answer, Espy reasoned, was that winds rushed towards the central column at the core of the storm as air in a room would be drawn in towards a burning fire – cool air from beneath replacing the warm heat travelling upwards. The science behind this idea was sound. In a powerful storm, Espy thought, the effect would simply be magnified.

Espy chose to unveil his ideas in the inconspicuous *Transactions of the Geological Society of Pennsylvania* in 1834. It was a cautious move, testing the water. Another two years passed until he was ready to bring his case before the wider scientific community, in April 1836. He now turned to his spiritual home, the *Journal of the Franklin Institute*. He aimed to impress. Full of confidence he announced:

> Gentlemen, I send you now, for publication, the first of a series of essays on rain, hail, and snow, water-spouts, land-spouts, winds, and barometric fluctuations, in which, I hope, it will be found that I have successfully traced these phenomena to their true causes . . .
>
> I promise the reader, in advance, that he will find developed in the following essays, a law in meteorology, which, founded on acknowledged dynamical principles, explains at once, with a simplicity which nothing but nature can equal, all the *seven* phenomena mentioned above.
>
> The importance of this law will be readily admitted, when it is understood that by it may be known whether there is a great storm raging at any time within four or five hundred miles of the observer, and also the direction of that storm, with the means of avoiding it, if the observer is at sea.
>
> Yours, James P. Espy[14]

It was a grandstanding announcement, the words of someone who did not doubt himself; not for a moment. Espy set out his case with a study of hail. He showed the evolving path of water vapour: rising from the earth, freezing in the atmosphere, falling over a specific area. He supported his argument with press reports and using his powers of rhetoric learnt from his twenty-five years of teaching.

So bold was Espy's theory that it instantly caught attention.

Admiring articles appeared in the press and were republished across the Atlantic. For his Theory of Hail he was awarded the prestigious Magellanic Prize by the American Philosophical Society. Suddenly the Pied Piper of American science, he turned his pen to a new purpose: to destroy all opposition to his ideas. This included Redfield and his concept of whirling winds. He resolved to take an axe to Redfield's theory and chop it down piece by piece.

Redfield first became aware of James Espy in early April 1835, though he did not quite know it. His attention was caught by an anonymous article, 'Notes of an Observer', that had been published the previous year in the *Journal of the Franklin Institute*. The piece was a critique of one of Redfield's articles for the *American Journal of Science*. Couched in the deferential style of the day, it had begun by praising Redfield's nimble prose as well as his 'laborious' efforts on behalf of science. Then it changed tack. The anonymous author revealed that he was disturbed by a number of Redfield's assertions. They were, he pointed out, 'so anomalous and inconsistent with received theories' that he could not put entire trust in them, 'and shall continue to doubt until I have the most certain evidence of the facts'.[15]

The 'Notes of an Observer' was a short article with enormous incendiary capacity. Issues like this were often chewed over in public but by several measures the piece fell short of the strict norms of the day. Firstly it was published in a journal unconnected to Redfield – making it likely that he would not see it and thus robbing him of the right of reply. Secondly it skipped over the central tenet of Redfield's argument. No mention, however brief, was made of his theory of circular winds. Such silence gave the piece a contemptuous air, bound to nettle. It was no surprise that Redfield remained entirely unaware of it for fourteen months until, in his own words, he chanced upon it. Aggravated, he sought to swat it away. Keen that his prolonged silence should not be misinterpreted, on 8 April 1835 Redfield sat down at his desk in New York City and dashed off a reply. The author of the anonymous piece was James Espy and this would be far from the end of the matter. It would mark the opening volley in a long war of attrition.

Redfield's response was curt. He sent it straight to the editor of the *Journal of the Franklin Institute*. He admitted that his claims about barometric variations in Edinburgh were wrong but brushed aside the other criticisms. Redfield was particularly scornful of the suggestion that his idea was speculative and 'inconsistent' with known science. Why should an idea be rejected or doubted for no better reason than being inconsistent or eccentric? 'Before we consent to give credence to new theories it may be well, perhaps, to inquire when and in what manner the "received theories" in meteorology have been demonstrated to be true.'[16]

Redfield reiterated his speculations about tides and his suspicion that, in some ways, they might be linked to the ebb and flow of the atmosphere. Though Redfield had not yet developed a complete theory he felt sure there was good ground to be made from this enquiry. He asserted, too, that he had read a sufficient amount to convince him of the folly of studying the flow of heat. He dismissed this as an 'error into which the whole school of meteorologists appear to have fallen'. Heat flow, Redfield wrote, might plausibly affect local weather – land or sea breezes – but it was far from being the prime factor in macro-atmospheric events. He repeated his assertion that storms were whirlwinds and, he added, this is caused by 'the rotative motion of the earth upon its axis'.

Redfield's thinking had progressed. No longer did he restrict his atmospheric investigations to storms: he was projecting them on to a bigger scale. He was beginning to see the sky as a theatre of moving parts governed by Newtonian laws of force and attraction. Essentially his ideas were drawn from his background as a mechanical engineer. He thought the atmosphere had a collection of great cogs and wheels like those of his steamers or George Stephenson's revolutionary locomotive *Rocket*.

There matters rested until April 1836, when Espy's series of meteorological essays started to appear. At first Espy concentrated on establishing his own ideas but by July he was ready to attack Redfield openly. In 'An Examination of Hutton's, Redfield's and Olmstead's Theories', Espy set about dismantling Redfield's ideas with the confidence of a man triumphant. No longer firing his arrows from the shadows, he was happy to attack his opponents directly. His critique of Redfield focused on the central claim of his theory: the evidence of the 1821 September Gale. 'Now it will appear by little reflection that all these facts agree with the idea of an upward vortex more consistently than with a horizontal whirlwind,' he pointed out. What would happen, Espy argued, when the winds that Redfield claimed rushed at each other met in the atmosphere? 'All the facts lead to the conclusion that in the storm at least the wind in the neighbourhood of the storm blew directly towards the centre and if so, it follows beyond all doubt, that there was an upward vortex in the middle of the storm.'[17]

Not only was Espy's prose bold, it also had a pompous edge. Academic discourse was governed by the same strict code of gentle-manly civility that dominated intellectual society, from politics to religion. At times Espy seemed waspish and condescending. None of his barbs would have cut deeper than the claim:

> If Mr Redfield should perceive that all the interesting facts which he had with such laudable industry collected, are fully explained by a theory which accounts also for the rain, I am sure he will not be very tenacious of his horizontal whirlwind; especially when he does not pretend to show that either the whirlwind is the cause of the rain, or the rain the cause of the whirlwind . . . I should be proud to enlist Mr Redfield under the banner of a true theory.[18]

But Espy was about to discover that Redfield was no timid oppo-nent. He hit back seven months later, in February 1837. Writing in the *Journal of the Franklin Institute*, he titled his piece bluntly, 'Mr Redfield in Reply to Mr Espy on the Whirlwind Character of certain Storms'. He announced that he had no wish to defer to Espy's theory of an atmospheric chimney. Instead he stood by the validity of his data and revealed that he had much more evidence than he had yet published. Redfield's reply covered fifteen scornful pages and stretched to around 10,000 words.

Already both men were entrenched. To Espy, Redfield's position was weak. He had no overarching theory, just a mere collection of nebulous observations about tides, and his great claim of whirling winds did not fit with any facet of known science. To Redfield, Espy was plain wrong. Enamoured with his idea he had grown too cavalier, too bombastic, and as a result he had closed his mind to the realities of fact. 'Mr Espy says that steam is the moving power in storms,' Redfield wrote, certain he had caught Espy out on a point of logic, 'but there is a much smaller supply of this moving power in winter than in summer and yet the greatest storms appear in winter.' How could it be?[19]

Espy ignored the point. And while a resolute Redfield vowed to produce more data to consolidate his position, Espy opted to take his ideas out on the road. The first of his lectures took place in November 1836 when he spoke before a 'large and respectable body of merchants, under writers, masters of vessels and others' at the Exchange in Philadelphia. 'The views of the Lecturer are not only new, but exceedingly ingenious and interesting, and calculated to convince any one who had not hitherto examined the subject,' reported the *New Bedford Mercury*.[20]

Espy's Philadelphia lecture was a prelude to a tour of the popular Lyceum circuit over the next four years that would take him to Harrisburg, New York City, Nantucket, Boston and a host of other locations. Standing on the stage with his charts and his nephelescope cloud chamber, he employed his gift as an orator and his natural charisma to make science accessible for a wide audience, logically explaining his arguments and citing his position as chairman of the Joint Committee as proof of his authority. It was a formidable combination. Glowing reviews followed in the local newspapers, and the appearances had the secondary benefit of boosting the size of his weather network. In January 1837 the Pennsylvanian State representatives were handed a petition that called for Espy to be appointed as official meteorologist of the state. This proved successful and, on 1 April 1837, the legislature allocated a $4,000 grant to a state-wide, Espy-led meteorological investigation.

His ambitions did not stop there. With his friend Professor Bache poised to depart on a European tour in 1836, Espy wrote, 'Please present to Dalton, Faraday, Brewster, Forbes, Airy, Apjohn, Daniell, Whewell, Scoresby or if not convenient to them to any others you may find interested in the science, those essays of mine already published on the Theory of Rain, &c.'[21] Europe was an exciting proposition. Espy knew that in America he had already gained an influential power base. But to plant his theory on an immovable footing he needed the backing of at least one of the great European men of science, a name whose reputation would dazzle. What such support could mean – funding from Congress, moral authority – he could only guess. Espy knew that a meeting of leading British scientists would take place in Newcastle upon Tyne in August 1838 and with Bache as his ambassador he hoped to have his ideas presented to the judges that mattered. But, unknown to Espy, his wily combatant Redfield would soon outflank him. Waiting in the wings Redfield had acquired an ally of his own who was about to tip the scales. His name was William Reid.

By 1838 Reid had been researching storms for six years. He had worked with military precision, exploiting contacts during his spell in the Caribbean to locate as many sources as possible. The notes had piled up, and by the start of 1838 he had started to draw them together. By 1 February he had progressed enough to introduce himself to Redfield. He began a letter to Redfield in New York, telling him he had read 'with good attention' his ideas on storms and that he had been 'impressed with the importance of the subject'. He said he was planning to publish a paper on storms for the Corps of Engineers and he sent him some example charts. 'I think what I am giving will be gratifying to you; and I beg leave to say it will give me great pleasure to hear from you on the subject.'[22]

It had taken two months for Reid's letter to reach New York, but when he read it Redfield was delighted. 'It affords me much satisfaction to find that inquiring into the true nature of storms is beginning to excite interest on your side of the Atlantic, and the observations which a residence in the West Indies has afforded must give you a peculiar advantage in pursuing the subject.' It marked the beginning of a long and warm friendship. During the spring and summer of 1838 Reid and Redfield corresponded often.

They shared data from ships' logs, newspaper cuttings and plans for publishing ideas. Redfield congratulated Reid on the quality of his charts, which he thought 'beautifully executed'.[23] When Reid mentioned that he had noticed some of Espy's essays on sale in London, Redfield cautioned him. 'I am informed that he is an ardent and at the same time an amiable man; but having examined his evidence in the cases on which he relies, and collated it with such facts as are in my possession, I feel justified in stating that not one of his cases can be sustained.'[24]

The friendship between the men was already blossoming by August 1838 when Reid decided to travel, by 'some degree of accident', to the annual meeting of the British Association for the Advancement of Science. Still in its first decade of existence, the British Association was the fashionable society of the day. Conceived as a more inclusive and radical alternative to the Royal Society of London, it managed to combine old authority with a new sense of purpose. Some of the pre-eminent minds of British science steered its executive committee, including Sir John Herschel, the exceptionally able, famous son of Sir William Herschel. Now forty-six years old, Herschel had already founded the Astronomical Society, served as secretary of the Royal Society, made huge contributions to both physics and astronomy and written his fabulously successful *Treatise on Astronomy*. Others on the committee included Sir David Brewster, the Scottish 'Father of modern experimental optics' and inventor of the kaleidoscope, and John Frederic Daniell, respected for the Daniell cell battery as well as his hygrometer. The Association meeting for 1838 was to be held in Newcastle upon Tyne – in an effort to get away from London a different provincial town was chosen each year – and more than 2,000 tickets had been sold to the lectures that were to be held on a range of topics.

Among the speakers was William Reid who, by chance, had been asked to present a lecture on his storm research. His lecture on Monday 20 August before a packed crowd turned out to be one of the highlights of the week. Reid introduced Redfield's theory of circular winds, explaining how reading the article had set him off on his quest. It was the first that many in the room had heard of Redfield and Reid explained how he had endeavoured to test the theory against as many historical examples as he could find. 'The more exactly that this was done,' Reid revealed, 'the nearer was the approximation of facts to a progressive whirlwind.'[25] One by one Reid presented a series of eight historical examples. 'My object is not to establish or support any theory,' he said, 'but simply to arrange and record facts.' It was an enterprise of pure Baconianism, free from faction and bias. His audience loved it. He left them with a final, tantalising finding: that while all the storms he studied in the northern hemisphere blew in an anticlockwise direction, he believed all those in the southern hemisphere revolved in the opposite way. His theories, he told the audience, were collated in his forthcoming book, *An Attempt to Develop the Law of Storms*.

Reid's *Law of Storms* turned out to be a highly practical work crammed with expert research. It included tips for mariners, accounts of exemplary storms, and cut-out diagrams, charts and real-time maps for reference in a storm. It was reviewed enthusiastically. The *Edinburgh Review* declared:

Having such impressions of the vast importance of the subject, we earnestly implore Mr Redfield and Col Reid, whose names will be forever associated with it, to continue their invaluable labours, and press upon their respective governments the necessity of some liberal arrangements for investigating more effectually the origin and laws of these disturbers of the deep. If we cannot bind them over to keep the peace, we may, at least, organize an efficient police to discover their ambush and watch their movements.[26]

When his copy arrived in New York in October, Redfield was thrilled. 'It is just such a work as I have long wished to see undertaken by one who comprehended the subject and had leisure and means to establish the subject in a detailed manner.' An unexpected hit, the book also managed to catch the attention of Whitehall's political elite. Even before the *Edinburgh Review* went to press, news had leaked that Reid was to be appointed governor of Bermuda, as it was 'a position peculiarly favourable for carrying on his valuable researches'.[27] More developments followed. On 3 January 1839, Reid wrote to Herschel revealing that Lord Melbourne's government was putting its might behind his project. Lord Glenelg, Secretary of State for War and the Colonies, had told British governors across the globe to keep weather journals and to record any unusual meteorological phenomena. Results were to be transmitted to the Colonial Office every six months. The Admiralty had acted also. As well as purchasing a number of Reid's books to distribute among intelligent officers, they had ordered ships' captains as well as ports, harbours and lighthouse keepers to begin observations, too.[28]

No longer would meteorology be a lonely meditation, as it had been for Beaufort, Forster, Howard, Redfield and Espy. The new meteorology was to be a subject of networks. This shift was noticed by John Ruskin, a wide-eyed undergraduate at Oxford University. In 1839 he published an ornate, optimistic essay, *Remarks On the State of Meteorological Science*, written against the backdrop of fresh interest in the subject. He realised that the meteorologist was 'impotent if alone'. He could not nurture his genius in isolation.

Let the pastor of the Alps observe the variation of his mountain winds; let the voyagers send us notes of the changes on the surface of the sea; let the solitary dweller in the American prairie observe the passage of storms, and the variations of the climate; and each, who alone would have been powerless, will find himself a part of one mighty mind, a ray of light entering into one vast eye.[29]

Reid's appearance at the British Association was a watershed moment. Years of patient research had combined to show Britain's scientific community what was possible. In the months before he left for Bermuda he became a familiar name in the newspapers. His election to the Royal Society was fast-tracked. His nomination slip read, 'Lieutenant Colonel William Reid, CB, of the Royal Engineers a Gentleman much attached to science and author of a Work On the Law of Storms'. Among his proposers were Sir John Herschel, future president Edward Sabine, polymath William Whewell and the old weather hand Francis Beaufort.[30]

Reid's work was destined to appeal to Beaufort for its detailed research and careful observation. Active as ever, Beaufort had retained an interest in meteorology. Twice he had served on Royal Council sub-committees on the subject and over the past years he had emerged as the Society's de facto meteorological correspondent, often reading reports at Society meetings on fog, wind or rain. There was something in Reid's work that captivated and impressed him. It seemed a thoroughly Beaufortian enterprise and perhaps, too, it was a spur for him to renew his interest in his own weather system. On 28 December 1838 – thirty-two years after it was first jotted down – the Admiralty officially adopted the Beaufort Scale for naval use.

After three years in the ascendancy in America, Espy's ideas had been swept away with barely a whisper in the halls of European science. News of Reid's triumph travelled across the Atlantic along with a copy of Reid's Law of Storms. Espy was shocked. Expecting to hear that his theories had been heard and praised, instead he learnt that even Sir John Herschel had sided with Reid and Redfield. Espy responded in indomitable style. He sought to discredit the data in Reid's book, twisting it to fit his own model rather than Redfield's.

On reading the logs of the several ships I kept the map of the particular storm open before me, and, drew my pencil across the point where the ship was, drawing an arrow so as to exhibit to the eye which way the wind was blowing at that time in that locality.

When several logs were read, and arrows made in every locality
– I was not a little pleased to see, in all the storms, decided proofs
of an inward motion of the air, if not exactly to one common
centre, quite as nearly so as any one had a right to expect; because
oblique forces are known to exist, which must vary the direction
of the wind.[31]

In this way Espy remodelled each of Reid's storms. He did not
tolerate conflicting facts and derided them as errors that needed
stamping out. This was too much for Redfield, who by now was
monitoring the *Journal of the Franklin Institute* like a hawk. For some
time he had hoped to settle the dispute. 'I have contented myself thus
far with an attitude towards him which is strictly favourable,' he wrote.
Another time he had even knocked at Espy's door in Philadelphia to
introduce himself. Unfortunately, Espy was out.

In the spring of 1839 Redfield's tone changed. He wrote apologeti-
cally to Reid, 'Mr Espy continues to busy himself with his theory of
aqueous condensation. His late amusements, as I am informed, have
given offence to the savants of Philadelphia to whose friendly support
he is mainly indebted.'[32] Another time he raged, 'It is a matter of
sincere regret with me that our friend Espy is not content with his
own conclusions relating to the storm of 1821, as you have seen in the
March number of the *Journal of the Franklin Institute*. To this I have
felt it necessary to reply and speak with freedom as well as candor of
his general course of proceeding in this matter, all of which I was
desirous to avoid.'[33]

It was not long before Espy felt the full force of his invective. In
July 1839 two waspish articles appeared in the *Journal of the Franklin
Institute*, the one true beneficiary of the brouhaha. Redfield expressed
his objections with caustic lucidity. He pulled Espy's analyses apart
storm by storm. He derided Espy's mission to 'explain nearly all the
physical phenomena of the earth by the theory of aqueous condensation' and regretted that he had been supported by 'plausible but
erroneous inductions' and 'friendly, though perhaps injudicious
support and announcements from highly respectable sources; and
aided also by the favor and guardianship of the Philadelphia press'.
He lampooned Espy's 'modest' announcement of April 1836 and
complained that he had spent far too much effort trying to fit the
facts to the theory rather than the theory to the facts. Espy's attempt

to connect every single atmospheric process to his ideas, Redfield complained, had become 'not unlike that of him who in essaying to climb should commence at the last and highest step in the ladder'.[34]

No longer friendly joshing, by now the furore was generating national coverage. *The Knickerbocker*, perhaps the most influential literary magazine in America, rejoiced in Redfield's assault, proclaiming Espy's ideas 'essentially demolished'.[35] Suddenly Espy's reputation was in danger of unravelling. Partly this was due to Redfield's exposé but it was also due to something even more improbable. Over the last few months Espy had formed a new pet idea and had started to advertise it at his lectures. He was claiming to be able to create rain.

Espy could have hardly chosen a more inflammatory path. Many had toyed with meteorological theory over the years, but few had aspired to command the elements themselves. But while the rain-making idea seemed ridiculous, to Espy it was merely a logical extension of his thinking. Now he had demonstrated how clouds and rain were formed, what better than a practical demonstration to prove him right? The 1830s had seen a craze for public experimentation most recently with early demonstrations of Daguerre's photographic process. For a country growing increasingly used to ground-breaking inventions – the decade had already seen the appearance of ether-based anaesthesia, grain elevators, a hand-cranked ice-cream machine, the first combine harvester, a steam shovel and a lockstitch sewing machine – the idea of creating rain tapped into the spirit of the age. Anything was possible.

Espy reasoned that if he produced a steady, controlled column of ascending air it would draw up the atmospheric vapour. When this vapour reached the ideal altitude it would condense into water. All that was needed was to produce a steady column of heat and the atmosphere could be modelled artificially. In some places this already happened, Espy told his audiences. He cited London – the most populous city on the planet – as an example, claiming that the hundreds of thousands of billowing chimneys combined to create a rainy microclimate in the Thames Valley.[36]

The American government was currently overseeing a vast process of deforestation in the newly settled states and Espy seized on this. If the government would let him start strategically placed fires in the west, he could exploit the prevailing winds to bring rains to the eastern states. The benefits would be immense. Canals could be filled with water to prevent drought. Farmers would be able to manage precipitation to produce perfect crops.

At the start of 1839 Espy approached Congress, proposing to test his idea. The plan excited and terrified in equal measure. The *Philadelphia Gazette* published a succinct appraisal of the dilemma:

> If our indefatigable friend Espy should succeed in establishing his theory with respect to the rains he will have effected a more wonderful improvement upon the weather, which, it must be confessed, is often sadly deficient in rectitude of conduct. The times of drought and 'some potatoes' will cease – for every farmer will have it in his power to burn a pile of wood, and grow his own thunderstorm. Think of taking out the copyright of a tempest or a patent for a whirlwind! The danger is, we think, if Espy succeeds, and the modus operandi of his system becomes known, that any one who likes will take possession of it, and create much mischief thereby – Naughty persons would cultivate storms from malice aforethought; – and they would come in time we fear, to conflict with the freedom of elections. The fires of party spirit would result in a deluge; and success, irrespective of the merits of measures or men, would crown the side that could make the most water! The progress of storms may yet be so great that a man may travel whithersoever he listeth by fastening to his carriage front a huge boiling tea pot, on wheels, and when this shall come to pass, and men will have it in their power to fling tornadoes or chain lightning at each other, we shall begin to tremble for the safety of our republican institutions.[37]

Elsewhere Espy's proposal elicited different responses. In February 1839 the *New Hampshire Sentinel* politely wondered whether 'Mr Espy would put his weather machine in motion and give us a good storm of snow – just enough to make good sleighing and no more – we would thank him in the name of the thousands of people who have not heard a set of sleigh bells this winter.'[38] In July the New Orleans *Times-Picayune* revealed, 'On Friday we prayed for a "forty horse Espy power" when lo! on yesterday morning, while yet in our gown and slippers, performing our diurnal ablutions, the rain descended in torrents, and with a force and velocity as if impelled by a power unknown to modern mechanics. Flash ! flash ! flew the lightning – and Boom ! boom ! went the thunder.'[39] Indeed some wit in the *Times-Picayune*'s newsroom had developed a special interest in the story. The writer tracked Espy's lecture tour, noting the strange frequency with which his arrival in a town was preceded by a sharp shower. The paper started to refer to Espy as 'Professor of Thunder and Lightning'. Once Espy even inspired the *Times-Picayune* to poetic rapture: 'I feel myself dissolving now – up in the clouds that's plain! Where's Mr Espy? In thin air, Diffused – evaporated, Without him, ne'er again I fear, To be precipitated.'[40]

His friends realised that Espy was risking his reputation. He seemed to have forgone the sure and steady ground of science for a hopeless, hubristic scheme. His rain-making claims certainly brought him attention – the English essayist William Hazlitt once wrote that 'The world loved to be amused by hollow professions, to be deceived by flattering appearances, to live in a state of hallucination; and can forgive everything but the plain, downright, simple, honest truth' – but was it the type of attention he needed? Bache ruminated on the 'strange course recently taken by my friend Mr Espy', while another of his circle, Joseph Henry, worried that he sometimes exhibited a 'want of prudence'.[41]

But Espy continued. In 1839 he petitioned Congress to back him publicly, asking for a grant of $25,000 in return for making rain over 5,000 square miles or $50,000 in return for an area over 10,000, 'or in such quantities as shall keep the Ohio River navigable during the whole summer'.[42] In the middle of a crippling financial crisis, it was no surprise that Espy's request was turned down. Instead he was left to debate with Professor Olmstead, Redfield's ally, at a series of public lectures in New York. The lectures were typically heated – both Olmstead and Espy passionate advocates of their ideas. Soon after, a letter appeared in the *Boston Evening Mercantile Journal* applauding Espy for his great contribution to American scientific life. 'The voice of ridicule is at length hushed,' it began. 'The name Espy may hereafter stand as high upon the list with Galileo, Harvey, Franklin and those other names precious to science and humanity.' Redfield kept Reid abreast of all this in his letters. 'Mr Espy . . . has lately given a course of lectures in this city, which was rather thinly attended. He maintains a thorough system of newspaper *puffing* and in this he has *full swing*, no one being disposed to enter the field as his antagonist.'[43]

While Redfield grumbled in private, Espy carried on. It was now high noon for the storm controversy: something had to give. Still smarting from the snub of the British Association in 1838, Espy resolved to extend his lecture tour with a transatlantic leg – a move that would allow him to plead his case to Europe in person. He sailed from Philadelphia on 6 June 1840 for Liverpool, bound for the annual meeting of the British Association, this time to be held in Glasgow. In the wake of Reid's storm book, meteorology was undergoing a mini renaissance in Britain. A detailed paper on the state of the science had been written for the Association by James Forbes, a professor of natural philosophy at Edinburgh University, and a succession of weather-related lectures had been prepared for the mathematics and physics sections. Two of them ironically enough focused on the effects of excessive rains while the third, by James Espy the rainmaker, was simply titled 'On Storms'. Brewster and Forbes sat among the audience that day. If Espy was to ever make his case, now was the time to do it.

He spoke for nearly two hours, carefully outlining his ideas. To appeal to a European audience he had selected a British storm from January 1839, like Reid before him using charts to illustrate his points, always stressing how the winds travelled in towards a central point. But the lecture did not go as Espy anticipated. When he finished he was questioned at length. Brewster produced a letter from William Reid which stated unequivocally that he had examined five waterspouts with a telescope, 'in all of which it appeared that there was a revolution of the particles of water in the manner of the hands of a watch, from left to right'. Forbes then presented Espy with several other difficulties. He thought it unlikely that the sheer mass of moving air in a storm could pass safely up inside a vortex. Equally he worried that if vapour was drawn up an ascending column of air, it would condense into water at a very low height.[44]

Espy laboured valiantly against these questions but he found that his audience had already been swayed by Redfield's idea and Reid's research. Espy left the meeting with the chairman's gratitude but little else. Writing to Redfield shortly afterwards, Reid gossiped: 'I hear from England that people's minds were satisfied with the revolving theory of Storms; so that few cared to listen at Glasgow, and elsewhere, to Mr Espy's explanations of his particular theory.'[45]

Espy found more support in France. Ever since the days of Descartes the French had maintained an interest in water vapour and Gay-Lussac's researches were a recent memory. The Académie des Sciences Committee listened to Espy and were impressed by what they heard. A commission, including such figures as François Arago, one of the stars of French science and Director of the Paris Observatory, and the physicists Pouillet and Babinet, was formed to scrutinise Espy's arguments and at length they produced a glowing assessment. 'The committee expressed then, the wish that Mr Espy should be placed by the government of the United States in a position to continue his important investigations, and to complete his theory, already so remarkable.' Another quote travelled home with Espy, possibly apocryphal, possibly accurate, from Arago himself: 'England has its Newton, France its Cuvier, and America its Espy.'[46]

Encouraged, Espy returned to America determined to capitalise on his success. He began by gathering together all his meteorological papers and publishing them in book form. The result was a repetitive tome with a succinct title – *The Philosophy of Storms* (1841). Espy began with a preface that set out with characteristic gusto his meteorological awakening in 1828 and ended with his triumphant appearance before the French Academy twelve years later, two moments that bookended what he clearly considered a stellar academic career. He lauded the French report as 'a beautiful analysis' and included his critical papers on Redfield, Reid and Olmstead. More a catalogue of Espy's published works than a coherent book, *The Philosophy of Storms* nonetheless manages to convey the author's enthusiasm. Page after page thrums with urgency. All that was now wanting for his complete triumph, Espy wrote, was time. 'After the controversy shall be terminated, and my system admitted to take its place among the acknowledged sciences, it will be time to write out a set of rules to assist the mariner in storms to the best advantage.'[47]

The Philosophy of Storms could not bring an end to the arguments but at least it helped Espy consolidate his reputation. It was now more than a decade since his meteorological quest had begun. All that time had been spent in an academic world of theoretical contention, and now Espy was ready to test himself in the practical sphere. He invented a 'Patent Conical Ventilator' – a conical chimney top, designed to maximise the upward flow of air – to help cleanse 'foul' atmospheres like that of a ship's engine room. Marketing it with characteristic aplomb, Espy even managed to have his invention fitted to the Capitol building and the White House in Washington – something he proudly noted in his publicity posters. This aggrandisement was beginning to annoy some. Benjamin Peirce, the Harvard mathematician, moaned about Espy and his 'air of self-satisfaction'. 'Even storm kings are intolerable in a republic.'[48] Espy succeeded in also irritating the old president, John Quincy Adams, who observed, 'The man is methodically monomaniac, and the dimensions of his organ of self-esteem have been swollen to the size of a goitre by a report from a committee of the National Institute of France, endorsing all his crack-brained discoveries in meteorology.'[49]

Such attacks did not seem to bother Espy, who had exchanged the familiarity of his Philadelphian base for the society of Washington. He aimed to turn the Academy's recommendations into reality and his efforts were soon rewarded. In August 1842 Congress set aside $3,000 for meteorological observations at military posts throughout the country. As the most prominent man in the field, Espy was chosen to oversee the project, making him, in 1843, the first ever official meteorologist of the United States of America. Espy's task was to write and circulate reports and to expand the network of weather-watchers.

Espy's appointment was a sign of the government trying to get a grip on the debate. By the early 1840s an overwhelming amount of contested research had been published. The appearance of Espy's long-winded book in 1841 had only added to the mass of half-digested facts. This deluge of data – between them Espy and Redfield had perhaps cited 100 storms – was half the problem. What was required now was not a new storm. What was needed was an objective figure and a method of corralling everything into a coherent whole. What was needed was something that is so common in today's society that we barely consider it an invention at all. What was needed was a weather map.

CHAPTER 6

Liquid Lightning

By the late Georgian age, mapping had become an important pursuit. An accurate map expressed the mastery of a terrain and fitted with the contemporary fashion for quantification. In 1791 the Ordnance Survey had been founded in Britain and ever since then triangulation surveys had been underway, reducing the varied peaks, vales, plains and glens of Britain into a set of homely symbols – information that could be absorbed at a glance.

While maps of the land and sea, like those commissioned by Francis Beaufort, had accumulated, few people had tried to do the same for the atmosphere. The earliest meteorological map was by Edmund Halley, who plotted the trade winds on the globe in 1686. But since then little had been done. In 1816 Heinrich Wilhelm Brandes, a professor at the University of Breslau, had apparently plotted a year's worth of historical weather data for 1783 on charts. None of Brandes' maps were published (some academics have doubted whether they were produced at all) and the only extant weather charts from Germany are from one of Brandes' students, Heinrich Wilhelm Dove, who had attempted to map weather fronts in 1828. These, along with some threadbare charts of Reid, Espy and Redfield's, were as far as meteorological mapping had progressed by 1840. It now seems like a historical anomaly, but at the time a good map expressed two things: precision and ownership. As the atmosphere was so imprecise, how could it adequately be caught on paper? Equally maps of lands or estates would often be commissioned by wealthy patrons: kings or politicians or dukes or squires, eager to chart the orbit of their powers. But who could possibly own the sky? What benefit was there in mapping it?

This attitude had stifled progress for years, but by the 1830s the European vogue for map making had crossed the Atlantic. One of those responsible for popularising the form was William Woodbridge, a Yale graduate. As a young man Woodbridge had taught in the Asylum for the Deaf and Dumb in Hartford, Connecticut, where he had discovered the power of visual information. After a European trip – when he had met Alexander von Humboldt – he had returned to America with an idea to display scientific information graphically. Woodbridge's plan to distil scientific data into maps challenged the orthodoxy that information was best learned by rote. But like a twenty-first-century infographic, it proved to be an effective way of arranging information. Woodbridge published a set of instructive maps in the 1820s, including his 'Isothermal Chart, or View of Climates & Production, Drawn from the Accounts of Humboldt & Others', which showed the boundaries of the equatorial, torrid, hot and temperate regions, by latitude, as well as the zones that produced various commodities: the finest spices, sugar cane, cotton, olives, wine grapes and peaches.

As Edmund Halley had with the trade winds before, Woodbridge showed that climate could be graphically depicted. But plotting weather seemed much more futile. As its movements were so swift what could be learned scientifically of a snapshot of a single atmospheric moment? Would a weather map not just be a record of complex arrangement of temperature, pressure, wind and rain never to be repeated?

It was another Yale graduate and a member of Espy's meteorological network who changed this belief. In 1840 Elias Loomis was twenty-nine years old. A professor of mathematics and natural philosophy at Western Reserve College in Hudson, Ohio, in 1840 Loomis was at the beginning of what was to become an illustrious career in American intellectual life. Already he had published research on shooting stars, comets and magnetism. Quietly watching the quarrels between Espy and Redfield, he had become interested in meteorology.

Loomis was a different character to Espy and Redfield. Reserved and intense with a devotion to Francis Bacon's ideal of objective science, he had used his inaugural address at Western Reserve in 1838 to call for a greater respect for the discipline in America. The country, he argued, should retain a class of men entirely devoted to science, and 'that such men, instead of being a dead weight on society, are to be ranked among the greatest benefactors of their race'.

Loomis' words resonated and were picked up by the national newspapers. The *American Quarterly Register* wrote:

We rejoice to behold in a youthful professor an ardour which men of phlegmatic temperament might condemn. Without it, no high eminence, no distinguished usefulness will ever be attained in the department of knowledge or of life. The address of Mr Loomis is crowded with interesting statements and illustrations, intended to show the practical value of the mathematical sciences. The unscientific reader may peruse it with the greatest interest.'[1]

Soon after taking up his post in Ohio Loomis had begun a weather diary, determined to involve himself in the storm debate. He had written to Espy, volunteering to keep a weather journal and his name was soon appearing in Espy's reports:

Western Reserve College, Hudson, Ohio. (NE Corner) – From Our Correspondent Prof Elias Loomis: March 15, dense drizzling fog, wind faint from NW, 16th Wind light from NW to NNW with some snow and drizzling. 17th wind fresh in the morning, strong in the afternoon from N varying from about NNW to NE (March wind). 18th perfectly clear and bright; wind light from NNW to N. The barometer nearly stationary on the 16th and 17th at about 28.86, on the 18th it fell to 28.79, and on the 19th to 28.47.[2]

It was a typically detailed report. By 1840 Loomis had resolved to extend his meteorological research. Believing that much of the existing data was tarnished by bias, he picked one specific storm, a memorable storm of 20 December 1836, and set out to study it in immense detail.

This was a shrewd choice. Vast and intense, the 20 December storm had stretched across the entire eastern side of North America. More importantly, it was well recorded. Aiming to produce a set of synchronous observations from places around the globe, Sir John Herschel had appealed to the scientific community to take thirty-six consecutive hourly readings of temperature, wind direction and barometric pressure. There were only a few who could have drawn the scientific community together in this way, but such was Herschel's

standing that people listened. The plan was set. At each equinox and
solstice a series of meteorological readings had been taken. As the
20 December storm hit just before the winter equinox, Loomis knew
there was a good chance that he could source data from right across
America.

Loomis worked diligently. He drew up a list of observers from the
New York Register 1837, petitioned Elisha Whittlesey, the Congressman
for Ohio, for access to the military weather readings, wrote to school-
masters and approached academies in New York. Within a few weeks
Loomis had established a network of 102 data points: diarists,
academics, judges and military men.[3]

His findings were read to the American Philosophical Society in
March 1840, impressing those present with the scope and depth of his
research. He gave an overview of the storm's characteristics and its
geographical extent. He did not think it was limited to the eastern
side of the country. He had uncovered data to suggest that it ranged
from the Rocky Mountains in the west, to the mid-Atlantic in the east;
south towards the equator, and an indeterminate distance north.
Loomis had even sourced a ship's log from Buenos Aires and European
data from Brussels, Milan and St Petersburg to compare what was
happening elsewhere.[4]

When studying wind direction he had formed another hypothesis.
In several places the wind seemed to be blowing, as Espy contended,
towards converging spots. But instead of concluding that air was then
drawn up an invisible chimney at the point of collision, Loomis felt
that it was more likely that the cold wind rushed underneath the
warmer wind like a door wedge – an idea he depicted with a diagram.
In any case, he wondered, the wind would be disturbed by any number
of local factors. He imagined the path of a river and how the particles
of water would be deflected on their way by pebbles or rocks:

An effect similar in kind, though much greater in degree, must
be expected from an elastic fluid like air. This is strikingly
exhibited in the narrow and straight streets of cities, with high
buildings on each side. The wind must here blow in the direction
of the streets, or not at all. So, also, a mountain gorge; straight
bed of the river with high banks; the shore of a lake, or the
ocean; or a mountain ridge might be expected sensibly to influence the direction of the atmospheric current.[5]

The wind, Loomis continued, varies 'not merely from day to day, and from hour to hour, but from minute to minute, and from second to second'. This made any study of it incredibly difficult and any analysis equally troublesome. To express this he added several charts at the end of his paper. They were drawn at six-hourly intervals, showing the progression of the storm and lines of equal barometric pressure.

Loomis' paper was well received. It was published the following year in the *Proceedings of the American Philosophical Society*, 'On certain storms in Europe and America: December, 1836', and lauded as the most complete study of a storm yet attempted. Loomis, though, was not satisfied. He only had good data for the southern flank of the storm, and the northern part of it had to be 'supplied by conjecture'. Nonetheless it had proved a useful starting point for Loomis and, as he later acknowledged, developed 'some peculiar methods . . . [which] had never been practised before'. Loomis was referring to his idea to map the storm at intervals of twelve hours, 'in such a manner that every important feature was made to appeal directly to the eye'.[6]

His framework now in place, Loomis waited for another opportunity. It soon came, in February 1842, when two storms blew through Ohio. Using the same network of correspondents that had supplied information about the 1836 storm, Loomis was able to harvest great quantities of data. He set to work immediately. By the spring of 1843 he was able to forward the result to Alexander Bache to read before the American Philosophical Society. The workmanlike title of the paper, 'On Two Storms which Occurred in February 1842', belied its significance. Where Loomis' first study had been applauded, his second created a sensation.

It was Loomis' maps that caught attentions. They were drawn with the same clarity and flair as Woodbridge's climate maps. Loomis had devised a simple colour-coded methodology that made the state of the atmosphere clear at a given moment. He used pastel colours: light blue for clear skies, violet for clouds, yellow for rain, green for snow and red for fog. Where he had an observation for wind direction he expressed it with an arrow, and he drew dotted lines through points of equal barometric pressure and temperature. With all this transposed over the eastern edge of the North American continent the viewer was able to see, in an instant, that on the evening of 3 February 1842 it was simultaneously foggy over New Bedford, cloudy over New

York City, raining over Harrisburg and snowing over Cleveland. To show the advance of what was clearly a moving block of weather, Loomis drew real-time maps, one for every twelve-hour split of the storm.

Like Eadweard Muybridge's photographic studies of a galloping horse later in the century, Loomis' maps provided a sequence of snapshots never before seen. The emphasis was on process. A judiciously picked freeze frame of the atmosphere could be used to support any idea as Redfield and Espy had proved. Loomis' mapping, though, was different. He had not attempted to create any theory, he simply wanted to invent an objective medium that could be interpreted by anyone. In doing so Loomis dramatised weather on paper, inventing what we today call the synoptic map.

Loomis' 'On Two Storms' would later be celebrated as one of the most important contributions in the history of meteorology. His biographer, H.A. Newton, would write: 'The method [of making a map] seems so natural that it should occur to any person who has the subject of a storm under consideration . . . But the greatest inventions are oftentimes the simplest.' Loomis, unable to present his own ideas to the Society directly, wrote a persuasive accompanying piece, arguing that if two daily weather charts were made each day for a year, 'some settled principles' would soon emerge and 'settle the law of storms'.

> No false theory could stand against such an array of testimony. Such a set of maps would be worth more than all which has been hitherto done in meteorology. Moreover, the subject would be well nigh exhausted. But one year's observations would be needed. The storms of one year are probably but a repetition of those of the preceding. Instead then of the guerrilla warfare which has been maintained for centuries with indifferent success; although, at the expense of great self-devotion on the part of individual chiefs, is it not time to embark on a general meteorological crusade?[7]

Loomis had a plan for his meteorological crusade. It involved a centrally controlled network of observers: well informed, dependable and accurate. They would span the twenty-six states, a territory especially suited to the study of atmosphere. The vast expanse would, he argued, allow for even the biggest storm to be studied. Should the Europeans attempt something similar they would have to over-come all the problems of politics, language, rivalry and practicality. In the United States these difficulties would be diminished. 'Here is one where the advantage is in our favor,' Loomis wrote. All that was needed was five or six hundred observers to complete the plan. It would be a weather experiment like no other before. 'If private zeal could be more generally enlisted, the war might soon be ended, and men would cease to ridicule the idea of our being able to predict an approaching storm.'[8]

Loomis' use of the word 'predict' here had the potential for contro-versy. It was a step beyond just understanding. Yet Loomis was confident that this was no spurious ambition. Rooted in a system of minute observation and conducted by a respected and approved governmental body, weather prediction became a distinct possibility if the process of recording, transmitting and interpreting data from disparate geographic positions could be undertaken quickly. Should such a process be possible then all the theoretical progress meteorology had undergone over the past few decades – an understanding of storm paths, the likely direction and strength of wind – could be turned to the advantage of the American people. The system could bring warnings of an imminent storm or even of a transatlantic hurricane. It was a tantalising prospect.

Loomis' paper was written in the spring of 1843, and though he did not appreciate it at the time an intriguing new technology was being trialled that would one day become central to his vision of a national weather system. In February 1843, after much delay, President John Tyler had signed a bill that granted $30,000 to a professor from New York called Samuel F.B. Morse for a test of his new machine, an updated version of Edgeworth's old idea, that he was calling 'The Electro Magnetic Telegraph'. Morse's invention harnessed the power of electricity to send messages through insulated wires. The news-papers had taken an interest in the progression of Morse's plans. They did not seem to know what to expect. Was it a chimerical folly? A costly vanity project? Nobody could be sure.

Throughout the summer and autumn of 1843 Samuel Morse had overseen the construction of a prototype line between Washington and Baltimore. There had been problems. Trenches dug for the wires had proved unsuitable and Morse had taken the decision to suspend them on poles in the air instead. The change of tack had brought him into conflict with one of his partners, the latest in a series of quarrels. On 18 December 1843 Morse had written to his son revealing that 'Troubles cluster in such various shapes that I am almost overwhelmed.'[9] It was a frustrated plea from a man at breaking point. From the start Morse's telegraph project had been a struggle against adversity. He had suffered financial ruin, the collapse of a career, the treachery of a friend and one political snub after another. He complained of 'snares' in his path, of 'dark and discouraging days'.[10] He was haunted by the words of a friend who had talked about the fate of the 'great inventors, who are generally permitted to starve while living and are canonized after death'.[11]

Years of effort had brought Morse to this point. His promise to the government that they would soon be able to send messages up and down the east coast weighed heavily upon him. But Morse had kept faith with his machine throughout – a series of wires and magnets fuelled by what some called 'liquid lightning'. His invention had the potential to revolutionise communication. It was a vision that had come to him on an Atlantic voyage, eleven years earlier.

For a week at the start of October 1832 winds cut over the Normandy port of Le Havre from the south-west. They penned in a fleet of merchantmen bound for India, South Sea whalers and the packet ships that operated a weekly relay, carrying cargo, mail and passengers between Europe and New Orleans, New York and Boston. Among those already embarked and waiting for the wind to veer round was forty-one-year-old Samuel F.B. Morse, who had boarded the packet ship *Sully* to complete his homeward route to New York. After days of delays, late on 6 October 1832 Morse scribbled a quick letter to his friend, James Fenimore Cooper. 'We are getting under way. Good bye.'[12]

Morse was tall, lean and dark-eyed; the last of his youth was fading as he stood on the *Sully*'s deck, watching the Normandy coast slip into the autumn night. With the prospect of a six-week voyage before him, Morse had time to meditate on his second great European trip. In many ways he had repeated the arc of his first tour – visiting galleries, painting, soaking up the culture – which he had undertaken twenty years before. Back then Morse had been full of artistic ambition, but for all his talent his career had not progressed as he had

hoped. As a young man, living in London in the same artistic community as John Constable, he had written home to his parents: 'My
ambition is to be among those who shall revive the splendour of the
15th century, to rival the genius of a Raphael, a Michael Angelo, or a
Titian; My ambition is to be enlisted in the constellation of genius
which is now rising in this country.' Twenty years later, on the cusp
of middle age, Morse's promise had faded. He was known as an artist,
but not celebrated. In his native New York he had risen to a position
of some note as president of the National Academy of Design. But
the inescapable feeling was of talent wasted and opportunities missed.
His best days already seemed behind him.

Morse's second European tour was an attempt to rekindle his artistic
career. He had spent three years visiting Britain, France and Italy. At
the Louvre he had embarked on a new project, a portrayal of the
interior of the grand gallery. The result was charming, chocolate-box
art – a nineteenth-century equivalent of the modern-day 'Best Of'
music album, that included microscopic depictions of forty-one paintings by masters like da Vinci, Claude, van Dyck and Poussin.

Morse hoped to exhibit *The Gallery of the Louvre* in New York on
his return. It was an engaging composition likely to appeal to cultured
Americans who had neither the time nor funds to make the lengthy
crossing to Europe. The painting had been stowed in the *Sully*'s hold
and the ship had begun her cruise into the Channel, crossing the
stretch of water that in the past year had seen Reid pass on his way
to Barbados and FitzRoy and Darwin shortly afterwards on their
voyage to South America. Joining Morse on the *Sully* for the passage
was an unusually intellectual blend of passengers: W.C. Rives, a senator
from Virginia and future American minister to France; Dr Charles T.
Jackson from Boston, a bespectacled twenty-six-year-old Harvard
graduate who had already made something of a name for himself in
medical research, and twenty-five or so others – businessmen,
academics and politicians, making the journey back from the old world
to the new.

This unfamiliar group were bound together in the close confines
of the *Sully*, and each evening they dined together. It was an unusual,
claustrophobic but highly sociable environment – a perfect place for
swapping stories. One of the chief raconteurs was Dr Jackson, who
delighted in telling his fellow travellers about the scientific lectures he
had attended in Paris. One evening after dinner Jackson had begun a

conversation about the curious properties of electricity. Morse had been at the table, opposite Rives, and he had listened to Jackson eagerly. Jackson recounted his experience of a memorable experiment at the Sorbonne where he had watched as an electric spark fizzed four hundred times about the great lecture hall in an instant.

Morse's recollection of this conversation was to remain with him for the rest of his life. He was sitting opposite Dr Jackson as Jackson gave his account of the lecture. Another passenger then asked whether a lengthy wire would obstruct the passage of an electric current. Jackson replied, 'No, Benjamin Franklin has demonstrated long ago in London that electricity travels at once through any length of wire.'

'I then remarked,' Morse remembered, 'if the presence of electricity can be made visible in any part of the circuit, I see no reason why intelligence may not be transmitted instantaneously by electricity.'[13]

It was as he spoke this sentence, Morse later claimed, that an idea appeared in his imagination. He excused himself from the table and went out on deck. As an artist Morse was used to being seized and carried along by ideas. Now he was projecting this mode of thinking in a different direction altogether, into the formation of a machine capable of transmitting intelligence by electricity. Such was the speed of electricity that messages could not just be sent five miles from one station to the next, but tens or hundreds or thousands of miles.

In fact this idea had occurred to Morse before. On 17 August 1811, when a young art student in London, he had written to his parents:

> I think you will not complain of the shortness of this letter. I only wish you now had it to relieve your minds from anxiety, for, while I am writing, I can imagine mam wishing that she could hear of my arrival, and thinking of thousands of accidents that may have befallen me, and I wish that in an instant I could communicate the information; but three thousand miles are not passed over in an instant and we must wait four long weeks before we can hear from each other.[14]

If the idea had lain dormant in Morse's imagination then Jackson's conversation had released it. For nights afterwards he was unable to sleep for excitement. During the daytime he dedicated his many free hours into developing the idea. It was the perfect environment. Free from distraction and with weeks of the crossing remaining, Morse could quietly work on his project in his artist's sketchbook. With no access to literature, his plans were incubated in an environment of intellectual independence. All Morse had was his existing knowledge of electricity – gained from a series of scientific classes at Yale College and a set of public lectures in New York by James Freeman Dana in 1828 – his natural ability for free-thought and the help of the other passengers. He would later claim, 'I cherished it as an antidote to ennui, maturing my invention principally in the sleepless hours of the night.'[15]

At length the visions in his imagination became drawings. If electricity could travel instantaneously then his challenge was to use this to make a signal. He realised that one way of signalling was to open and break the electrical current. In his sketchbook he drew the series of incisor-shaped jigsaw pieces that would become a key feature of his first design. He thought the current could be made to flow or stop to a rhythm dictated by the shape of these little pieces. Morse also considered using electric sparks to chemically decompose a strip of paper. He asked Dr Jackson, who was well-read in chemistry, whether this would be possible. Jackson replied that electric charge should leave brown marks on a roll of paper stained with turmeric and coated with sulphate of soda. Excited by the idea, the two men agreed to experiment in Jackson's laboratory in Boston to find out which chemicals worked best.

By the beginning of November Morse's project had progressed so far that he had begun to share news of developments and drawings with the other passengers.

> I devised a single circuit of conductors from some generator of electricity. I planned a system of signs, consisting of dots or points, and spaces, to represent numerals, and two modes of causing electricity to mark or imprint these signs upon the strip of ribbon and paper. One was by chemical decomposition of a salt which should discolour the paper; the other was by the mechanical action of the electromagnet, operating upon the paper by a lever, charged at one extremity with a pen or pencil. I conceived the plan [as] moving the paper ribbon at a regular rate, by means of clockwork machinery to receive the signs.[16]

The first person Morse showed his sketches to was Senator Rives of Virginia, 'explaining with ardour their entire feasibility'.[17] By the time the *Sully* was nearing New York everyone on board was aware of Morse's pet idea. When the ship docked on 15 November, Morse bade the ship's captain farewell, saying, 'Well, Captain Pell, should you hear of the telegraph one of these days as the wonder of the world, remember the discovery was made on board the good ship *Sully*.' Met by his brothers Sidney and Richard, Morse told them at once of the 'important invention – one that would astonish the world'.[18] 'He was full of the subject of the telegraph during the walk from the ship, and for some days afterwards could scarcely speak about anything else,' Sidney remembered.[19]

Sixty-five years after Richard Lovell Edgeworth conceived his idea to transmit intelligence across space Samuel Morse was experiencing the same intoxicating thrill of invention. Morse's idea could be considered a grandchild of Edgeworth's, an heir to the optical system that had been established by Claude Chappe in 1793. For the past three decades the semaphore – as it had become known – had held sway. The world's leading telegraphers, in France, had built five great lines that connected Paris to Calais, Strasbourg, Brest, Toulon and Bayonne. Britain had built a line between Portsmouth and London and Russia had just unveiled its first ever semaphore. The optical telegraph, though, remained an unreliable technology. These telegraphs were useless in bad weather or at night, as Edgeworth's struggles in Ireland had proved. But fitful as the European signal network was, the old world was still far ahead of the United States, which had yet to invest in any telegraphic infrastructure. Mail sent on the enormous journey from Boston in the north to Covington in the south would take more than a week to arrive. The results of presidential elections in Washington frequently took a week to fan out to all corners of the Union. Morse realised at once that there was great scope for his invention. His challenge was to make it work.

After this initial explosion of energy there was a prolonged lull. Morse's ever precarious domestic arrangements and perilous financial position ruined any chance of progress. Between the end of 1832 and the summer of 1835 Morse moved home three times. His focus remained on his art and the promotion of *Gallery of the Louvre*, and his time was further taken up by a foray into politics that saw him run for mayor of New York. Neither this nor the *Gallery of the Louvre* was a success – the *Gallery* failed even to cover its exhibition expenses. It

was only in July 1835, when Morse was selected as professor of literature of the arts and design at the University of the City of New York,
that his situation became settled. The post came with a suite of rooms
at the new building, enough for a residence, an artist's studio and a
workshop. Delighted with this change in fortune, Morse moved in
before the building work was completed. His son would later write:
'The stairways were in such an embryonic state that he could not
expect sitters to attempt their perilous ascent. This enforced leisure
gave him the chance he had long desired and he threw himself heart
and soul into his electrical experiments.'[20]

Morse immediately turned back to his old plans. He refined the
drawings from his *Sully* sketchbook and began to work them into a
prototype. Without money to buy apparatus, little of which was available anyway, Morse improvised. He took a canvas stretching frame,
the wheel of an old clock, a counterweight, carpet binding, wooden
rollers and a wooden crank and fashioned them Heath-Robinson-style
into a first design. The prototype had taken Morse two months, and
it resembled a printing press in form. The design had two main
components. There was a transmitting section that Morse called a
port-rule. This was a three-foot-long wooden rail with grooves that
ran along its length. Into the grooves he planned to insert different
metal 'types' cut with teeth at the top – the jigsaw pieces he had
sketched on the *Sully*. The pattern of each of the different teeth would
correspond with a number from one to nine. To transmit a message
the operator had to assemble the type in the grooves in a sequence
that signified a word in an (as yet unwritten) dictionary. The operation
would be completed when the operator cranked a handle, running a
lever over the teeth that, as it rose or dipped, opened or broke an
electrical circuit. The telegraphic signal was then sent through a wire
to the receiver: a wooden frame adapted from a painter's canvas
stretcher. Here an electromagnet would pulse back and forth depending
on the electric signal, a motion that Morse exploited to press a pencil
upon a paper tape that was moved by clockwork.

Few of the great inventions in human history could have looked
so clumsy and crude as Morse's port-rule telegraph of September
1835. Even Stephenson's *Rocket*, Trevithick's 'Puffing Devil' or Fulton's
Clermont steamboat were beautiful in comparison. Essentially a creative not a technical thinker, Morse desperately needed help. He
turned to Professor Leonard Gale, a scientific colleague who taught

at the New York College of Pharmacy and the University of the City of New York. Gale had just published a book, *Elements of Chemistry*, and he became the perfect foil for Morse over the months and years to come.

Gale introduced Morse to Professor Joseph Henry's revolutionary 1831 paper on electromagnetism and these ideas were soon applied to the telegraph. Gale also urged Morse to replace his old battery with a new, updated one from Britain invented by John Frederic Daniell. With Daniell's forty-cell battery and the strengthened electromagnet, the infant telegraph began to gain power in Gale's lecture room. In the quiet hours at the end of the day, Morse and Gale were now able to transmit simple messages over ever-expanding distances: 200 feet, 300 feet and then 660 feet through coils of bundled wire. To counter the problem of a weakening current, Morse invented a relay system of circuits using the electromagnetic charge of the current to open and close subsequent circuits. It was a stroke of true ingenuity – 'a wonderful invention'.[21]

Progress was slow. Morse's finances were stretched and his artistic career seemed to be in terminal decline. Then in March 1837 a circular arrived at New York University. Written by Levi Woodbury, Secretary to the Treasury, it was an appeal for information. The government was at long last ready to commit to a state telegraph line and Woodbury had been charged with eliciting suggestions from the commercial and academic worlds. This circular caught Morse and Gale flat-footed. Their telegraph was far from complete. Morse would later admit that in 1837 his invention 'existed in so rude a form, that I felt a reluctance to have it seen'. A bundle of wires tied to a painter's canvas stretcher was unlikely to impress. A month later the dilemma would intensify. In April 1837 news reached Morse that a French duo had arrived in New York with a telegraph that could send information at astonishing speed – 'A hundred word despatch might be sent from New York to New Orleans in half an hour.'[22] Morse was thunderstruck.

It was Edgeworth's fate all over again. Through indecision and reticence Morse had been outpaced. Worse was to come. Rumours of the French telegraph were followed shortly afterwards by reports of a British equivalent designed by William Cooke and Charles Wheatstone. In despair, Morse had no choice but to act.

He asked his brother Sidney to write to the *New York Observer*, revealing that

A gentleman, of our acquaintance, several years since, suggested that intelligence might be communicated almost instantaneously, hundreds if not thousands of miles by means of very fine wires, properly coated to protect them from moisture, and extending between places thus widely separated. It is well known, that the electric fluid occupies no perceptible time in passing many miles on a wire; and if it is possible, by connecting one end of the wire with an electrical or galvanic battery, to produce any sensible effect whatever at the other, it is obvious that if there are TWENTY-FOUR WIRES, each representing a letter of the alphabet, they may be connected with the battery in any order; and if so connected in the order of the letters of any word or sentence, that word or sentence could be read or written by a person standing at the other end of the wires.[23]

The summer of 1837 was spent in Gale's lecture room perfecting the machinery. By the end of August the apparatus was constructed and ready for a public demonstration. Better news, too, had emerged. The French telegraph had turned out to be only an improved semaphore, not an electrical device. Relieved, Morse had been invigorated by his shock. Mindful that speed was now of the essence he began to plot his own launch. He invited friends to Gale's lecture room on 2 September 1837 to witness the machine in action. He also drew up a circular for his fellow passengers on the *Sully*. 'My object in writing you [is] to ask whether you remember my conversing on the subject of the Electric Telegraph as my invention when a passenger with you on board the ship *Sully* in the month of October, 1832.' The letter was sent to five men he hoped would bear witness: W.C. Rives in Virginia, Captain W. Pell, J. Francis Fisher, Charles Palmer and Dr Charles T. Jackson in Boston.[24]

On Saturday 2 September 1837 a small audience filed into Gale's lecture rooms for a first glimpse of the telegraph. Among those present was Professor Dalby of Oxford University who was also a member of the Royal Society. This was a coup for Morse: a respected objective figure who could carry news back across the Atlantic. During the demonstration Morse and Gale managed to send a message through 1,760 feet or one third of a mile. Morse described the success of the demonstration for the *New York Journal of Commerce* in an enthusiastic note written two days afterwards. To enliven the piece he attached

an engraving depicting a specimen of 'telegraphic writing', showing how the sequence of numbers 215 / 36 / 2 / 58 / 112 / 04 / 1837 could be transformed into the sentence, 'Successful experiment with telegraph. September 4 1837.'[25]

In private things were even better. A passionate new advocate had emerged. Alfred Vail was a thirty-year-old ex-student of the university. He had watched Morse and Gale demonstrate the apparatus and had been so impressed that he promised Morse his full support. Alfred Vail had arrived at the ideal time. His father, Judge Stephen Vail, owned an iron and brass works at Morristown in New Jersey. Promising to work for free for a share of the invention, and making his father's premises a base for development work, Vail solved a trinity of problems – time, money and machinery – at a stroke. Morse agreed to grant a 25 per cent share in the invention to Vail and in return Vail was to work on developing the project at Morristown using his own means.

While improvements were under way at Vail's Speedwell iron foundry and further demonstrations were planned for the new year, Morse was rocked by a quite unexpected development. Three weeks after sending his circular letter to his fellow passengers on the *Sully*, he received a letter from Dr Charles T. Jackson, now working as a state geologist in Maine, Rhode Island and New Hampshire. Jackson's reply, dated 10 September, though congratulatory in spirit, contained an astonishing claim.

S. F. B. Morse, Esq.

Dear Sir; Mrs Jackson has forwarded to me your favor of the 28
ultimo, in which you give me some account of the success of
our Electric Telegraph. I have seen several notices of it in the
newspapers, but observe that my name is not connected with
the discovery. I am greatly rejoiced to learn that you have been
successful in the trials of its power. This, I felt confident, would
be the result, as there are various ways of marking at any distance
required.

    I suppose that the reason why my name was not attached to
the invention of the Electric Telegraph, is simply that the editor
did not know that the invention was our mutual discovery. It is,
I suppose, an accidental inadvertency of the editors. I trust you
will take care that the proper share of credit shall be given to
me, when you make public all your doings.[26]

Morse was shocked. In a year that had seen his artistic ambition
destroyed and his telegraph almost eclipsed by a foreign rival, the
prospect of Jackson wresting away his claims to invention were too
much. He immediately drafted a reply.

To Dr Charles T. Jackson,

My dear Sir; Yours of the 10th instant from Bangor, I have
received, and I lose no time in endeavouring to disabuse your
mind of an error into which it has fallen, in regard to the Electro-
Magnetic Telegraph. You speak of it as 'our Electric Telegraph'
and as a 'mutual discovery.' I am persuaded that when you shall
recall the circumstances as they occurred on board the ship, and
shall also be informed of the nature of the invention of which
I claim to be the sole and original inventor, you will no longer
be surprised that your name was not connected with mine in
the late announcement of the invention.[27]

Morse's letter swept on with earnest lucidity, documenting in
forensic detail the dinner conversation on the *Sully*, how Jackson
had been absorbed in geology and anatomy during the Atlantic
voyage as he, Morse, had laboured on the telegraph in his sketchbook and how they had agreed to try a chemical experiment in

Boston but had never managed to find time. 'You were always otherwise busily and necessarily engaged, and the experiment was never tried.'

There was no invective in Morse's reply. Instead he reasoned with Jackson, setting down a crisp, even-handed account of the matter. Even so, there was a defiant undertone. 'The plan of making by my peculiar type. And the use which I make of the electro magnet was entirely original with me; all the machinery has been elaborated without a hint from you of any kind, in the remotest degree. I am the sole inventor.'

If he was disturbed by Jackson's letter, at least there was better news from elsewhere. In quick succession he heard from Senator Rives and several others who remembered his telegraphic excitements on the *Sully*. He also heard from a sanguine Captain Pell, 'I am happy to say I have a distinct remembrance of your suggesting, as a thought newly occurred to you, the possibility of a telegraphic communication being effected by electric wires . . . I sincerely trust that circumstances may not deprive you of the reward due to the invention, which, whatever be its source in Europe, is with you at least, I am convinced, original.'[28]

Cheered by these letters, Morse turned his attentions to the government. It was now six months since Levi Woodbury had sent his circular requesting suggestions and he had set a deadline of 1 October for replies. Morse seized the opportunity to pitch his electro-magnetic telegraph directly to the government. In a long, flowing letter he stated his case. It was far more efficient than any semaphore, he argued. There were five distinct benefits of an electro-magnetic system, Morse contended. First – that the information was sent instantaneously. Second – messages could be sent by day or by night equally well. Third – the whole apparatus was compact (scarcely six cubic feet). Fourth – the intelligence was recorded permanently on paper. And fifth – communications were kept secret from all but the persons for whom they were intended. Morse ended with a patriotic flourish. The Electro Magnetic Telegraph should be justly known as the American Telegraph.

The same day Morse applied for a caveat at the Patent Office, a move that protected an incomplete design from imitation for up to a year. His letters written and his pleas made, Morse spent the next few months shuttling between New York City and Vail's workshop in

New Jersey. There the telegraph was undergoing a radical transformation from its crude prototype into something quite different under Vail's diligent eye. One improvement in particular involved the telegraphic language, which Morse had originally intended to be numerical, just as Edgeworth had envisaged. At some point during the New Jersey redesign, this system was changed entirely. Instead of using a numeric code, either Morse or Vail, or a combination of the two of them, invented a new language of dots and dashes, the system we know today as Morse Code. Just which of them should be credited with the language has been the subject of some controversy. Whatever the truth of the matter, it seems that Vail made important contributions not just to the language but also to the telegraph machinery in these months. Indeed the new design was taking shape and Morse, Vail and Gale had begun to make plans for a demonstration in Washington on New Year's Day.

Stressed and confined to bed in New Jersey, Morse could have done without a second letter from Dr Jackson – a man who was fast becoming not so much a thorn in his side as a gaping gunshot wound. Jackson's reply was written on 7 November but Morse did not receive it until early December when he returned to New York. The letter brought bad news. Jackson had written a thousand-word epic, recording his 'surprise and regret' at what he considered Morse's mischievous conduct. 'I have always entertained the highest opinion of your honour and fairness, and should be very sorry to have any reason to change my opinion of your character,' he began.

Jackson's recollection of happenings on the *Sully* were formed from what he called 'a strong and retentive memory as to the facts'. Jackson proceeded to give a quite different account of the *Sully* dinner conversation. He had been enthusing about electricity, 'one of my favourite studies from boyhood to the present hour'. 'The company were all listeners.' No one knew anything of electricity but him. He had told a story about Benjamin Franklin, who made a spark travel twenty miles up along the Thames through a wire. Morse had confessed to Jackson that he had never heard of this experiment and had asked Jackson many questions about the electric current, which Jackson had answered. His mastery of the subject was in sharp contrast to Morse's complete ignorance and he had gone on to work on the idea in Boston. Jackson reached a devastating conclusion:

Hence, since I had performed all the experiments in detail, and had here brought [them] together for a specific purpose, I was, so far as they are concerned, the true inventor, and I do claim to be the principal in the whole invention made on board the *Sully*. It arose wholly from my materials, and was put together at your request, by me.[29]

Whatever restraint Morse had exercised in his first reply to Jackson would vanish in his second. He wrote back scathingly on 7 December, 'It is with the deepest regret after the attempt I made in my last to disabuse your mind of its errors on the subject of the telegraph, that I perceive the danger of a collision with you to be more imminent than at first.' He began by rubbishing Jackson's claim that Franklin had sent an electric current up the Thames, an anecdote that after much enquiry Morse had discovered to be complete fiction. This, Morse argued, was the first of Jackson's errors. Over five vicious pages he pulled apart Jackson's last letter, accusing him of labouring under 'a most serious delusion':

You very well know that you never entertained, until lately, since my invention has been publicly announced, and since the subject of telegraphs has excited the attention of foreign nations, any interest in it, while I always pressed the subject of the experiment to be tried when I met you in Boston. You were full of other engagements, and by your manner of always dismissing it as soon as you could, whenever I introduced it, you showed to me that you esteemed the matter as of little importance.

Now, Sir, I not only deny that all the materials were furnished by you, but I deny that I am indebted to you for any single hint of any kind whatever which I have used in my invention. I go further, Sir; I assert that all the consultation I have hitherto had with you on the subject has had only the effect to retard my invention, by holding out expectations that you would try an experiment which you never tried, but which was necessary to be determined one way or the other, before I could advance a single step.[30]

Morse ended his barrage with the wish, seemingly futile, that the quarrel could still be resolved in private for the sake of Jackson's reputation. And there, for a time, matters rested.

Thanks to Alfred Vail's industry and skill, by the new year the telegraph was at last ready. Yet his father, Judge Vail, had grown sceptical after months of slow progress. One day in early January 1838 Morse and Alfred enthusiastically invited him to the workshop. With Morse at the distant end of the wire, Alfred asked his father to jot down a message on a sheet of paper. Judge Vail had written out 'A patient waiter is no loser'. He then stood in silence as Alfred tapped out the message. Moments later, Morse approached him with the words written in duplicate. So amazed was Vail senior that he immediately offered to ride to Washington and petition the government for a national line.[31]

Judge Vail's reaction to the telegraphic demonstration set the tone for what was to come. There was an innate theatre to this process of sending and receiving messages. It had all the power and thrill of a magic trick. There was potential for audience participation, there was the allure of the message which could either be profound, comic or moving. Thereafter the audience had to endure silence, anticipation and tension as the demonstration commenced. If the machinery did not work the effect could be equally negative and Morse knew he might face the backlash of an audience who could be angry or scathing. Morse's confidence was growing, though. Before he and Vail embarked for Washington, he ran a series of successful test demonstrations. He wrote to his brother Sidney on 13 January 1838, 'The machinery is at length completed and we have shown it to the Morristown people with great éclat. It is the talk of all the people round, and the principal inhabitants of Newark made a special excursion on Friday to see it. The success is complete.'[32]

And the success continued. A week later they were in New York, where Vail and Morse sent a coded signal through ten miles of wire before a 'scientific' audience at the university. In an impressive burst of diplomacy Morse invited a grandee, General Cummings, to compose the message. Cummings played his role exquisitely, carefully noting down a phrase, concealing it and then handing it over to Morse. 'The assembled company was silent,' Morse's son would later write, relishing the drama of the moment, 'and only the monotonous clicking

of the strange instrument was heard as the messages were ticked off
in the dots and dashes, and then from the other end of the ten miles
of wire was read out this sentence pregnant with meaning – "Attention,
the Universe, by kingdoms right wheel".'[33]

Once again the whole triumph was reported in the *Journal of
Commerce*, which was now becoming something of an ally. 'The great
advantages which must result to the public from this invention will
warrant an outlay on the part of the Government sufficient to test
its practicability as a general means of transmitting intelligence,' it
enthused. The piece could almost have been written by Morse himself,
for now he had his sights on the conquest of Congress. On 8 February
the telegraph was exhibited before the Committee of the Arts and
Sciences of the Franklin Institute in Philadelphia. They were impressed. From there Morse and Vail travelled to Washington where
they secured permission to display the apparatus in the chambers of
the influential Committee on Commerce. Morse planned a test signal
of ten miles. He had waited years for the opportunity, it was 'a
supreme moment in his career'. Throughout February figures of
rank, repute and influence came to see Morse's curious machine.
Even President Van Buren watched a test signal, as did his Cabinet.
'The majority saw and wondered,' wrote Morse's son, 'but went away
unconvinced.'[34]

The demonstrations in Washington would set the tone for what
was to come. Morse and his telegraph were admired, but treated with
suspicion. Onlookers saw the telegraph as more of a scientific curiosity
than a machine of any practical worth. Over the months and years
that followed Morse would travel to Britain and France to advertise
his invention. He would receive compliments from such men as
François Arago and Alexander von Humboldt, and at one point he
very nearly secured a lucrative contract with Czar Nicholas. But all
came to nothing – the Czar's fear that the telegraph might be used
by enemies was a typical response. Morse's invention was either too
hot to handle or too complex to work properly. His fortunes at their
lowest ebb and with Vail and Gale's attentions wavering, Morse gave
up on his invention. For a while it seemed as if his electro-magnetic
telegraph would be consigned to history like Erasmus Darwin's
horizontal-axis windmill or Edgeworth's velocipede. Morse, however,
never quite gave up hope.

★   ★   ★

In 1842 Morse resolved to have one final shot. With America at last emerging from an economic depression that had crippled the economy since the Panic of 1837, he hoped the time might at last be ripe. By now his telegraph had undergone a series of improvements. It was powered by a stronger battery and the wires were better insulated. Morse had also had his case bolstered by Professor Joseph Henry of Princeton University. Henry was perhaps the most influential American scientist of the day. He had written to Morse in February 1842, encouraging him to give his telegraph another go. 'About the same time with yourself Professor Wheatstone, of London, and Dr Steinheil, of Germany, proposed plans of the electro-magnetic telegraph, but these differ as much from yours as the nature of the common principle would well permit,' Henry had observed. 'And, unless some essential improvements have lately been made in these European plans, *I should prefer the one invented by yourself.*'[35]

By the end of the year Morse was in Washington and making progress. Congress was long familiar with his ideas and for once seemed ready to back him. After a favourable report by the Committee for Commerce, he had been forced to wait in the capital for three torturous months as a bill of governmental support crawled through one administrative hoop after another. Finally on 3 March 1843, more than a decade after Morse had first set foot on the *Sully*, he received a grant of $30,000, enough to fund a test line between Washington and Baltimore. Tottering on the verge of complete poverty, Morse could barely believe it.

The line to Baltimore took a year to complete. As ever, the process was characterised by conflict – Morse falling out with one of his major partners – and issues with the wires and machinery. By degrees though the line progressed on its thirty-six-mile route, running parallel to the path of the railroad. It was officially opened on 24 May 1844. A crowd had been invited by Morse to see the first public message transmitted. They stood in the chamber of the United States Supreme Court to watch as a message – selected by Miss Annie Ellsworth, the daughter of one of his political allies – was flashed along the line. The words Annie Ellsworth chose have gone down in history. Taken from the twenty-third verse of the twenty-third chapter of Numbers, they rang prophetically: 'What hath God wrought.'[36] In the aftermath of the opening, one of Morse's great congressional opponents approached him and admitted, 'Sir, I give in. It is an astonishing invention.'[37]

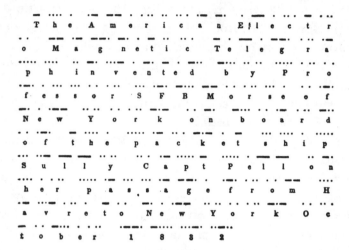

There followed a blizzard of publicity. By chance the opening of the line coincided with the start of the Democratic National Convention and within days Morse was able to flash the news of James Polk's shock presidential nomination to Washington from Baltimore, a fabulous publicity coup and perhaps the first example of breaking political news. Eleven minutes after Congress heard of Polk's success they were able to reply telegraphically: 'Three cheers for James J. Polk.' The world had suddenly shrunk. 'Information is communicated with *lightning speed*,' cheered the *Pittsfield Sun*. 'Locomotives go at a snail's pace compared with it.'[38]

What followed was a true telegraph craze, equal to that which had swept Britain in the autumn of 1794. In June the *Berkshire County Whig* enthused:

Here we are, in one part of the capital corresponding with a city forty miles distant, by means of the lightning, and in another

part taking miniatures of the sun. What shall we not compel
the elements to do for us yet?[39]

The *Albany Journal* was even more excited. 'Puck undoubtedly
thought that he was promising great things when he said to Oberon,
"I'll put a girdle round about the earth in forty minutes;" – But that is
nothing to what Professor Morse can do. They could give Puck an
hour's start and beat him easy at that.'[40] Others were creatively exploiting
the new technology in ways Morse could never have imagined. In
November a 'gentleman' from Baltimore challenged another from
Washington to a game of long-distance chess. The whole contest was
played out telegraphically, on duplicate boards. Over seven games 666
moves were transmitted from one city to the other and not a single
mistake was registered. (Unfortunately there is no record of who won.)
So long derided, Morse's liquid lightning was the toast of the day. At
Walden Pond, Thoreau was growing sick of the enthusiasm. 'Men think
that it is essential that the Nation have commerce, and export ice, and
talk through a telegraph, and ride thirty miles an hour,' he grumbled.[41]

A year after the line's opening Alfred Vail published his personal
history of the telegraph. There was only affection for Morse in its
pages. He dubbed the invention a 'Labour of scientific genius'. In the
preface he sketched the year's progress:

> The experimental line from Washington to Baltimore has been
> in successful operation for more than a year, and has been the
> means of conveying much important information: consisting of
> messages to and from merchants, members of Congress, offices
> of the government, banks, brokers, police officers; parties who
> by agreement had not met each other at the two stations, or
> had been sent for by one of the parties; items of news, election
> returns, announcements of deaths, inquiries respecting the
> health of families and individuals, the daily proceedings of the
> Senate, and House of Representatives, orders for goods, inquiries
> respecting the sailing vessels, proceedings of cases in the various
> courts, summoning of witnesses, messages in relation to special
> and express trains, invitations, the receipt of money at one
> station and its payment at the other, for persons requesting the
> transmission of funds from debtors, consultation of physicians,
> and messages of every character usually sent by mail.[42]

Later in his book Vail speculated about the other uses the telegraph could be put to. It might, for instance, he argued, be used to send secret or urgent messages. For an example of this he picked a scenario off the top of his head: 'A thunder storm is rising in the west.'[43]

This is perhaps the first recorded suggestion for using the electro-magnetic telegraph as a weather warning device. So far the idea had not occurred to anyone, but if Loomis' vision of a state-based warning system was to be implemented, then what better machine than the telegraph?

It would be another year before Vail's suggestion was repeated in more forceful terms by a scientific man. Commonly acknowledged as the first to lobby for the telegraph to be harnessed by meteorology was William C. Redfield, who wrote in the *American Journal of Science* for September 1846:

In the Atlantic ports of the United States, the approach of a gale when the storm is yet on the Gulf of Mexico, or in the Southern or Western States, may be made known by means of the electric telegraph, which will probably soon extend from Maine to the Mississippi.[44]

Science, innovation and progress, nineteenth-century watchwords, were combining with ever greater speed. The old, localised, agrarian world was being swept away and in its place a different society, built on a platform of statistics, rationality and industry, was appearing. Morse's telegraph in America and Cooke and Wheatstone's in Britain were to be the scientific agent to knit this new world together in ways unimaginable just a decade before. From now news would fly in a flash – or a series of earnest taps. News of death, news of war, news of hope, and, as William Redfield so rightly speculated, news of storms.

Midday

At midday the cumulus cloud is drifting on a current through a pale blue British sky. This shade of blue is unmistakably British. At southern latitudes, in Italy or Spain, the sky has a different hue. In Lima or Cairo or Sydney, it is different again.

The sky has no physical reality. Its blue tint is nothing more than a scattered light show that reaches the earth eight minutes and twenty seconds after being emitted by the sun. When sunlight arrives in our atmosphere it exists as a combination of colours which make up white light. It then travels onwards through the five principal layers: the exosphere, the thermosphere, the mesosphere, the stratosphere and, finally the troposphere, where almost all organic life exists and weather occurs.

The blue we see is the scattering of sunlight as it strikes against air molecules or other particles as it passes through the atmosphere. Each collision splits the white light into its constituent colours – red, orange, yellow, green, blue, indigo and violet – but it does not do so evenly. Every colour has a distinct wavelength – the highest, red, at 710 nanometres, the lowest is violet at 400 nanometres. The molecules in our atmosphere are more effective at scattering low-wavelength light like violet, indigo or blue. As the human eye has not evolved strong sensitivity to violet, which is the dominant tint in the sky, we see a vaulted blue space above us. The process can be imagined as billions upon billions of distinct explosions in the atmosphere, each a firework of bluish light.

The different blues of skies across the world are the product of the atmospheric composition in that particular place. From sea level the sky is bluer near the equator. The bluest skies on earth are said to be over Rio de Janeiro in Brazil. At northerly latitudes when the atmosphere is thinner the skies have a whiter tint. Altitude has an effect too. Two thousand feet up on the plains of Castile, Madrid has distinctive hard blue skies. Even darker hues, lapis or Prussian blue, are seen by Himalayan climbers or balloonists. In every case the deepest blue is seen at an angle of 90° from the eye to the sun.

Other combinations of colour can be generated by pollution. Smoke particles are much larger in size than air molecules and they intensify the scattering of light at all wavelengths. Polluted cities can have off-white skies with barely a trace of blue in them. This same colour can be seen if you look dead ahead. You are now looking through a much greater thickness of atmosphere, and as a result the blue has disappeared completely to be replaced by the familiar white haze of the horizon.

The pale blue British sky is a product of our humid, or damp, atmosphere. The greater concentration of water vapour particles increases the scattering of sunlight, weakening the tint. Sometimes a rain shower changes everything, clearing the atmosphere of dust and vapour and reinvigorating the skies: making them a lucent, fresher blue.

PART THREE

Experimenting

CHAPTER 7

Steady Eyes, Delicate Skies

'Singularly hot day. Them. 88. Thunderstorm & cool afternoon,'
Francis Beaufort scribbled in his pocketbook on 5 July 1846. It had
been like this for weeks and it was not what he needed. His daughter
Emily was suffering from 'a little cholera' and was 'low and weak'.[1]
Beaufort like most others worried that bad air lay at the root of such
afflictions, and reasoned that London with its heaving, perspiring
population became a stew of disease in the dog days of summer.
Those with enough time and money had already fled town for the
brisk, invigorating air of the coast.

No such luck for Beaufort. He was still lashed to his oar at the
Admiralty where he was working on the fine details for a survey
of eastern Australia and the Great Barrier Reef – still a hazard for
ships seventy years after Cook had sailed HMS *Endeavour* straight
into it. Twice over the next weeks he would work twelve-hour
shifts. It was not bad going for a man of seventy-two. Long past
the usual age for retirement his work rate had barely slackened.
Wanting to escape the Westminster bustle he had taken the lease
of a house at 11 Gloucester Place, near Baker Street, on the edge
of the city. Never the healthiest and still riddled with war wounds
from half a century before, Beaufort had adopted a strict regime
of brisk walks in Regent's Park – a minute's walk away – and cold
morning baths. As he sweltered in the July heat, Beaufort busied
himself with work, and breakfasted with friends like the Irish scientist Edward Sabine. Then on Sunday 26 July an old acquaintance,
Robert FitzRoy, knocked at his door. It was FitzRoy's first visit since
his return from New Zealand a month before. It was an excellent
chance for them to catch up.[2]

There was much to tell. It was now a decade since FitzRoy had
returned from his circumnavigation in HMS *Beagle*. For both men the

voyage had been a triumph. In a report for the House of Commons, Beaufort had referred to 'the *splendid* survey of Captain FitzRoy'.[3] He had returned to Britain with eighty-two meticulously drawn charts and eighty maps of harbours and coasts, all annotated and labelled with sailing instructions. Never again could South America's coastline be considered a wilderness. Now charts of the whole region were available from Beaufort's office for any commander with just a few shillings. It was an admirable achievement; FitzRoy's work would remain in circulation through the century. During the voyage he had kept up a lively correspondence with Beaufort. FitzRoy had paid his respects to his superior, even naming a pair of Chilean bays in his honour.

If the charts were one success then Darwin's vast haul of botanical specimens – the shells, bones, rocks and insects, tangible wonders plucked from distant lands – were yet another. Happily for them all, FitzRoy and Darwin had turned out to be a good match. But for a few spats, they had thrived in each other's company, FitzRoy cheerfully christening Darwin his 'fly-catcher' or 'Dear Philos'.[4] Both had returned heroes and had set to work writing up their experiences. Darwin had revelled in the task. Still enamoured with memories of his voyage, his pen had flowed with eloquence and ease. FitzRoy, by comparison, had laboured, struggling under the weight of detail. Once during the preparation of what was to become the *Narrative of the Surveying Voyages of H. M. Ships Adventure and Beagle*, Darwin had glanced over a passage, originally written by Captain King, that FitzRoy was trying to knock into shape. 'No pudding for little schoolboys ever was so heavy,' he concluded.

But FitzRoy had plenty to keep him busy. He was in high demand for months after their return – a dashing, aristocratic officer, back from a circumnavigation of the globe. His custodianship of the *Beagle* had been impeccable. One of his men – a mate of four years' standing – wrote, 'If one had told me that I was not a seaman when I joined the ship, I should have been greatly offended, but now I know that I never knew what real seamanship was until I saw this vessel.'[5] This was high praise, but it was deserved. In five years at sea, not a single mast or yard was lost, a sail split or a man blown overboard. FitzRoy's decision to fit lightning conductors to the masts and booms had also been vindicated. Although they were hit several times no damage was ever recorded. The main triumph in the minds of the scientific community, though, was FitzRoy's attention to his instruments. The logs were diligently kept, the readings unwaveringly accurate. The ultimate test came when he rated his chronometers – watches used for determining longitude at sea. Having circumnavigated the entire globe he was able to compare his calculations with local British time and he discovered that in five years his cumulative deviation was just thirty-three seconds. It was a terrific achievement.

The Geographical Society of London, later to become the Royal Geographical Society, was among the first to honour him. Only established in 1830 – with Beaufort one of its principal figures – it awarded FitzRoy its Founder's Medal in 1837, its highest honour. Then in short succession FitzRoy had been appointed an Elder Brother of Trinity House, the body responsible for lighthouses, and Conservator of the Mersey, ensuring that the River Mersey in northwest England remained navigable. Both were positions of wealth and prestige. But FitzRoy's ambitions had not stopped there. Soon after he entered political life, winning a seat in the Commons at the 1841 general election representing the constituency of Durham. A man of influence and ambition, FitzRoy had married the beautiful and devout Mary O'Brien, daughter of an army major. They seemed the perfect match. Robert, Mary and their young children had settled at the upmarket Lowndes Square in London's Belgravia where they were tended by a bustling army of domestic staff.

Yet FitzRoy's ascent to the lofty echelons of British life had not been all smooth sailing. First there had been his election to Parliament. With such influential support it should have been straightforward, but FitzRoy's boisterous behaviour at the hustings had led to a quarrel with his fellow candidate. A blaze of venomous letters had flown between FitzRoy and his opponent, Mr Sheppard, and a duel had narrowly been avoided. Polling day had come and FitzRoy had been elected, but that was not the end of it. In Westminster FitzRoy had been accosted by Sheppard who was lying in wait outside the United Service Club in Pall Mall, brandishing a horsewhip. 'Captain FitzRoy!' he had bellowed. 'I shall not strike you but consider yourself horsewhipped!' This was too much for FitzRoy. He had set upon Sheppard with the only weapon he had – his umbrella – and the confrontation had ended up deliciously for the press with FitzRoy knocking Sheppard to the ground.[6] It was the type of undignified event that Westminster outwardly despised but inwardly relished. It was reported in the newspapers, a bad end to one of the most acrimonious election campaigns in Westminster's long and varied history.

Yet FitzRoy's parliamentary career recovered from this sorry start. He had taken his seat on the back benches behind Robert Peel and struck up a friendship with Sir John Gladstone, father of the future prime minister. He played a key role in drafting legislation to improve the quality of merchant seamen's education through a system of examination and was selected to accompany Archduke Frederick of Austria on a tour of Britain. Fancied for a Cabinet position, his Westminster career had come to a halt in 1843 when Lord Stanley asked him to become governor of New Zealand. It was a dilemma for FitzRoy. Leaving the country meant forfeiting the financial security of his seat and his position at Trinity House. Yet 'however distant and ill-remunerated' he knew the position to be, FitzRoy accepted nonetheless, out of a spirit of duty. It turned out to be a terrible mistake. 'A more thorny path he could hardly have chosen,' one journalist later wrote.[7] Having sailed halfway around the world FitzRoy arrived to find the thinly populated island near bankruptcy and torn apart by a bitter conflict between the Maori tribes and Western settlers. For two years he worked at the problems but everything he tried backfired. In 1845, with matters worsening by the day, he was recalled by Stanley. A sad end to a noble project.

To set things straight FitzRoy had written a pamphlet on his experiences that was doing the political rounds – William Gladstone was currently reading it – but his problems were not over. He had sailed home in a merchant ship called the *David Malcolm*, a vessel commanded by the lax and lazy Captain Cable. They had sailed east from New Zealand across the Pacific and Cable had cut up through the Strait of Magellan. For FitzRoy his days in the whaleboat or scrambling up mountains must have seemed an age ago. But the waters still held powerful memories. 'The Strait of Magellan is proverbially stormy, wet and dismal,' he wrote,

Yet, in the rare intervals of fine weather, grander scenery, more striking combinations of high, snow-covered mountains, *extensive* glaciers, forests with every tint and shade, immense precipices, numerous waterfalls, and *deep blue sea* at their base – cannot be found in all the world.[8]

The weather remained fractious. Captain Cable had sailed almost the full length of the Strait to Mercy Harbour at the eastern end. There he had thrown out the shortest scope of cable and lightest anchor, and retired below. FitzRoy was unimpressed. Luckily enough he was carrying a pair of his sympiesometers. He later wrote:

Anchored thus, with all yards and masts aloft, as arrived from the sea, [Cable] would have gone to sleep, as usual, on April 11, though the writer's two sympiesometers then told *him* a storm approached, and by *great* exertion he did induce this Captain, *Cable*, by name, to send down light spars, point yards, and veer chain; besides getting a second anchor ready. Then the skipper made himself happy, in his own peculiar way, below – and was soon too sound asleep, to be seen again that night.[9]

FitzRoy's vigilance was to prove decisive. After Cable had disappeared FitzRoy had remained on anchor watch. His anxieties were not just for the ship: his whole family – Mary and their three children – were sleeping below. FitzRoy watched as the pressure continued to dip. He knew what was coming. Assuming control, he had a second anchor dropped. 'The night was beautiful, clear and still moonlight,' he recalled. Everyone thought he was wrong. But at two o'clock in the morning the weather turned, as it so often did. A roar came out of the west and 'a white dense cloud of driven water as high as the lower yards' buffeted the ship. For a few desperate minutes the *David Malcolm* was pitched at a terrifying angle, a stone's throw from granite rocks. 'Had that ship been taken unprepared,' FitzRoy declared, 'not a soul would have been saved, in human probability; only God's providence could have rescued any one in so desolate, wild, and savage a country.'[10]

The *David Malcolm* had emerged from the Strait of Magellan intact but with no thanks to Captain Cable. For FitzRoy it was a narrow escape and he later wrote about his family's deliverance by Providence.

In the years since the *Beagle*'s voyage, FitzRoy had experienced something of a Christian rebirth. Always a believer, he had nonetheless

been timorous in his conviction in his youth. 'I suffered much anxiety
in former years from a disposition to doubt, if not disbelieve, the
inspired History, written by Moses,' he revealed in his *Narrative*. But
by the mid-1840s, FitzRoy was ardent in his faith. He wrote about
his previous 'wavering' of mind, his 'fancy' that the Old Testament
'might be mythological or fabulous'. His friends had noticed this
shift. He had become increasingly intolerant of dissenters and
eschewed theories like those of the geologist Charles Lyell, in
Principles of Geology, who argued that the earth was of ancient, not
recent, origin.

His friends attributed this to Mary's influence. As he had once
with phrenology, FitzRoy was demonstrating his tendency to
embrace an idea wholeheartedly. 'When once convinced,' a journalist
later wrote of FitzRoy's character, 'nothing could shake or remove
that conviction.'[11] And with the ire of the convert he had used the
final chapters of his *Narrative* to substantiate the claims of the Old
Testament. He sought to link what he had seen on his travels with
what he had read in the Bible. He declared that there were at least
twenty-three distinct races on earth, as neatly stratified as different
types of rock. The black, red and brown tribes he encountered were
the cursed descendants of Cush, the grandson of Noah, while the
handsome, prosperous, white races of Europe were the favoured
sons of Shem and Japheth. A scriptural literalist, FitzRoy traced
everything back to the Old Testament. One of FitzRoy's arguments,
later lampooned by Darwin, was that the dinosaurs had been wiped
out because they were too large and ungainly to clamber up the
gangplank of Noah's Ark.

In 1846, though, Darwin and FitzRoy remained on good terms.
Learning that his old friend was back in London, Darwin rushed
off a note. 'I cannot resist writing to congratulate you on your
safe arrival after your bad passage home,' he began. 'I hope that
your health has not suffered and that you are as strong & vigorous
as formerly . . . I am aware how little chance there is of your
having time to spare, but if ever when in Town Mrs FitzRoy &
yourself should feel inclined to spend a few days in the country
– it would give my wife & myself real pleasure; – we have a toler-
ably comfortable house in a very quiet, retired, airy part of the
country.'[12]

★   ★   ★

The prospect of visiting Darwin's house at Downe in Kent must have appealed to FitzRoy as July wore on. Temperatures had soared even higher. The grass in the parks had turned golden yellow, the ground beneath scorched hard as a ship's biscuit. London sweltered in the heat. Many escaped the sun's glare by staying indoors or sheltering under the elm trees that lined the Mall and Birdcage Walk. Now by population the biggest city on earth, London had doubled in size since the frost fair of 1814 from one to two million. But in late July 1846 that frozen Regency frolic of jingling sleighs, ice skating, gingerbread, gin and steaming punch seemed impossibly distant. Instead people bathed in the Serpentine and waded in the Thames at low tide. It was the kind of weather that Luke Howard, the erstwhile cloud classifier, called a 'Coup de soleil'.[13]

Over the years Howard had continued with his weather research, publishing his influential *Climate of London* and *Seven Lectures on Meteorology*. In these works Howard had claimed, for the first time, that cities had grown to such a size that they were capable of forming their own 'micro climates'. Howard argued that the cumulative heat created by the density of smoking fires and belching chimneys in the capital combined to keep urban temperatures $1.579\,^{\circ}$F higher than in neighbouring rural areas. He also believed the sheer mass of human bodies crammed together also acted to raise the temperature further. He posed an analogy:

> Whoever has passed his hand over the surface of a glass hive, whether in summer or in winter, will have perceived, perhaps with surprise, how much the little bodies of the collective multitude of Bees are capable of heating the place that contains them: hence in warm weather, we see them ventilating the hive with their wings, and occasionally preferring, while unemployed, to lodge, like our citizens, about the entrance.[14]

For Howard, London was a hive on a majestic scale and by 31 July 1846 Londoners were like the bees, flapping at its entrance. The month ended with thunderstorms growling like yard dogs on the southern horizon. Beaufort was among the sufferers. On 1 August he woke early with a 'smart of cholera'. Unable to sleep he had his doctor summoned, who prescribed a pill of opium and calomel which 'completely quieted' him. Sedated, Beaufort slept through the morning haze which cloaked the city. Only at 10 a.m. did the sun break through, and when it did it shone with intense power. Temperatures shot up to about 90°F. At Greenwich the skies were alive with cirrostratus. Scud clouds darted over rooftops. A keen wind blew up, rustling trees, rattling windowpanes. Thunder was heard. Then at about three o'clock, as Beaufort woke from his slumbers, the skies darkened. Minutes later a summer storm broke in spectacular style.

London had experienced nothing like this for years. The rain fell with great force. Jets of lightning forked out of the gloom. On and on the storm went, gathering momentum. The water soon overpowered the drains, and streams went dashing down streets and rising in hollows. It was as if the city was under attack. On the Thames a steamer was struck by lightning. Electricity shot through the vessel, blowing off the starboard paddlebox and narrowly missing the captain on the bridge. A separate strike hit 17 Mornington Crescent, the electricity travelling down the chimney flue and knocking a female servant to the floor. In Norwood, to the south of London, a gang of reapers in the meadows did not escape so lightly. All four were killed in an instant.

At Green Park the weight of the water burst open an iron sewer. A wave went cascading down the Mall into St James's Park where a flock of sheep were swept off their feet. As the waters rose a second flank of the storm arrived, bringing 'a hail storm hardly remembered by man living'. As alarming as the ferocity of the hail was the size of the stones. Larger and heavier than a marble and some of them as big as a halfpenny, the hailstones resembled lethal shards of ice. *The Times* later reported that one stone had weighed in at an ounce and a half. It was almost as if all the shingle had been gathered up from Brighton beach and had rainied down on the capital. Sheltering at home, people listened to the reverberating echo of ice on glass. For two hours the rain and hail continued. It was not until a quarter past six that residents were able to venture out. Everywhere was confusion. Neighbours were bailing water out of windows with milk pails.

The damage was reported in the London papers over the next few days: 'Terrific Storm of Thunder and Hail', 'Destructive Thunder Storm' and 'The Great Storm of Saturday Last'. These articles gave panoramic descriptions of the havoc caused by the combined force of electricity, wind, water and hail. Every pane of south-facing glass had been smashed along a two-mile stretch of Wands-worth Road. At the new Houses of Parliament 7,000 panels of old crown glass had been destroyed, along with 2,736 windows and skylights at the fashionable Burlington Arcade and 14,000 more at a factory in Millbank. At Buckingham Palace a cultural calamity of the first magnitude had been narrowly averted when the skylight over the Picture Gallery was broken, letting rain pour through the gaps. In an hour it had risen to a height of several feet in a room containing masterpieces by Cuyp, Parmigianino, Steen and van Dyck.

Countless reports of damage surfaced in the papers in the following week. Beaufort's house had avoided the worst of it, and within forty-eight hours he had had 'our little damages' fixed up by the glazier. It was a lucky escape. He was nonetheless sensible to the suffering of others. 'Sad account of the devastation of the storm in broken greenhouses, flooded floors, burst sewers &c &c', he noted in his pocketbook. A week later he was still ruminating over the matter. 'It is said that Saturday's storm of hail or rather of lumps of ice destroyed £100,000 worth of glass in London,' he wrote.[15] It was an enormous sum, almost as much as Isambard Kingdom Brunel had just spent building SS *Great Britain*, the world's first screw-driven iron ship.

The storm was reported eagerly by all the press, but in particular by the sensational new weekly, the *Illustrated London News*. Still only four years old, the paper had made its name through its nimble features of recent news events, accompanied by evocative illustrations. The storm was the perfect quarry for the paper and it devoted two pages to the story. The real draw was the accompanying woodcut depicting the storm from an elevated viewpoint on Blackheath near Greenwich. Below, London spread out in miniature: streets, roofs and spires, the dome of St Paul's standing – like Herbert Mason's iconic photograph in the Blitz – defiantly in the distance.

The woodcut, by the artist Frederick James Smyth, is a superb depiction of extreme weather. The London skyline is black, streaked with a bolt of lightning. Rain hammers down at a sharp angle. Smoke billows sideways from a chimney. In the foreground of the composition Smyth places a figure in the chaos to lift the piece and give it

LONDON, FROM BLACKHEATH, DURING THE STORM.

focus. Bent to the wind, the man clamps a hand to his head, stopping his hat from being blown into oblivion. A dog slinks along on his right-hand side. They hurry down what must have once been a gentle path but has now transformed into a mass of mud. A minute or so away are the outskirts of the city. Their first hope of shelter.[16]

In 1846 the *Illustrated London News* had a man from Blackheath as its meteorological correspondent. For weather advice, reports and features it increasingly turned to a young, earnest superintendent of the Magnetic and Meteorological Department from the Royal Observatory in Greenwich called James Glaisher. Glaisher lived at 13 Dartmouth Terrace, a few minutes from the scene in Smyth's woodcut. Could it be Glaisher in the woodcut? If it is, it transforms the illustration. It becomes a picture of Glaisher darting home from a morning's work at the Royal Observatory, under a wild, uncertain atmosphere he would devote his life to understanding.

In 1846 James Glaisher was thirty-seven years old: tall, thin as a reed, as sharp-eyed as Loomis, as industrious as Beaufort. He had been at the Observatory on the morning of the storm. He had seen the cirro-stratus and low-flying scud clouds. Glaisher had recorded how the rain had started to fall at ten past three, accompanied by thunder and lightning. He had studied the hailstones at the Observatory – 'not especially big' – and later at home in Blackheath. There Glaisher had measured more stones, or fragments of ice, and reasoned that the average specimen was about the size of a large hazelnut. Later he had felt a tremor as nearby lightning shook his house. In the evening, after the rain subsided, he had watched an 'eerie fog' settle over the city. 'So dense was the mist,' he wrote, 'that lamps at Blackheath could not be seen at a distance of a few yards.'[17]

Always the observer, James Glaisher had other traits that made him an excellent nineteenth-century scientist. He was rationally minded, punctilious and formidably productive. In August 1846 he was putting his skills to good use, writing up the conclusions of a mammoth meteorological research project. This project examined the formation of dew and the passage of terrestrial radiation. Titled 'On the Amount of Radiation of Heat, at Night, from the Earth, and from Various Bodies Placed on or Near the Surface of the Earth', Glaisher's was one of the most intensive studies of dew since W. C. Wells had published his classic 'An Essay on Dew' in 1814, demonstrating for the first time that dew was produced by vapour condensing on objects instead of falling from the sky. Now Glaisher wanted to analyse this process in greater detail and for three years the project had consumed his spare hours. Before he had even started he had invested months in finding the best thermometer, trying all imaginable lengths, sizes, shapes of bulb and colours of finish before settling on a suitable model. His apparatus ready, his framework drawn up, Glaisher had spent night after night in the grounds of the Observatory with thermometers, watching the subtle transfer of heat from earth to air. Everything was measured: the height of the individual blades of grass, air pressure, air and ground temperature, wind direction, humidity, the dew point, the proportion of cloud cover. Glaisher at work reminds us of P. G. Wodehouse's description of his magnificent arch-villain, Roderick Spode, who had 'the sort of eye that can open an oyster at sixty paces'. If Spode could open an oyster at sixty paces, Glaisher could at seventy.

Glaisher's devotion to his experiment was total. He considered every variable. He held his breath when taking readings to avoid contaminating the thermometer and kept his reading lamp at a distance to prevent interference. Months turned into years as he stockpiled data. Gradually he expanded the scope of his research. He examined long grass, short grass, soil beneath the grass, changing every variant to test its effect. He then started covering the grass with materials: wool, flax, lead, blackened tin, charcoal, glass, rabbit skin, chalk, cotton. This done, he moved on to colours to see how they absorbed heat differently. Black absorbed the most, he discovered, followed in turn by yellow, scarlet, orange, white, green, crimson, dark blue and light blue. Glaisher spent hours stretched out in the grass at Greenwich. One of his joys was to watch dew form on the individual blades.[18] It captivated him, but made him ill. The damp brought on a rheumatic complaint that would bother him for years.

For Glaisher it was worth the hazard. Now in August 1846 he was putting the finishing touches to a paper that was to be read before the Royal Society in the new year. Still the elite forum for any aspiring scientist, for Glaisher this was a first opportunity to impress. The paper was to be his intellectual début and, as was customary for non-members, it would be read on his behalf, in Glaisher's case by his superior at Greenwich, the Astronomer Royal George Airy.

At forty-five, Airy was eight years older than Glaisher and as illustrious a superior as could possibly be. He was short of stature but colossal by reputation: few in Britain, perhaps in the Western world, could match his intellectual pedigree. Born in Northumberland and raised in Suffolk, Airy had startled his schoolmaster at the age of fifteen by repeating 'with no great distress' 2,394 lines of Latin verse. It was the first flash of brilliance from a mind destined to dazzle. Quiet and studious with a relish for rambles over the Fens, for folk songs and poetry, Airy excelled as a mathematician. As soon as he was old enough he had gone up to Cambridge where he had won a hatful of prizes and surprised absolutely no one by graduating as Senior Wrangler – the top mathematics graduate – for 1823. Thereafter he had shot up through the academic ranks: Lucasian Professor of Mathematics at twenty-five, Plumian Professor of Astronomy and Director of the Cambridge Observatory at twenty-seven. Long coveted by the London political elite, Airy had eventually been prised away from his Cambridge fiefdom in 1835 to replace elderly John Pond as Astronomer Royal at Greenwich. There he had set to work with characteristic verve, refreshing the Observatory's outlook. Ruthlessly efficient, if one trait defined Airy above others it was his mania for order. A biographer would later write:

In everything he was methodical and orderly, and he had the greatest dread of disorder creeping into the routine work of the Observatory, even in the smallest matters. As an example, he spent a whole afternoon in writing the word 'Empty' on large card, to be nailed upon a great number of empty packing boxes, because he noticed a little confusion arising from their getting mixed with other boxes containing different articles; and an assistant could not be spared for this work without withdrawing him from his appointed duties.[19]

To Glaisher, Airy was a formidable superior, but by agreeing to read his paper he had become a significant intellectual ally. It presented Glaisher with the latest opportunity to elevate himself into even higher circles of influence. It was to be the latest triumph in a life that had started in quite different circumstances.

James Glaisher was born on 7 April 1809 on the south bank of the Thames in Rotherhithe. When he was still a young boy his family moved to Greenwich, the site of the old royal palace where the Tudor monarchs Henry VIII, Mary I and Elizabeth I were born. The town was a blend of pretty Georgian streets and naval schools that fanned back from the riverside. The scientific beacon of the place, though, stood high up in Greenwich Park. Now one hundred and fifty years old, the Royal Observatory had become an established icon in the intellectual landscape. Built in the seventeenth century, the Observatory rose over the treetops like a scientific castle of red brick and glass domes. It is at the Observatory that we hear the first of Glaisher as a bright-eyed visitor of twenty, roaming among the instruments: the magnificent quadrants and sextants.

Glaisher's apprenticeship would continue in different climes. Appointed an assistant on the Ordnance Survey he exchanged east London for the wet, windswept hills of western Ireland. For two years, in 1829 and 1830, he climbed hills in Counties Galway and Limerick, carrying theodolites to take angles for the triangulation. Like Beaufort on Croghan Hill and FitzRoy in Tierra del Fuego, Glaisher found inspiration at this elevation. Later he traced his atmospheric awakening to this time:

I was often enveloped in fog for entire weeks, first on the mountain Bencor in Galway, and afterwards upon the summit of the Keeper Mountain, near Limerick . . . in the performance of my duty I was often compelled to remain, sometimes for long periods, above, or enveloped in cloud. I was thus led to study the colours of the sky, the delicate tints of the clouds, the motion of the opaque masses, the forms of the crystals of snow.[20]

The wet climate, though, took its toll. Within two years of his departure, Glaisher was back in England recovering from exposure. On the lookout for a new position, Glaisher found work at the Cambridge Observatory as a mathematical computer for Airy. He began work in 1833, starting an association between the two that would stretch for almost forty years. He worked on a mural circle – a telescope fixed to a bracing wheel – plotting the position of stars as they shifted across the sky, as part of a team of three assistants in the Octagon Room. But even during these years his affection for the atmosphere was never completely lost. 'Often between astronomical observations,' he would write, 'I have watched with great interest the forms of the clouds, and often, when a barrier of cloud had suddenly concealed the stars from view, I have wished to know the cause of their rapid formation, and the processes of action around them.'[21]

Glaisher's success depended on his attention to detail. Notoriously fastidious, Airy was a difficult man to please. At length, though, Glaisher worked his way into his confidence. When Airy was appointed Astronomer Royal in 1835 one of his first acts was to have Glaisher brought down from Cambridge. He was to be a part of a bold, rejuvenating plan for Greenwich, a place Airy decided was in a 'queer state'. Four years later with the project in full swing, Airy appointed Glaisher superintendent of the newly established Magnetical and Meteorological Department. It was a decisive moment for Glaisher that turned his talents towards meteorology. Airy had passed over a field of better-qualified candidates for the role and it crowned, for Glaisher, a fabulous rise in status.

As superintendent, Glaisher became Britain's first governmental meteorologist under Airy's supervision. Airy and Glaisher – two more finely wrought meteorological names you would struggle to find – had free rein to innovate. They began by devising a timetable for daily observations. It was to be data collection as never before. Glaisher set about sourcing the most accurate instruments. Airy meanwhile drew up the list of different observations required. His plan was to be followed to the letter. Except on Sundays, readings were to be taken every even hour of Göttingen mean time, night and day, by Glaisher or one of his assistants. They were to note wind direction, cloud cover, type of cloud, variety of currents, wind strength and the power of the sun's rays. Airy and Glaisher even devised a special stand for their instruments to ensure consistency which stood four feet above the ground in the shade. From the day the observations started in the early winter of 1840, everything ran with metronomic accuracy. But collecting the raw data was just the start of it.

Glaisher wrote in 1844:

Harassing as the observations are, requiring as they do the most vigilant care in regard to the state of the instruments, they demand afterwards such massive calculations, that the observation is, in comparison, a mere trifle. From them are deduced the mean, daily, monthly and yearly positions of each instrument, with an accuracy that can be obtained in no other way than fully reducing regular observations, taken at whatever time of the day or night it may be necessary to observe them.[22]

The routine produced an ever-increasing quantity of atmospheric data. A flavour of Airy's approach and his fanaticism about reductive science can be found in his habit of calling Glaisher, and his assistants Edwin Dunkin and John Hind, not by their surnames, as was the convention, but by the abbreviations, A, B and C.

It was in the mid-1840s that Glaisher began his study of dew. Forced to check instruments throughout the night, this side project was a way to turn his spare hours to fruitful purpose. There was also the incentive of writing his reports up into a paper. After more than a decade as one of Airy's subordinates, Glaisher was ready to establish his own reputation. He already had two astronomical papers to his name and since 1843 the words 'Mr Glaisher' had been appearing sporadically in the papers. His journalistic break had come in March 1844 when the *Illustrated London News* published a long feature about the Magnetical and Meteorological Department at Greenwich.[23] The piece, written by Glaisher, was a sober tour through the day-to-day operation of the department. Glaisher wrote with pride about their 'novel means of obtaining and recording scientific results of the utmost interest and importance'. It seemed harmless enough, but the feature ended up causing a spat between Airy and the newspaper. Airy pointed out that all official communications from Greenwich should come from him and not his deputies. Glaisher's name should not have appeared at all.

It was a trivial affair but very revealing. Since childhood Glaisher had displayed an ability to seize opportunities when they presented themselves. Getting his name in the *Illustrated London News* – its circulation was about 50,000 weekly – was a coup. It began a long association between the ambitious superintendent and the aspiring newspaper. Soon Glaisher became its unofficial meteorological correspondent. He offered comments and quotes when needed and was asked to draw up the copy for the spin-off *Illustrated London Almanac* in 1846. Everything was coming together. Glaisher had a distinguished job, the room to innovate and a public profile. He had risen from computer to assistant, superintendent to correspondent. And he was far from finished yet.

Before the 1830s meteorological reportage had been the domain of the almanacs: *Old Moore's* or *Murphy's*, titles that sold in the hundreds of thousands each year, crammed with meteorological nonsense scripted for a forgiving and forgetful public who loved to be dazzled by predictions for boiling summers, mild winters, frost fairs, storms or fogs. Just occasionally one of the predictions would come true. It happened famously in 1838 when 'Zadkiel', exotic pseudonym of the more orthodox-sounding Richard Morristown, correctly predicted 20 January 1839 as the coldest day of the year. It was a lucky strike. In an instant Zadkiel was famous, a weather wizard, and Morristown was rich, magnificently rich. The prediction made his career.

Glaisher's entrance into weather journalism, though, represented a shift. His outlook was informed by the trends of his time, and by the 1840s there was a growing faith in numbers and rational quantitative truth. 'No sound knowledge can exist,' the *Edinburgh Review* had claimed in 1838, 'but that based on observation and experiment, which either rests immediately on facts, or is reduced from them by mathematical reasoning.'[24] At last society seemed on the verge of realising René Descartes' vision of *mathesis universalis* – the total understanding of nature through mathematics. One of the most influential books of the 1830s had been Mary Somerville's *On the Connexion of the Physical Sciences* (1834). For Somerville, mathematics was a pure subject, so formidable in its application that it was like a spiritual pursuit. 'This mighty instrument of human power itself originates in the primitive constitution of the human mind, and rests upon a few fundamental axioms which have externally existed in Him who implanted them in the breast of man when He created him after His own image,' she wrote. Later she would dwell on the moment she first realised the allure. 'I could hardly believe that I possessed such a treasure when I looked back upon the day I first saw the mysterious word Algebra.'[25]

In the 1820s a Belgian astronomer in Paris, Adolphe Quetelet, had begun to apply the error law, typically used for determining the position of stars in the night sky, to different purposes. Collecting statistics, he applied his maths to all sorts of social phenomena from crimes to health, and over time the error law acquired its modern name 'standard deviation'. By the 1840s numbers were everywhere. They were immutable facts. In 1841 the first full-blown census of the British population was undertaken, and plans were under way for a more ambitious survey in 1851.

In 1846, a French geometer called Urbain Le Verrier had thrilled the scientific community with an unprecedented feat. For some time he and John Couch Adams, a Cambridge mathematician, had been working separately on 'perturbations' in the orbit of the planet Uranus. It was thought these irregularities suggested that an interfering mass lay somewhere beyond the planet in the solar system. Le Verrier and Adams had issued predictions based on their calculations. Then, in September 1846, Le Verrier sent a letter to the Berlin Observatory asking them to examine the night sky for a possible planet at a specific location. Johann Galle, an assistant in Berlin, picked up the note, and soon found an uncharted star within a degree of the predicted position. It turned out to be a planet, later christened Neptune. Stunned by the achievement, Arago famously declared that Le Verrier had discovered a planet 'with the point of his pen'.

Le Verrier's triumph showed the power of mathematics. It could be harnessed to inform spending by the government, to track criminals, to conquer disease. There were more and more public displays of precision. Charles Babbage, who was teased mercilessly by friends for reducing every problem to a mathematical formula, drew excited audiences to see his Analytical Engine solve logical problems. In 1833, one of the last orders of Airy's predecessor, John Pond, was to have a large metal ball attached to a pole at the Royal Observatory. At

1 p.m. each day it dropped to earth with an echoing thud so that ships in the Thames could calibrate their chronometers. Pond's time ball became a familiar part of the Greenwich soundscape.

Glaisher, too, fell under this spell. Everything he planned, every experiment he executed, every paper he published was saturated with numbers. A perfect specimen of the Victorian *scientist* – a term coined by William Whewell in 1833 on the suggestion of Coleridge, Glaisher encapsulated the idea that with a sheet of paper, a pen and a table of carefully sourced numbers, truth could be found. In the 1840s scientists were building a fortress of knowledge, each stone of which was a beautiful line of mathematics.

Glaisher, of course, concentrated on meteorological matters. His efforts throughout the 1840s marked a massive improvement in the way that atmospheric information was harvested. It was only twenty years before that John Frederic Daniell had written a bad-tempered introduction to his *Meteorological Essays*, taking the opportunity to swipe at the Royal Society for their methods of data collecting. He had been horrified that the foremost scientific society had overseen such a calamitous process. Bad instruments, inattentive observers, measurements taken at all times of day:

> The carelessness exhibited in this department has, for a long time past, been the subject of serious and public complaint; and there is scarcely a person who has had occasion to consult the records, who is not declared to be unworthy of confidence. Mr Dalton, Mr Thompson, and Mr Howard, have recorded their dissatisfaction; and the latter gentleman has been compelled to ask – 'if this learned and highly-respectable body feels the subject of the weather no longer worthy its notice, would it not be better, at once, to dismiss the register from its transactions?'[26]

Daniell's critique ran on and on. Readings were 'regulated by nothing but the observer's night cap', the mercury in the barometer had never been boiled and no effort had been made to remove the residual air or moisture from the tube. The Society relied on a weather vane from a neighbouring building, and so lax were the estimates of wind speed that out of a total of 730 observations he had studied, 669 had given the same result. Most alarming of all, Daniell revealed, was his discovery that the rain gauge was jammed beneath the cowl of the chimney, meaning that almost no precipitation was measured on windy days.

Airy and Glaisher's efforts at the Observatory were reflective of a new, clinical approach to measurement that banished these methods to historical anecdote. For decades, whenever the newspapers had wanted to emphasise extreme weather, they had always fallen back on the refrain, 'In the memory of the oldest person living'. But from now on such vague platitudes would just not cut it. Numbers were needed. No argument could survive without statistics. No theory without formula.

Some time in 1846 Glaisher came across an error in the quarterly report of the Registrar of Births, Marriages and Deaths. In its brief meteorological section the report claimed that the mean temperature in York was 5° warmer than that of London. Puzzled, Glaisher had written to the Registrar General, George Graham, explaining that this was a 'physical impossibility'. In due course a letter returned, thanking Glaisher and confessing that nobody in the department understood meteorology. Glaisher pounced. He fired a letter back, proposing that all the regional weather reports be sent to him instead. Arrangements were put in place. Airy allowed Glaisher to work on the statistics in his spare time.

Glaisher set to the task with typical gusto. He wrote to the existing regional correspondents and was unimpressed by the response. They were 'utterly impractical' he complained. 'They did not have the facts of nature', held 'peculiar views' and 'but few of them attended either to the character of their instruments or to the application of the corrections necessary to reduce their observations'. Glaisher did precisely what Airy would have done. He sacked the lot of them. At the beginning of 1847 he set out afresh, exploiting his position at Greenwich to contact members of the leading learned societies. He approached fellows of the Royal Society, members of the Royal Astronomical Society and he even persuaded Airy himself to fill in his newly fashioned skeleton forms from his home at the Observatory – a delightful twist in their relationship.

Glaisher conducted his new hobby with the sort of enthusiasm that Loomis had described across the Atlantic when he had written of a 'meteorological crusade'. Glaisher travelled to all corners of the kingdom in his spare hours to meet correspondents and fired off letters to potential conscripts whenever he could. Little by little his network grew, from Southampton in the south to Glasgow in the north, from Lowestoft in the east to Liverpool in the west. His travels were made possible by the extension of the railways. Just two decades earlier, the kind of journeying that Glaisher was now undertaking would have consumed weeks. As it was, he could dart into one of the London terminuses in a hansom cab, jump onto a train and be at his destination within hours. There he could meet a contact and check their instruments, calculate the longitude and latitudes of their station, demonstrate the process for taking readings, shake hands and be back in Greenwich the same day. In many cases Glaisher brought improved equipment along to replace inadequate models and steadily his network started to expand.

Initially these forms were used merely for crafting quarterly reports for the Registrar General. But in August 1848 an opportunity arose and Glaisher, as ever, leapt at it. In 1846 Charles Dickens had thrilled Fleet Street by establishing a newspaper, the *Daily News*, and appointing himself its first editor. It was an exciting prospect. The most famous writer of the time unleashed on a daily basis, able to square up to the Peelite *Morning Chronicle* and champion social issues as he did in his novels. Though Dickens had soon passed on the editorship so that he could concentrate on the latest serialised numbers for *Dombey and Son,* the *Daily News* retained its taste for innovative content. In the late summer of 1848, with 'rainy and inclement weather' threatening the harvest, an editor at the office had written to Glaisher to see if he could deliver a weather update. Glaisher did even more than that. With his network of meteorologists fully operational, he had the ability to offer an experimental service of weather reports. It was quite unlike anything ever done before.

The key was the British electro-magnetic telegraph. Different in design to Morse's model it had nonetheless generated a similar level of excitement. The *Illustrated London News*, always eager to showcase technological advances, had kept a close eye on progress as the telegraph, designed by William Cooke and Charles Wheatstone, was strung out over the British landscape. In May 1844 the paper had published a feature on the new Slough terminus, illustrated with woodcuts of the stationhouse and pictures of its earnest operators hard at work. Nine months later the Slough station had become a sensation itself when it became embroiled in a murder case. Under the headline, 'Murder at Salt Hill', the *Illustrated London News* described 'an extraordinary instance of the working of the newly-supplied power of electro-magnetism'. As the suspect fled from Slough a message had been telegraphed along the line to Paddington:

A murder has just been committed at Salt Hill, and the suspected murderer was seen to take a first-class ticket for London by the train which left Slough at 7h, 42m p.m. He is in the garb of a Quaker, with a brown great coat on, which reaches nearly down to his feet; he is in the last compartment of the second first-class carriage.[27]

The train arrived at Paddington where the suspect was seized by policemen. It was a revelation, as far away from the old hue and cry as possible. No man could outpace electricity, however fast he ran. As in America a telegraphic craze ensued. A long-range chess game was played, between a Mr Staunton and his partner Major Kennedy at the Portsmouth terminus against the *'celebrated'* Mr Walker and a friend at Vauxhall. At the end of its report the paper concluded:

Our forefathers would have ascribed much less astonishing effects to witchery. The invention is not astonishing only, however, for it is brought into the most useful operation. It is employed to telegraph trains and to convey messages to and from the servants of the company. On payment of a small sum the public may avail themselves of it for matters of business. It is hardly possible to contemplate all the purposes to which it may be applied, but it is at once evident that the Electrical Telegraph can be devoted to the conveyance of all sorts of important intelligence, and that it may hereafter, to a certain extent, supersede the present mode of Post-office communication.[28]

Three years later, in the August of 1848, the *Daily News* and Glaisher agreed to harness electric power to fuel a series of weather reports. The newspaper bore the cost of the plan and Glaisher curated the data. At nine o'clock in the morning simultaneous readings were taken at twenty-nine telegraph stations across the country and forwarded to London where they were processed by Glaisher and published the following day in the *Daily News*. The reports were simple. They included the name of the place, the direction of the wind and the general state of weather. The first bulletin appeared on 31 August 1848, detailing the weather conditions across the country the previous day. It is the first recorded use of telegraphed weather reporting anywhere in the world.[29]

State of the Wind and Weather

[The State of the weather for the next two months must have such important consequences that we have made arrangements with the electric telegraph company for a daily report.]

At Nine o'Clock Yesterday Morning the wind and weather at the undermentioned places were as follows

Chelmsford	. . .	W	. . .	Fine
Colchester	. . .	WSW	. . .	Fine
Derby	. . .	NE by N	. . .	Fine
Gloucester	. . .	ESE	. . .	Fine

etc.

The *Daily News* intended these reports to reassure farmers throughout the harvest. By autumn they had seemingly run their course and were dropped. Unwittingly, though, the newspaper had hit on something – weather reporting. This new idea appealed for both its practicality and its novelty. It turned out that readers in London had enjoyed studying tables of weather as they ate their morning toast. It had allowed them to discover whether a friend or family member had been caught in the rain at Brighton or basked in the sunshine at Deal. For weeks following the end of Glaisher's reports the newspaper received letters asking if the bulletins would return.

One enquiring note arrived from none other than Airy himself. Whether Airy wrote on Glaisher's behalf is uncertain, but he told the editor that he had found the information useful. He had started to plot it on a map and had been on the point of writing to praise an idea that was 'likely to lead to results of great scientific value' when the scheme had abruptly ended.

Airy's intervention was significant. A keystone in British scientific life, his opinions carried weight and the *Daily News* could hardly brush him aside. Instead they wrote back with a resolve to start the project afresh. This time, though, the scheme was modified to work over a longer period of time. They decided to replace the telegraph with the railways. Wondrous as it was, the telegraph was too expensive for daily use and the network was still limited. Railways by comparison were a better alternative. By 1848 about 5,000 miles of track had been laid across Britain, far more than the amount of telegraphic wire, and the *Daily News* managed to elicit the support of the Great Northern, Great Western, South Western, South Coast, Lancaster & Carlisle and York, Newcastle & Berwick railways, who agreed to carry reports to London for free.

It was an exciting development. Airy and Glaisher drew up a list of fifty railway stations spread across Britain, and from there Glaisher took on the task of communicating with the stationmasters. Within a few weeks Glaisher was once again on his travels, this time carrying a simple skeleton form, including the place, date, wind direction, strength (calm, gentle breeze, strong breeze, hard wind, storm or heavy gale) and type of weather: cloudless, partially cloudy, overcast, foggy, scud, rain, heavy rain, snow, hail or thunderstorm. Communication between the governmental departments seems to have been poor; had Airy talked to his friend Beaufort in Whitehall he would have learnt about his more advanced wind scale and weather code. But the departments remained fractured, ploughing their own distinct furrows. Still, at least it was a start.

By the next summer the operation was under way. The first bulletins were published in the *Daily News* on 14 June 1849 under a heading 'Meteorological Table – Showing the State of the Weather at each of the following places at Nine o'clock yesterday morning'. Meanwhile back in Greenwich Glaisher was doing more than just archiving the old data.

On receiving the returns, I first examine every one by itself; second, I divide them into groups, including the observations from one known good observer, and then I compare every result in every return with the corresponding result in the standard return, taking into account difference of elevation, etc.; next I form groups according to latitude, and another according to longitude; by these means I usually detect any errors, and I believe very few escape. After this I proceed to their combinations.[30]

As Loomis had done in America, Glaisher also began to plot the data on an outline map of the British Isles. He used a long, narrow-headed arrow to show the direction of the wind and other symbols – he doesn't elaborate – to indicate the barometric pressure, and the temperature at the various stations. Glaisher started these maps in July 1849 and continued to experiment with them over the next year, day after day. Although they were never published, these are thought to be the first weather maps drawn from contemporary data. Coming so soon after the earliest telegraphed weather reports, they cement Glaisher's position at the beginning of weather reportage, not just in Britain but in the world.

No longer Airy's assistant, by 1849 Glaisher had a reputation of his own. In January 1849 he was elected a fellow of the Royal Society, with Airy, Herschel and Charles Babbage signing his nomination form.[31] Cited for his meteorological papers in the *Philosophical Transactions* and for his eminence as a meteorologist, for Glaisher the achievement crowned a remarkable decade. Quick to capitalise on his status as a fellow, Glaisher was soon petitioning the Royal Society for a grant to help him expand his annual register and weather network for producing the governmental statistics. He had been heading this project for four years and shouldering the costs himself. Meteorology, Glaisher argued in his letter to the Council, was undergoing a revolution. It had been overshadowed by the more solid grounds of physics, chemistry and biology, but not any longer. 'Within the last two years there has been a remarkable increase in the attention to this much neglected science,' he pointed out.[32]

There could be no quibbling with this claim. All of a sudden the scientific world seemed abuzz with meteorological activity. The 1848 meeting of the British Association for the Advancement of Science had been dominated by news of weather-related schemes and experiments. Three of the keynote speeches were given over to meteorology, including an address by a German scientist called Heinrich Dove on conflicting air masses. The same year a sailor in the East India Company, Henry Piddington, had published *The Sailor's Horn-Book for the Law of Storms* on the theories of Redfield and Reid. Lucid, clear and compelling, it was positioned as a reference work for mariners rather than academics and it sold well. He made another contribution, too. Eager to standardise vocabulary, he aimed to replace the various words for storm winds – sirocco, helm wind, simoom, Sumatras – with a catch-all standard. He proposed that sailors adopted 'cyclone', an old Greek word for 'the coil of a snake', to describe circular storms. The idea would catch on. Back in America there was no sign of an end to the storm controversy: Espy and Redfield were still lobbing missiles at each other in the scientific press while a further combatant, Robert Hare, had entered the debate in belligerent style with ideas about electrical force.

Elsewhere there was more activity. In 1841 the British Association set up a meteorological observatory in Kew. Superintended by Francis Ronalds, the Kew Observatory was as industrious as its rival in Greenwich, harvesting all manner of atmospheric data, using kites and electric apparatus. In 1849 William Reid published an updated version of his original handbook entitled *Progress on the Law of Storms*, while the same year, in America, Espy finally had his chance to make rain. He made his experiment in Fairfax County, Virginia, where he set fire to twelve acres of pine forest and waited for the rain. It never came. Thereafter little more is to be found of Espy's rain-making ideas in the newspapers although in private he explained away the failure of his experiment as the result of interfering factors.

At the same time a different American scientific scheme was getting under way. In August 1846, as the hailstorm struck London, President Andrew Jackson had opened the Smithsonian Institution in Washington, DC. The Smithsonian was a new 'establishment for the increase and diffusion of knowledge among men', funded by the surprise bequest of around $500,000 by an English philanthropist. Appointed as first secretary of the Institution was Joseph Henry, the star of American science, and soon Henry announced the Smithsonian

Meteorological Project – the first major scheme of the new body. It was an ambitious affair. Advised by Loomis and Espy, Henry drew up plans for a bigger network of weather-watchers than ever. A thousand dollars was allocated to the project and soon results were streaming in from all across the States by telegraph. The data was rich enough for a weather map to be displayed each day in the Institution's lobby. The map soon became a tourist attraction for its scientific novelty.

At last meteorology was in vogue. No longer was it an eccentric plaything; the biggest problem it now faced was co-ordinating all the disparate work. The occasional sub-committees of the Royal Society in London were not equal to the task and all previous attempts to create a meteorological society, including one by Thomas Forster in 1823, had failed. It was a glaring problem and one that was at last addressed in April 1850 when Glaisher joined a meeting of fellow meteorologists at Hartwell House in Buckinghamshire. The meeting lasted three days, and ended with the establishment of the British Meteorological Society. Announcements of the Society were soon circulating in the press and a first meeting was arranged for 7 May. 'Gentlemen desirous of promoting the science might be admitted members on signing their wishes to Mr Glaisher, of Blackheath,'[33] announced *Jackson's Oxford Journal*. Glaisher himself did not take the position of president – he left that to Samuel Whitbread, a fellow of the Royal Astronomical Society. Instead he was appointed secretary. It was the perfect role for Glaisher: pulling strings, in the thick of the detail.

Glaisher's reputation was no longer confined to Britain. When Joseph Henry wanted to find out about the customs of collecting weather data in Britain, out of the whole scientific community in Britain he wrote to Glaisher. Glaisher was overjoyed at the honour. He replied the same day, full of enthusiasm. He gave a full account of his successes over the years and forwarded to Henry some sample skeleton forms for data collection, adding:

> With the papers I shall send you will find a few copies of an Address of a new Society, which myself and a few gentlemen have formed . . . At the meeting of the Council of the Society, held a few days since, I did myself the pleasure of reading the letter with which you have favoured me, and it was resolved that a form for collecting observations, drawn up by myself, and now

in the printer's hands, should be sent to you, and the Council expressed a wish to co-operate with the Smithsonian Institution as far as possible.[34]

Throughout the spring of 1851 Londoners looked forward to May Day. Always a cause for celebration, this year it would be extra special: it was the date long fixed on for the opening of the Crystal Palace of Industry in Hyde Park and the beginning of the Great Exhibition. When the day finally arrived the capital was charged with anticipation. The *Illustrated London News* described the scene:

> Never dawned a brighter morn than on this ever-memorable 'May Day'; the sky clear and blue, the sun coming forth in undimmed splendour, the air crisp, cool yet genial, as a poet's spring morn should be. London with her countless thousands, was early afoot; by six o'clock, the hour fixed for opening the park-gates, streams of carriages all filled with gaily-attired company, came pouring in from all parts of the metropolis and the surrounding districts, while whole masses of pedestrians marched in mighty phalanx towards the scene of action.[35]

The Palace of Industry ran along the southern edge of Hyde Park, a shimmering building of crystal walls and iron pillars. It stood, one journalist wrote, as one of the most 'splendid monuments of human progress in every quarter of the globe'. Inside, a maze of boulevards and avenues waited to be explored by 'the wealthy to whom money is of no importance and the respectable to whom shillings are a grave concern'. The avenues were bedecked with sculptures, jewellery, fine art, pottery, steam engines, photographic equipment, musical instruments and everything else imaginable. Never before had there been such an outpouring of collective creative genius. Erasmus Darwin and Richard Lovell Edgeworth would have loved strolling amid the blaze and bustle, examining tents that popped up like umbrellas and admiring rocking beds that lulled their occupants off to sleep.

Always in demand for his organisational skills, William Reid was ever present at Crystal Palace as Chairman of the Executive Committee. Glaisher was there, too, appointed as reporter to the Section of Philosophical Instruments. He stalked the avenues, inspecting each of the specimens with his usual care – and deciding that very few were of the required standard. On 8 August Glaisher would make his own mark on the Exhibition, publishing the first British Daily Weather Map from data sourced by telegraph. Visitors could glance at the weather visually: it was another first. Glaisher's experiments were changing the way people understood the atmosphere.

One visitor impressed by Glaisher's work was Prince Albert, who asked him to present a lecture the following year. Glaisher would begin his address, 'For years I have pursued the science of meteorology and have long been convinced that a widely spread and universal system of simultaneous observation, uniformly reduced, must be the groundwork of its establishment as a science.'[36] How right he was.

CHAPTER 8

Beginnings

Catching sight of Robert FitzRoy across a west London street in early 1852 was to catch sight of a man in the prime of life. FitzRoy looked every inch the distinguished Victorian gentleman. Although he was firmly in middle age his hair had kept its rich chestnut colour and his face still shone with boyish vigour. Clad in a dark frock coat, waistcoat and necktie, he had retained something of his taste for fashion. The previous summer he had been elected a fellow of the Royal Society – 'eminent in scientific navigation' – his nomination paper signed by the recently promoted Admiral Beaufort.[1] He had followed this with a social triumph, securing his election (without ballot) to the discerning Athenaeum Club, skipping its sixteen-year waiting list. A gentleman of consequence, FitzRoy's appearances at society events were noted by the newspapers. Though he lived a little out of town in Norland Square in Holland Park with Mary, their four children and three servants, he was often in the thick of action. Known and respected, admired and connected, FitzRoy was in perfect health. Yet all was not well.

On Monday 15 March 1852 at his desk FitzRoy headed a sheet of paper in his flourishing fashion: 'Private and Confidential: FitzRoy's Curriculum Vitae, his past service, his health and readiness for future duties. Memorandum.'[2]

It was the start of a thousand-word summary of his professional career. FitzRoy outlined his experiences in succinct paragraphs. He sketched all his triumphs: the first medal, promotion to lieutenant of the *Thetis*, captain of the *Beagle*, Gold Medal of the Geographical Society, election to Parliament.

Then he reached more difficult territory.

In April 1843, Lord Stanley proposed to Captain Fitz Roy to go out to New Zealand as governor; and he thought it was his duty to undertake the onerous office, however distant and ill-remunerated. He gave up his seat in Parliament – and other employments, (though tenable for life by remaining in England) and went, with his family, in a merchant ship to New Zealand.

His proceedings there gave such umbrage to the New Zealand Company – then very powerful – that they occasioned his recall in 1846.

It was plain and accurate, an unvarnished account of a painful memory. It is not hard to detect the undercurrent: that by accepting Stanley's offer he had forfeited a stable position at home. This sense of events gone awry permeates FitzRoy's Memorandum, a document that is more complex than first appears. What was the point of writing it? Why should a man as well known as FitzRoy be asked to write a CV? Reading it a hundred and fifty years later you are left with a sense of a man seeking to convince himself of his worth. His use of the third person, although a common period device, is also revealing. Does it enable him to be objective? Is it a way of confronting perplexing questions?

FitzRoy had reason to be perplexed. Once so wealthy he could have bought, built or maintained a stylish house in the shires and had plenty left to splash on a social life, by 1852 much of his money had been frittered away. Not a man given to frivolity, the majority of FitzRoy's losses had come in the course of public duty: the New Zealand project, or buying extra support ships during the *Beagle*'s voyage. FitzRoy's faith in the government's spirit of *noblesse oblige* had been his undoing. Having fully expected them to reimburse him, in October 1834 he had been shattered to find out they would not return a shilling. This unexpected blow – not only to his finances but to his moral code – had cast FitzRoy into a depression. He had locked himself away in his cabin and drafted his resignation. For days he had protested to the *Beagle*'s surgeon that he was about to go the way of his uncle, Lord Castlereagh, who had slashed his throat with a penknife. A few dreadful days slipped past before he recovered, talked into sense by his officers. Soon he was at his place on the quarterdeck exhibiting, Darwin noted in a letter, his 'cool inflexible manner'.[3]

But the episode was not forgotten. Ever since Lord Castlereagh's dramatic and very public suicide when FitzRoy was a seventeen-year-old midshipman, a terror had haunted him. His entire personal philosophy had been built on the strength of his family pedigree. It was this that had brought him his station, his connections, his judgement and his poise.

His drive and his intellect were also inherited, talents passed down through his illustrious line of kings, lords and prime ministers. Yet by this logic he was prey to the same madness or fragility as Castlereagh. FitzRoy would have known Shelley's famous poem, *The Mask of Anarchy*, that fixed Castlereagh's reputation as a complex figure who concealed his true feelings behind a mask of political suavity. The second stanza ran:

I met Murder on the way
He had a mask like Castlereagh
Very smooth he looked, yet grim;
Seven bloodhounds followed him

It was often noted that FitzRoy resembled his maternal uncle, both in appearance and character. For FitzRoy it must have been an unsettling comparison. The possibility that he had inherited a weak emotional core rattled him. Beneath the brilliance a restless demon lurked.

In recent years this demon had reared again. After his failure in New Zealand, disillusioned with politics, FitzRoy had returned to the Navy. Well regarded from his *Beagle* days, he was soon put to work as superintendent of the Woolwich Dockyard and he wasn't long there before a ship had come his way. HMS *Arrogant* was a different beast to the *Beagle*. She was a hybrid that combined a screw-driven propeller with a traditional sailing rig. Relishing *Arrogant*'s novelty, FitzRoy had enthusiastically had her fitted out. He had been appointed captain and under his careful eye *Arrogant* had sailed on her maiden voyage to Lisbon. It was a pedestrian affair compared to the Strait of Magellan – a mere breeze across the Bay of Biscay – but it meant leaving Mary and the children in London. Years before, Darwin had counselled against this. 'I am astonished to hear that you have any thoughts of taking a ship, considering the sacrifice of leaving your family, but you are an indomitable man.'[4]

His words were prescient. In his Memorandum, FitzRoy detailed what happened next:

In February, 1850, after having proved the *Arrogant* in every way – and fairly tired himself out – Captain Fitz Roy was obliged to yield to the effects of fatigue and anxiety about home affairs conjoined which had unnerved him for a time.

At Lisbon he consulted with Commodore Martin, and gave up his ship, in order that he might settle his domestic affairs, and regain his usual uninterrupted health. A week's change of air only, with absolute rest, sufficed to make him feel himself a different person; and a few months in England, after arranging difficulties that had harassed him, entirely recruited his health, which since that time has been, as it always was throughout his whole previous life – remarkably good.

FitzRoy was recasting in far more palatable terms an uncomfortable memory, and he only tells half of the story. At Lisbon he had suffered another neurotic attack, sparked by a fretful letter from Mary about money. This sent FitzRoy spiralling. A naval man, steeped in the values of the Service, he must have known that abandoning his command was a terminal step. A long queue of replacements waited on the reserve list, all able and ready to fill his place.

In his Memorandum he skirts the truth, avoiding mention of the *Beagle* debacle, keen to cast the Lisbon affair as an aberration. But the thought was inescapable. Was FitzRoy, the famous navigator who had thrived in some of the harshest environments on earth, not in command of his own temper?

It was a temper that had long unsettled his friends. Darwin had noticed FitzRoy's strange manner early during the *Beagle*'s voyage, writing home to his sister, 'I never before came across a man whom I could fancy being a Napoleon or a Nelson.'[5] He had grown accustomed to FitzRoy's 'most unfortunate temper', and had learnt to stay out of his way in the mornings when it was at its worst. Darwin found being close to FitzRoy an anxious experience. He could charm or inspire, but equally he could terrify and attack. It was like having a lion locked up in a London town house and never knowing when it was going to roar. The worst of their arguments had come, after the *Beagle*'s return, during publication of the *Narrative*. In FitzRoy's view Darwin had failed to give due credit to the ship's crew in his account. On a visit FitzRoy had exploded at Darwin, who later ruminated: 'I never cease wondering at his character, so full of good & generous traits, but spoiled by such an unlucky temper. – Some part of his brain wants mending: nothing else will account for his manner of seeing things.'[6]

For Darwin this temper was an indelible blot on an otherwise sparkling personality. Those who knew him best were well aware of this paradox. Among his crew he was admired for his 'zeal, energy and self-sacrifice'. They were 'most struck with his extraordinary nerve and seamanship on small trying occasions when the safety of the vessel and all on board depended on him'. But the same man who could withstand so much could also crumple at a trifle. To Darwin he seemed a strange mix: gifted, masterful, tragic. In two decades he had risen from star pupil to respected captain, from celebrated navigator to rising politician. Yet at forty-six years old, when he might have hoped to be nearing the zenith of a glittering career, he was unemployed and uncertain of his future course. He ended his Memorandum:

> Captain Fitz Roy's health is now perfect. He is free to undertake any service. His age is forty six. He speaks French, Italian and Spanish – has learned Latin and Greek (as dead languages) and has studied various scientific as well as professional subjects.

Matters, though, were about to take a terrible turn. Unbeknown to FitzRoy as his pen scratched over the paper, Mary's health was ailing. For sixteen years she had been his emotional lynchpin but now she was sinking fast. In March 1852 he wrote to his sister. Mary had a chill, he explained, and 'a pain about the heart, left side and back'. She didn't improve and by the start of April it was clear that she was gravely ill.

Her death was announced in the *Morning Chronicle*. 'On Monday the 5th April in Norland Square, Mary Henriette, the beloved wife of Captain Robert FitzRoy, R.N.' Friends rallied round. Among them was Sir Thomas Gladstone, heir to the Gladstone baronetcy at Fasque in Aberdeenshire. He comforted FitzRoy and gave him a book, *The House of Mourning*. FitzRoy replied a week later. It is the most touching and revealing of all his letters:

> I do not grieve as one without hope. My beloved Wife was so sincere and consistent a Christian that I know she is safe – and permanently happy – while I am just as certain that if I do my

best to follow her blessed example and earnestly work out my own salvation I shall rejoin her sooner or later. How far her spirit may be permitted a cognisance of earthly things we cannot yet tell – but He who made and governs all is unlimited in power – We wonder at the mystery of the Electric agency – How minute a fraction of the Almighty's infinite creation is that which alone is to us incomprehensible.

The reliance of my dear departed blessing, on the Mediation of our Saviour was invariable. She had a critical knowledge of His holy Word exceeding that of many Divines – and her entire esteem for it was proved by her undeviating adherence to truth – consistency – and Christian purity of life.

FitzRoy revealed to Gladstone that Mary had known about her illness since the previous autumn and that she had kept it secret. Since February she had also, without his knowledge, started to write 'some most touching letters' to their children, 'as from beyond the grave'. The only one to know about these letters and Mary's illness was their nursemaid. FitzRoy ended his letter with a wish:

May God grant that her children and her bereaved husband may so profit by her example that they may all join her after and that not one head may be missing of those she so tenderly loved.[7]

The letter is powerfully charged with emotion. In writing just nine days after Mary's death to an intellectual and social equal, FitzRoy's cloak is torn away. The shocking discovery that Mary had kept her illness secret was devastating proof of her high moral character, innate goodness and religious purity. Instead of calling for doctors, she had worried about his career and put the children before herself even in the face of death. It was the ultimate sacrifice. The final passage of the letter is the most revealing. So tender is his hope, 'they may all join her after and that not one head may be missing' that he feels the need to apologise for his frankness.

Darkness shrouded FitzRoy. Months passed. He was now a widower responsible for four children and for a while he let his career slide. He rejected an offer from Beaufort to superintend a British tidal survey. A year went by before he took some employment, this time as principal private secretary to the Commander-in-Chief of the British Army,

Lord Hardinge. It was a bureaucratic, suffocating role for someone of FitzRoy's talent. Furthermore it came through a nepotistic network of family loyalty – Hardinge was FitzRoy's maternal uncle – rather than recognition of his ability.

But at least it was work. FitzRoy oversaw Hardinge's timetable like a glorified Bob Cratchit. Darwin, after meeting him in early 1854, recorded: 'I saw FitzRoy rather lately, & he looked very well & was very cordial to me. Poor fellow, I fear besides his other misfortunes, he is rather poor; at least he has given up House-keeping.'[8] Darwin's verdict was clear-headed as usual. At least FitzRoy was keeping the wolf from the door. Soon he was to find a level of domestic stability, too, by marrying his cousin Maria Isabelle Smyth. Whether it was a practical arrangement or a loving relationship is hard to judge, though the union did bring a solution to his household woes. He could now focus on finding a more stimulating job. Something was bound to turn up. Sure enough, something soon did.

At the end of the summer of 1853 a maritime conference had been called in Brussels. It was organised by a lieutenant in the American army called Matthew Maury who had been agitating for the meeting for some time. Maury was not the kind of man to be ignored. He had bombarded European governments for months with news of his scheme. A first-rate organiser, Maury had spent the past decade pioneering a new brand of practical meteorology in the United States; now he wanted to extend his system and needed help. His work concerned wind charts and was based on good bookkeeping. Desk-bound in Washington, DC, he had noticed how ships' logs had been completely overlooked as data sources. Vast piles of information on wind and currents had been forgotten in pigeonholes and offices and he set out to see if he could curate it into a meaningful, useful shape. It was exactly the idea that Beaufort had suggested four decades before to Edgeworth.

Maury's wind charts had become famous in the United States. They showed in any given part of the ocean the typical wind speed and strength at a specific month in the calendar. For navigators they were a vital breakthrough. For years sailors had simply sailed in a direct line towards a destination. Maury's charts exposed the folly of this. Just as an overland traveller making a journey from Manchester to Sheffield would navigate around the Pennines to avoid going up and down hills, a sailor from Rio to Montevideo might avoid zones Maury had marked as having contrary winds or, even worse, no wind. For years sailors had been forced to rely on experience – what they called navigating by Guess or God. But now with Maury's maps they were able to plan intelligent routes tailored to the time of year, just as overland travellers might do with an Ordnance Survey map. The result was reduced sailing times, and sharply increased profits. It was a cocktail bound to appeal to European nations and Maury was a superb advocate for his own work. Full of ambition, he realised that his future success relied on gathering even more ships' data and to achieve this he had lobbied all the maritime nations of the Christian world. No longer, he argued, was it right for results to be scattered aimlessly. It was time to draw them together and turn them to good use.

Maury's plan had crystallised in the conference in August and September 1853. Ten nations in all – Britain, France, Belgium, Denmark, the Netherlands, Norway, Portugal, Russia, Sweden and the United States – had taken part. Britain had been represented by Frederick Beechey, an old naval captain, and Henry James, leader of the British Ordnance Survey. Reluctant to be led by an American, the British government had told the duo to commit to nothing. But despite their pessimism, events had progressed smoothly. In what was as much an exercise in diplomacy as anything else, Maury's chivvying enthusiasm had worked. All ten participants had agreed on a skeleton log for 'co-operating merchantmen, whatever be their flag'.[9] At set times each day captains of the vessels were to take their position and readings of air pressure, temperature, wind force and direction. Beechey and James left satisfied and reported back to the Admiralty in Whitehall that someone was needed to undertake this new initiative. As was customary, a committee of the Royal Society had been asked to draw up a report and various individuals had been approached for their opinion. Beaufort was consulted, as was Airy, the Hydrographic Office and the Royal Observatory, two prime candidates to take on responsibility.

Allocating Maury's plan to a department was not straightforward. Beaufort's Hydrographic Office was the natural home of anything related to the sea, but it was already overworked. Equally busy was Airy at Greenwich, and the prospect of farming the project out to bodies like the Royal Society, the British Association or the Meteorological Society meant ceding government control. When

Sketch of the young Robert FitzRoy about the time of his meeting with
Charles Darwin in 1831. 'As far as I can judge: he is a very extraordinary person',
Darwin wrote to his sister; 'I never before came across a man whom I could
fancy being a Napoleon or a Nelson.'

John Constable, 'Study of Cirrus Clouds', *c.* 1821/2. The back of this oil sketch is inscribed with the word 'cirrus', evidence of Constable's knowledge of Luke Howard's cloud classification system.

John Constable, *Spring: East Bergholt Common.* 'One of those bright and animated days of the early year, when all nature bears so exhilarated an aspect.'

The violent weather Robert FitzRoy encountered in Tierra del Fuego while captain of HMS *Beagle* would leave a lasting impression on him. In this watercolour the *Beagle* is pictured in front of Mount Sarmiento in the windswept Strait of Magellan.

By 1850 Francis Beaufort had risen to become one of Britain's most influential civil servants. Here he is pictured in familiar guise, at the heart of the action, during a meeting of the Arctic Council.

The rival storm theorists of the 1830s advocated their contrasting ideas
at scientific gatherings like this meeting of the Royal Society in *c.* 1844.

By the 1850s America was beginning to forge its own scientific identity,
set apart from what Ralph Waldo Emerson dubbed 'the courtly muses of
Europe'. Here Christian Schussele depicts nineteen American inventors.
In the centre sits Samuel Morse with his Electro-Magnetic Telegraph.

Robert FitzRoy, the 'Clerk of the Weather', in decline, *c.* 1864.

James Espy, the 'Professor of Thunder and Lightning'.

James Glaisher, at the time of his ballooning fame in the 1860s.

Francis Galton, Charles Darwin's half-cousin and FitzRoy's nemesis.

Scientific ballooning of the 1860s allowed meteorologists to experience the atmosphere in new ways. This image of a halo was seen by the French aeronaut Camille Flammarion in 1868. James Glaisher saw something similar in 1862. 'On looking over the side of the car the shadow of the balloon on the clouds was observed to be surrounded by a kind of corona tinted by prismatic colours.'

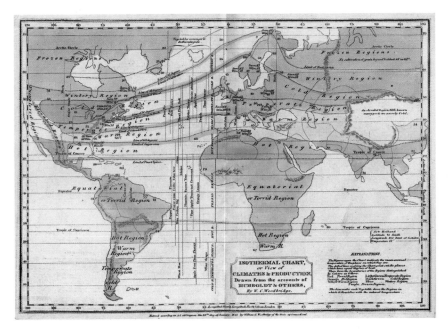

In the 1820s William C. Woodbridge's 'isothermal charts' demonstrated how scientific data could be portrayed on geographical maps. His ideas paved the way for synoptic weather charts.

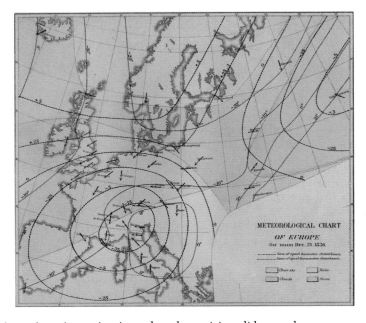

Elias Loomis, an American scientist and mathematician, did more than anyone else to pioneer the art of weather mapping. This 1859 chart shows the progression of a famous snowstorm over Europe on Christmas Day, 1836.

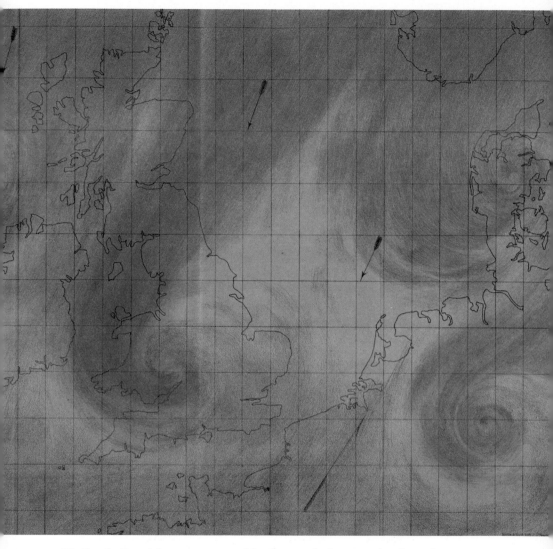

FitzRoy believed that great storms like the Royal Charter Gale of October 1859 happened along the unstable fault lines between hot and cold air masses. This revolutionary idea is depicted here in FitzRoy's *The Weather Book* (1863).

James approached Airy for his opinion he replied, blunt as usual, 'In regard to the digest of and deductions from extensive meteorological observations, I think it totally unimportant to what Department of Government this is attached, its work will be different from that of any official bureau . . . the *man* will be the only important selection.' Coming to the nub, he added, 'As regards myself abstractedly I have no objection to undertaking it, and under certain circumstances I might be willing to do so, but it would require arrangements on some points which cannot be settled this moment.'[10]

FitzRoy, too, replied to James and Beechey. If Airy had been diffident, then the letter FitzRoy returned was enthusiastic. A draft of his reply, written on 5 November 1853, is kept at the National Archives. It is loaded with crossed-out sentences, underlined words and scribbled notes. FitzRoy thought that a naval officer *might* be suitable, but worried that such a man would have insufficient education. He argued that a candidate would need a breadth of skills and practical experience. He would have to understand 'such diverse subjects as Tides – Currents – Winds – Temperature – Magnetism – Electricity and the Atmosphere. [It] would require an Arago a Whewell – a Rennell – a Reid – a Sabine – and a Faraday and a Herschel combined under Humboldt.'[11]

FitzRoy's scribbles then take a pointed turn. 'You might select a zealous trustworthy officer . . . who is able to devote himself for years to come to the great object you have in view.' Finding his theme, FitzRoy gathers pace: 'This officer would have to pull the laborious car steadily . . . to have more experience and have been more accustomed to think for others than lieutenants or masters in the Navy.' You get the sense FitzRoy's pulse is quickening as he writes.

A captain of considerable length of service should be selected . . . an actively disposed officer living in or near London – might collect materials by day and reduce them in the evening – at his own residence – he might work alone finding his own clerical assistants when required, provided that his salary was fairly considerable and so far encouraging for him to give up other remunerative employments.[12]

Although FitzRoy suggests some names, it is clear there was really only one that he had in mind. His own.

Indeed FitzRoy seemed peculiarly qualified for this new, as yet unspecified, role. His days on the *Beagle*'s foredeck gave him an advantage no landlubber could replicate. He had set foot on Horn Island, run with the wind across the boisterous Bay of Biscay, charted the Strait of Magellan. Few knew weather as he did. Years later when he heard a dramatic account from a Mr Rowell in Oxford, who had seen a broad-wheeled wagon lifted from a road and thrown over a hedge by a gust of wind, he was able to reply, 'I have known such things myself. I have known the wind to lift a boat into the air and shake it to pieces.'[13]

FitzRoy was not just a practical man but a scientific one too. His talent with instruments and commitment to accuracy had already been demonstrated and over the winter of 1853 his friend Admiral Beaufort was lobbying on his behalf. As time wore on and proposals were debated in the Commons, it became increasingly clear that FitzRoy was the best, if not the perfect candidate. He knew the systems of government. He knew the ports. He had the respect of sailors. By the spring of 1854 everyone agreed that it was FitzRoy's job.

But first the government had to find the money. Eventually £3,200 was set aside by the Admiralty. It led to a memorable Commons exchange when John Ball, a newly elected member for Carlow in Ireland and a science enthusiast, stood to record his hopes for the new body.

I trust the Vote will be increased in future years, as no branch of science has of late years, progressed so rapidly as that of meteorology. I hope that the observations made by different persons upon land as well as upon sea will be collected, as, if that is done, I anticipate that in a few years, notwithstanding the variable climate of this country, we might know in this metropolis the condition of the weather twenty-four hours beforehand.

[Interruption – Laughter.]

[Mr J. Ball, continuing] Science has in modern times achieved even more astonishing things than this, and therefore my anticipation is not so ridiculous as some Honourable Gentlemen appear to suppose.[14]

It is hardly surprising that Ball's colleagues laughed. This department was not established to predict the weather – that would have been a gross folly. Instead it was to be a chart depot charged with producing maps to reduce the cost of shipping. Ministers in London had noticed Maury's success in America. The idea that a quarter to half a million pounds could be shaved off the annual expenditure was an enormous temptation and that is what they were appointing FitzRoy to do.

Across the Channel there were more ambitious schemes in mind. After the Brussels conference the Dutch had appointed a mathematics professor, C. D. H. Buys Ballot, to set up a Royal Meteorological Institute at Utrecht. This had been established by 1854 and was soon producing meteorological maps. Even more forward-looking was France and the new chief of the Paris Observatory, Urbain Le Verrier. Le Verrier, renowned for his discovery of Neptune in 1846, had been selected by Napoleon III to replace Arago, who had died a month after the Brussels conference. Autocratic, irascible, ambitious, Le Verrier saw this as a unique opportunity to refresh the Observatory's outlook and to bring pride to France, a country with 'an eminently scientific spirit'.[15]

One of Le Verrier's chief concerns was meteorology. He realised that Paris had fallen far behind Britain and America, where Joseph Henry continued to innovate from the Smithsonian. In December 1854 he published his 'grand proposal'. He set out a template: the Observatory would be supplied with better instruments, it would publish weather records daily in the press and he would establish a telegraphic network supplying weather data to the Observatory from the provinces every day. 'By joining with telegraphic lines,' he declared, 'various stations at which meteorology observations are made, it should be possible to know from moment to moment the direction and velocity of propagating storms, and to announce several hours in advance on the coast high winds, especially the most dangerous ones.'[16]

This, at the very least, hinted at a bold new policy in France. Less than a decade before Arago had written tartly, 'those learned men who are honest and careful of their reputation, will never venture to predict the weather, whatever may be the progress of science'.[17] Yet within months of stepping through the Observatory door Le Verrier was advocating just this. A French storm-warning system would be an international first, another feather in Le Verrier's cap.

The technology at Le Verrier's disposal was far beyond what Arago had envisaged back in 1846. By the time of his appointment Europe was strung with a true spider's web of telegraph lines. Britain was the most advanced of all, where fewer than 10 per cent of the population lived more than ten miles from a station. France, Belgium and Italy were all close behind, and busily investing in the infrastructure. On a continental walking tour in 1854 Charles Dickens had written home, 'Few things that I saw, when I was away, took my fancy so much as the Electric Telegraph, piercing, like a sunbeam, right through the cruel old heart of the Coliseum at Rome. And on the summit of the Alps, among the eternal ice and snow, there it was still, with its posts sustained against the sweeping mountain winds by clusters of great beams – to say nothing of its being at the bottom of the sea as we crossed the Channel.'[18] This submarine Channel line had been opened in 1850. Thirty miles of wire, 'one tenth of an inch in diameter and encased in a covering of gutta percha [an insulating natural plastic] the thickness of a little finger'.[19] The wire ran along the bottom of the Channel between Dover and Cap Gris Nez, meaning messages could be fired from London to Paris, Berlin or Milan in seconds.

FitzRoy would keep a close eye on Le Verrier, his French rival. But in Britain there was certainly no appetite for storm warnings as he officially began his job on 1 August 1854. He was to be the Meteorological Statist at the Meteorological Department of the Board of Trade, a true mouthful, and a job title that sounded as dry as one of the *Beagle's* biscuits. A year earlier, Airy had written to James, 'In the whole world there is no science so overwhelmed with undigested facts as meteorology,'[20] and FitzRoy's job was to bring order to this mess. He was given a salary of £600 and by the end of the year he had assembled a staff of three – Pattrickson, a draughtsman, and Babington and Townsend, two clerks. To begin with they had no fixed office and, briefly, they set up shop at 15 Regent Street. It would be the next spring before they settled at their permanent home, 2 Parliament Street, a minute's walk from the Palace of Westminster. Poised, FitzRoy was back on course.

A short distance along Whitehall from Parliament Street, Francis Beaufort was coming to the end of his tenure at the Hydrographic Office: a time of industry and expansion, wisdom and foresight. Beaufort had spent twenty-six years projecting his intense curiosity on to a global canvas. In figures alone Beaufort's output was striking: 1,437 charts, more than a hundred different surveys. Every foot of the British coastline had been surveyed. Distant shorelines – New Zealand, Australia, Greece, Argentina – had received the same care. Everything had been done with a scrupulous attention to detail: the initial orders, the correspondence, the debrief, the chart-making and publication. In government Beaufort's industry was renowned almost as much as his breadth of knowledge. Over the years he had been consulted like a nineteenth-century *Wikipedia*, a quick and accessible store of enormous amounts of information. When Prince Albert wanted to identify a location for a new house on the Isle of Wight (later Osborne House), he wrote to Beaufort. When the headmaster of Rugby School wanted to check the topography of the Bay of Tunis, he wrote to Beaufort. When Airy wanted to map the path Agathocles took in the Punic Wars, he wrote to Beaufort. He seemed to know everyone. Once while strolling in Hyde Park with his family he had bumped into Darwin, who 'introduced them to some curious facts about the bumble bee'.

With the gash in his leg from Turkey and the shotgun pellets in his chest from Spain, Beaufort was a physical relic from a hostile age. He had even suffered a suspected heart attack at his desk, but still walked home afterwards. His age, however, was beginning to tell. His hearing was going, he suffered from frequent back spasms. On 27 May 1854 Beaufort turned eighty. That March he had offered his resignation but it had been turned down – with the country at war with Russia in the Crimea he had been deemed indispensable. He had carried on throughout the year and for a brief spell had overlapped with his friend and protégé Robert FitzRoy. It must have been a joy for Beaufort to see FitzRoy's talents being turned to productive use. He knew more than anyone, though, how steep the challenge was ahead. For years Beaufort had complained about dunderheaded politicians as they shaved departmental budgets. In 1851, learning that his funding had been slashed again, he had written: 'I will not trifle with your time by repeating here any hackneyed truisms about the comparative expense to the country in the cost of surveys or the loss of ships and cargoes, but I will entreat you to weigh the small sum which you propose to save against the large amount of mischief which may result.' It was a reality FitzRoy would have to grapple with himself. Beaufort's famous words, 'The natural tendency of men is to under-value what they cannot understand,' would speak to him as much as anyone.

New Year's Eve 1854 finished in colourful style. According to the *Manchester Guardian*:

> A superstitious or fanciful age might have seen in the grand effects of light and colour which adorned the last sunset of the old year – in the ragged clouds which seemed to shower down fire, the very rainbow bathed in crimson, and the delicate water tints of the far west – in these, and afterwards in the calm moonlight through which the bells rang out announcing the birth of 1855, and the grey and stormy morning – tokens of the mysterious sympathies of inanimate nature with events present and to come.[21]

It was a time of beginnings and endings, circles closing, enterprises begun. As FitzRoy got to work in his suite of rooms at 2 Parliament Street in January 1855 Beaufort was writing the last of his communiqués to his surveyors around the world, telling them to report back to his successor, Captain John Washington. By 30 January the last letters had been sent, last charts drawn, final plans finished. Beaufort left his office, shut the door and walked into Whitehall feeling 'rather lost'. No longer able to hare through the streets as he once had done, he trampled over the snow. For a fortnight it had been falling. London was a wonderland again. Catching sight of Beaufort's tiny frame inching up the street, few would realise what a colossus he had become as he walked away into the haze.

The snow had been falling since 13 January, gently at first but with increasing vigour as the days passed. By now Britons had developed a romantic taste for snow. The spectre of London in winter, an idea seized on by Charles Dickens in his Christmas numbers and by the *Illustrated London News* in their Yuletide specials, had distilled an idea of White Christmases, Victorian gentlemen slipping on the cobbles, rookery-boys lobbing snowballs across the street: red noses and wringing hands, the seductive contrast between the icy street and the drawing room with its roaring fires, an idea that tapped into the powerful division between private and public. It was a potent narrative. But stacked against all the other frozen winters of the Victorian age, there was an additional bite in 1855. The temperature hovered at freezing for days. Darwin estimated that four-fifths of the birds in his garden were destroyed. The snow lay perfectly preserved on the ground. To Glaisher it was a seductive sight. The crystals had not fallen in a flurry as they would in a blizzard. Instead they had dropped at rare intervals, from a calm, cold sky. 'Never do I recollect such an infinity of crystals as have lately fallen beneath my observation.'[22]

But for all Airy's diffidence about the Brussels conference, Glaisher might well have got FitzRoy's job. He was Britain's most visible and perhaps most able meteorologist. By the start of 1855, however, his workload had expanded to such an extent that even he might have struggled to cope. In addition to his work at Greenwich he was now co-ordinating a vast range of supplementary tasks: his work for the British Meteorological Society and voluntary collection of data for its Annual Register. During the past year he had been approached by the General Board of Health to compose a paper on the link between air quality and recent cholera outbreaks. In the few spare hours that remained he had plunged into the pursuit of two new scientific hobbies, photography and microscopy. These interests were about to propel him into a fresh, invigorating project.

It began one day at the start of February. Glaisher had been hiking through Abbey Woods to the east of London, a good walk from his home in Blackheath. He had noticed the sparkling brilliance of the snow and had knelt down to examine it with his Coddington lens, a lightweight and powerful magnifying glass popular with beetle-hunting naturalists. The ground was dusted with fine crystals, each seemingly isolated, shining like jewels. To Glaisher they looked 'like ravellings of fine white cotton, knotted here and there'. Staring through his lens he saw that each snow crystal looked like a six-rayed star, with beautiful arms that stretched out from a central hub or nucleus. Suddenly it began to snow again.

The air was calm, the snow lay upon the ground, and the sky was overcast. They were certainly 4-10ths of an inch in diameter, and could be readily distinguished. The temperature at this time was at or near 32°. After falling about a quarter of an hour, they became intermingled with a variety of very complex crystals. Some of these last exhibited all the rigidity but harmonious proportions of geometric figures; others the fanciful luxuriance of the fronds of the Lady fern; others, again, exhibited an arrangement of trefoils, and some there were, with pinnae of unequal size, three being large and fully-developed, of fern-like character, and three being little more than spicules. The air, during the continuance of the shower, was considerably cooled, and was at its coldest when these beautifully-varied figures were falling. Towards the close of the shower, fleecy groups of stars were again prevalent, the air was less cold, and half an hour after, when the shower ceased, the sun was endeavouring to penetrate the gloom which before prevailed.[23]

Many had been intrigued by snow crystals before Glaisher. In 1662 Robert Hooke had presented a series of ink-wash drawings of them at an early meeting of the Royal Society. Hooke believed these were imperfect shards of ice, the ruins of perfect forms of crystal that had split from the clouds. Another series of drawings had been published in 1834 by George Harvey in his *Treatise on Meteorology*. Now in February 1855 Glaisher decided to turn his clinical ability in observation to the task.

Back in Blackheath Glaisher used a Coddington lens to magnify the crystals, and sketched what he saw. On 8 February he 'secured drawings of some of the most remarkable' and watched them continually through the day to see whether individual crystals endured in the same state or if they mixed and melded together. After dark he was still working. 'At long past midnight when I was out of doors, the crystals sparkled in the snow, like mica in a piece of granite, and every cobweb every leaf and knotty projection was laden with countless myriads of crystals which seemed to defy every effort to individualise their character, or group them into classes.'[24]

Glaisher's new scheme revealed the hidden beauty of a world invisible to the naked eye. During the days that followed he returned again and again to his task, recording, classifying and sketching individual specimens from his house window. Just as he had sought to demonstrate that there was order in nature with his studies of dew a decade earlier, so he attempted to prove that the snow that was banked up against the houses, slipping from rooftops and clinging to the branches of trees was not a haphazard heap of particles but a true and perfect arrangement of geometrically precise specimens. On 16 February he was out again. Each distinct snow particle was like a miniature Crystal Palace, with avenues, arches and structural architecture that enabled the light to deflect and reflect as it collided with its edge. He captured what he saw in nimble drawings. Of one snowflake he wrote, 'its diameter was about 0.05-inch. It glistened brightly, and was highly crystalline. Its general effect was similar to the drawing, and the clusters of prisms round the outer boundary of the figure chiefly arrested the attention; the nucleus appeared a glistening speck.'[25]

Glaisher worked with the relish of a fossil hunter. But with no way of storing the physical crystal he was forced to sketch it as best he could. He marked the maximum and minimum temperatures, the snow cover, the place of observation. As he progressed he found a rich variety of form and pattern and an array of geometric shapes: hexagonal, rhomboid, rectangular prisms, darting shafts of crystals. Far from being haphazard, they displayed an astounding symmetry. Of one crystal, 'as graceful a combination as can possibly be imagined', Glaisher wrote:

[It] exhibits a nucleus composed of two hexagons, so centred as to present a double set of angles; from six of these spring the main radii of the figure, surmounted by crystalline plates of laminae of the greatest transparency. From these laminae spring leafy tufts. Which as it were, crown the structure. The lower part of the ray, near the nucleus, serves as an axis for an elongated prism, near the apex of which spring on either side, leaflets of graceful form. Immediate and springing from the angles of the under hexagon, are a set of shorter rays, the axes of similarly elongated prisms, surmounted on the top by three others, the one of similar, the other two of dissimilar figures. It is hardly possible to imagine a more graceful composition, which is greatly enhanced by the delicacy and admirable execution of the drawing.[26]

At home Glaisher was joined in his experiments by his twenty-six-year-old wife Cecilia. A talented artist, Cecilia would establish a reputation of her own later that year with photographs of British fern leaves. She worked with a natural flair and sensitivity hitherto unknown to the subject. Soon her drawings mounted up in Glaisher's study: a variety of crystals in breadth, beauty and microscopic detail to rival any great Victorian collection. Glaisher would write up his fruitful researches several times. First for the Greenwich Natural History Society, then for the *Illustrated London News* and later for the British Meteorological Society under the title, 'On the Severe Weather at the Beginning of the Year 1855, and on Snow and Snow Crystals'.

With Cecilia's support, the snow-crystals project became a success. Reading his report now you can detect his sense of loss as the thaw set in at the beginning of March. Like a friend at a deathbed he watches as the crystals melt one into another, deliquescing, tumbling, falling – towers and castles crashing down.

Owing to the high temperature, the figure of the crystals continued rapidly to change: collapsing in the most curious and kaleidoscope manner possible, the upper groups of prisms collapsing first, the next in order next, and so on. When I say collapsing, I mean the sudden dissolving of three or more prisms into one, a change effected with instantaneous rapidity . . .

At noon the snow had all but ceased. The temperature attained to 37°. Cocks crew as anticipating a change; the birds, which for six weeks previously have been silent, answered each other from the trees; icicles two feet in length, which I had noted for sixteen days previously, were fast melting away: all nature but the birds seemed motionless, as waiting the advent of a change; and, what is rarely seen, the trees were dripping moisture while the snow lay like a rime upon their branches and bended stems. Half an hour after the thermometer rose to 38°, and a complete thaw set in.

At 3 o'clock, the thermometer was at 35.5°, small and fine snow was falling, water was dripping everywhere, the birds were singing joyously, and a dead calm prevailed.[27]

*

Across London in Westminster FitzRoy was setting to work with the energy he had poured into the *Beagle* preparations more than two decades before.

He began with the ports. He knew that establishing a line of communication with merchant captains was essential and one of his first acts was to send a circular to the marine boards and shipping agents. He advocated an economic argument. 'It has been shown,' he pointed out, 'that Lieutenant Maury's charts and sailing directions have shortened voyages of American ships by about a third. If the voyages of those to and from India were shortened by no more than a tenth, it would secure a saving, in freightage alone, of £250,000 annually.'[28]

FitzRoy's small office was not encumbered by bureaucracy and it was soon developing into an industrious arm of government. He sourced instruments, sent them over to Kew for testing and then co-ordinated their dispersal through the Mercantile Marine or the Admiralty. By mid-1855 fifty merchant ships and two men-of-war were already fully equipped. A succession of others soon followed. He travelled widely, meeting captains, agents and naval officers. Not one to wait around for results to come in, FitzRoy had his draughtsmen work on historical data in the meantime. Dividing the oceans up into 10° squares, he plotted what he termed 'wind stars'. There was one star for each square, and at a glance the chart showed the set of average wind data for a three-monthly period. FitzRoy's wind stars were a simple innovation on an old idea and a variation of Maury's wind charts. They transformed the seas of the world into a chequer-board of easily interpreted data. Faster routes to India, America and China instantly became visible. It was exactly the work the Lords of the Admiralty had envisaged in 1854.

FitzRoy's departmental notes from this time fizzle with intent, and his magnificent looping handwriting hops over the pages. FitzRoy's jottings are of the moment: exhortations to work hard, be neat and turn up at a punctual hour. They give a flavour of the man who had once strode across the quarterdeck of a ship of the British Navy, bellowing at his crew:

Mr Pattrickson –
I am sorry to find that in consequence of my own attention not having been given sufficiently to the progress of filling data books

– some of them are very <u>discreditable</u> – in their appearance –
however exact we may hope they are in their contents.

. . .

Mr Pattrickson –
Adverting to former office arrangements and recent changes!

. . .

Mr Pattrickson,
Adverting to my remarks on office arrangements – I have now
to say, that I, for my absence, for more than a day, either you or
Mr Babington ought to be here.[29]

FitzRoy was as puritanical as ever. In letters he complained that
'the office has only three young men, not sailors, so merely a minimum
of work can be done'.[30] Happily for the government, a FitzRoy
minimum still meant an impressive output. He worked his charges
hard and often reminded them that they were undertaking a public
duty. Like Beaufort before him, FitzRoy considered his hours at
Parliament Street sacrosanct, devoting them to government work. 'So
scrupulously did he carry out this principle, that during that time, he
carefully avoided all personal matters, writing his own private letters
after five o'clock, before his return home.'

This was a cardinal tenet of FitzRoy's philosophy: his mission to
serve as best he could. As he had on the *Beagle*, he soon gained the
respect of his subordinates. He argued successfully to the Board of
Trade that they should have more time off and fought their corner
when they were bullied, unreasonably he thought, into taking civil
service examinations. At the end of 1855 FitzRoy jotted down the
routine of the infant department: a rigid hierarchy – FitzRoy then
Pattrickson then Babington then Townsend, all engaged on their
specific duties, churning through the enormous sea of data.[31]

A year into his job, in 1855, he was mulling over the suitability of
the Meteorological Register. It had been drawn up by Beechey after
the Brussels conference in 1853 and FitzRoy had decided it was restric-
tive. Ignoring instructions he composed a replacement. This provoked
a furious response from Beechey, who soon found out. Henry James

was drawn into the ensuing furore. 'I have read your proposed new edition of the Meteorological Register &c,' he wrote to FitzRoy, 'and if it had been proposed at the Brussels Conference I should unhesitatingly have adopted it.' As it was, James told him to continue with what he had.

The next day FitzRoy petulantly hit back, reeling off a list of grievances. Beechey's form was not fit for purpose: Glaisher thought so, Major Sabine and Washington the hydrographer, too. He sent James a second replacement form but a week later James wrote starkly back: 'I am clearly of opinion that your best and safest course is to revert to the Brussels form of log, pure and simple.' It was irritating news for FitzRoy, who scrawled, 'Why? If found defective?' in the margin.[32]

It was an early sign that FitzRoy was not going to just follow his orders without question. Accustomed to thinking for himself, FitzRoy was discovering that Whitehall bureaucracy could pour water on the best of ideas. Despite FitzRoy's cry to Henry James that 'I am not the Innovator', he was not acting the part of the sleepy civil servant and at times he grew devious in his attempts to sidestep his superiors. He encouraged Maury to write secretly, so that they could discuss plans in private. Their clandestine correspondence remained just beneath the radar, but in Whitehall the feeling that FitzRoy was not entirely to be trusted may have been the reason he was passed over for promotion when Beechey died in 1857. FitzRoy had coveted Beechey's position as chief naval officer to the Board of Trade – and probably the accompanying salary of £1,000 a year – and although he was ideally placed the role went to his old friend and subordinate from the *Beagle*, Bartholomew Sulivan. It was a slight that must have stung. Since starting his job FitzRoy had been busily rebuilding his social standing. He had moved his family from Norland Square to the elegant Onslow Square in South Kensington, a new development of white stucco-fronted town houses by the fashionable society architect Charles Freake. One of the most coveted addresses in London, a thousand a year was the very least he needed to keep his family in suitable style. It was an expense he could not really afford.

Still, people couldn't fail to be impressed with FitzRoy's productivity. As his network of captains continued to grow he turned his attention to the secondary aim of the department. While the production of time-saving maps was the driving rationale when money was set aside in 1854, there was also mention of storms. Sabine had written:

It is much to be desired, both for the purposes of navigation and for those of general science, that captains of Her Majesty's ships and masters of merchant vessels should be correctly and thoroughly instructed in the method of distinguishing *in all cases* between the rotary storms or gales which are properly called *cyclones*, and gales of a more ordinary character.[33]

FitzRoy had noted this brief passage. It extended the reach of his activities, broadening his remit into questions of science. It was an attractive proposition. If FitzRoy could examine the scientific aspect of weather it would be a perfect way both to satisfy his intellectual curiosity and to experience a sense of religious fulfilment. Bettering the lives of those at sea, explaining the Creation: these were excellent ministries. With Beaufort's example before him, too, he had a blueprint for an office that was an administrative hub as well as a hotbed of philosophic enquiry.

FitzRoy already had a grasp of meteorology. Years spent roaming the oceans had brought him into intimate contact with all types of weather: pamperos, St Elmo's Fire, lightning, williwaws, Channel gales and black squalls. Initially a lunarist, his belief in the moon's influence on weather had been destroyed in 1836 when, on the *Beagle*, he met Sir John Herschel at the Cape of Good Hope, who had reasoned him out of the idea. Over the years he had studied Espy, Redfield, Reid and Piddington. And since his appointment to the Meteorological Department he had plunged even further into the subject; one book in particular had impressed him. Heinrich Dove's seminal *Law of Storms*, which he had had translated into English.

In March 1858 FitzRoy renewed his correspondence with Herschel. He wrote to say that he had just read and enjoyed his article on meteorology for the *Encyclopaedia Britannica*: 'to my shame and sorrow I did not know of its existence'. In his letters to Herschel FitzRoy is at his most deferential. He calls himself merely a 'practical man'[34] and appeals to Herschel's good nature for titbits of information. He comes across as an eager schoolchild, lingering behind after class with a favourite master. After 1858 the letters increased in frequency. Herschel seemed to enjoy the correspondence and the two men traded private views. FitzRoy complained about the word 'meteorology', thinking it a complicated tongue-twister. 'I wish one might say "mitrology",'[35] he wrote. At another time he comes across as duplicitous, revealing his true opinion of Maury. 'Entirely do I subscribe to your (private and strictly confidential) opinion of Lieu Maury,' FitzRoy admitted. 'He has collected facts (or data <u>some</u> very doubtful) very industriously (aided by a <u>large</u> staff) and has set other people to do likewise.'[36]

The Herschel letters give an early indication of what would later become plainly obvious. By the late 1850s, despite protestations to the contrary, FitzRoy was trying to establish a scientific reputation of his own.

FitzRoy had the latitude to do this. He was in close correspondence with the Kew Observatory, which was responsible for supplying him with standardised thermometers, barometers and hygrometers. Taking an interest in the quality of these instruments, FitzRoy devised a new model of barometer, later to become known as the FitzRoy Barometer, that substituted the familiar numerical scale for one with descriptive words, intended to make reading the instrument as easy as possible. The barometers were soon in production. Of equal use was FitzRoy's twenty-five-page *Barometer and Weather Manual*, a typically straight-forward, clear and accessible booklet that took inspiration from Piddington's *Sailor's Horn-Book* and Reid's *Law of Storms*. Kept inten-tionally brief 'so as not to diminish the portability of the compilation or increase its price', it was a chance for FitzRoy to condense all he had learnt into a single place.

The book drew on FitzRoy's rare combination of scientific under-standing and real-life experience. Like Forster's *Researches About Atmospheric Phaenomena*, FitzRoy's manual speculated at length about the possibility of predicting weather. 'Difficult as it is to foretell weather accurately, much useful foresight may be acquired by combining the indications of instruments with atmospheric appearance.'[37]

This was no timorous work of science, and it was significant because FitzRoy was writing in his capacity as the governmental meteoro-logical statist. It was the first state foray into weather prediction – unsanctioned and probably unnoticed. FitzRoy's advice was clear and simple, easily accessible for captains at sea. His section on clouds could have been written by Thomas Forster himself:

> Soft-looking, or delicate clouds foretell fine weather, with moderate or light breezes; – hard edged oily-looking clouds, wind. A dark gloomy, blue sky is windy; – but a light, bright blue sky indicates fine weather. Generally, the *softer* clouds look,

the less wind (but perhaps more rain) may be expected; – and the harder, more 'greasy' rolled, tufted, or ragged, the stronger the coming wind will prove. Also a bright yellow sky at sunset presages wind; a pale yellow, wet; – and thus by the prevalence of red, yellow or grey tints, the coming weather may be foretold very nearly; indeed if aided by instruments almost exactly.

Small inky-looking clouds foretell rain, a light scud, driving across heavy clouds, wind and rain; but if alone wind only.[38]

You can imagine these thoughts racing through his mind on a brisk, bright morning in Tierra del Fuego, the crew rushing across deck, climbing rigging, shouting orders from the crow's nest. Some of the salt and spray of life at sea is captured by FitzRoy's documentation of several poems or 'saws':

> When the glass falls low,
> Prepare for a blow.
> When it rises high,
> Let all your kites fly.[39]

FitzRoy's booklet was well received. A copy reached Maury in America, who wrote back that it was 'a capital little weather guide'.[40] Three years into his meteorological work FitzRoy was sailing smooth and quick. He had established his network, issued wind star charts, designed a barometer and released a book. Whitehall might have lost Beaufort but it had gained FitzRoy.

There was no need for Beaufort to stay in London after his retirement. For better air and distant horizons he moved to Brighton on the south coast. Beaufort could gaze from his bedroom window out over the English Channel and, perhaps, relive his many voyages that had started and finished there. More than thirty years before, Constable had made the same journey from London to Brighton, 'a Piccadilly by the sea', for the sake of his wife's health. Its bright, fresh air, sloping topography, sweeping beach and bustling waters had become a subject for his art. Here he had painted a range of oils. *Seascape Study with Rain Cloud* shows a cloudburst, a torrent of water tumbling dramatically into the sea. Another depicts the seafront on a grey, brooding day. It is filled with chiaroscuro: patches of light burst through gaps in the clouds. In the distance a brig is anchored in the bay. Two figures crunch over the shingle arm in arm. The Georgian age Constable depicted had now faded into brash and busy Victorian Britain.

Strolling on the shore in his final years Beaufort would have identified with Constable's vision of the lonely brig. The ship in the bay could stand for one of a dozen – the wrecked *Vansittart*, the fighting *Phaeton*, HMS *Woolwich* of the wind scale, *Frederickssteen* of his tour to Karamania. Or it could represent one of Beaufort's own survey vessels: *Beagle* or *Sulphur* or *Rattlesnake* or a hundred others Beaufort chose, filled with men, charged with purpose and sent around the world. The weather was always with Beaufort. He had kept his weather journals for more than half a century – tallies of temperature, pressure, wind and rain that ran on and on. He kept up correspondence with his London friends and remained cheerfully lucid. On 27 May 1857 he turned eighty-three. By the end of the year, though, he was weakening fast. As Christmas approached his family was called to Brighton to bid him farewell. There they found him weak in body but bright in mind, with a 'wonderful clearness, memory, spirit and animation in the discussion of a book he was reading'. On 15 December he held a conversation with his doctors on the merits of a good historian. The next evening he told his children he was tired. He gave each a kiss and settled down to sleep. He died quietly in the night.[41]

The tributes to Beaufort were fulsome, kind and true. His legacy was profound. With no formal education he had risen right to the very top of the British administrative system to be ranked alongside friends such as Airy, Sabine, Whewell and Herschel, and he was celebrated for his sagacity. A fund was established that became the Beaufort Testimonial Prize, an honour to be awarded to the highest-ranking candidate in navigation in the annual lieutenant or master's exams. To reflect Beaufort's personality the prize would not be a Gold Medal but 'some instrument or work of a professional character that may be practically useful to a Naval Officer'. The award was narrowed down to one of the following:

1. A set of the best telescopes, by Ross, Dolland, or other optician of established character
2. A sextant, by Troughton, or Cary, or some other well-known maker
3. A gold lever watch, or pocket chronometer

4.   A geographical atlas, as Keith Johnson's *Political and Physical Atlas*
5.   A set of books of standard authors chiefly on subjects connected
       with the Naval Service[42]

Three trustees were appointed to administer the prize, one of them
Beaufort's old protégé: the star of the Naval Examinations himself
and the recently promoted Rear-Admiral, Robert FitzRoy.

Other members of the old generation were fading away too. William
C. Redfield died on 12 February 1857, a distinguished man in American life, one of the founders and first president of the American
Association for the Advancement of Science. The news shocked his
friend William Reid. For twenty years the two friends had been in
constant communication, writing hundreds of letters, passing on
observations about storms right to the end. On hearing the news,
Reid wrote to Redfield's son John: 'I have felt all day as if a heavy
loss has befallen me, and I am sensible that it is not an illusion.'[43]
Reid himself would die the following year, bringing to an end a life
which had seen him serve as governor of Bermuda, the Windward
Islands and Malta. 'His efforts for the improvement of agriculture
and education and for the promotion of circumstance were unremit-
ting,' declared Redfield's son, 'and well-earned for him the title of
"The Good Governor" applied to him in Dickens' *Household Words*.'[44]
    Reid and Redfield's deaths were followed by those of Thomas
Forster and their great foe James Espy in 1860. Espy's contributions
to the infant science in the 1830s had been vital and profound, yet his
ambition to be seen as a Newton of the air would never be realised.
For all his brilliance, Espy had been hampered by pugnacious perse-
verance in a theory that was too bold in its extent, and too colourful
in its application. More than anything else he would be remembered
as Espy the rainmaker – a slight that is unfortunate, but of his own
making. Luke Howard, the father of modern meteorology, would
outlive them all. An elderly man in his eighties, his memory failing
and his body frail, he retained his interest in the skies. His son Robert,
who nursed him at their family home in Tottenham, north London,
preserved an image of his elderly father, gazing up into the skies
'morning, noon and night', watching the clouds glide towards
Hampstead even after he could no longer remember their names.
Howard would die aged ninety-one in 1864.

Howard's contributions to meteorology over half a century were profound. He had not limited himself to clouds but had looked at radiation, urban heat islands and wind flow. Fifty years before, at a lecture in Tottenham, he had even suggested that the rotation of the earth might deflect winds off course. He had explained to an audience that as the air travels north or south the earth is forever 'slipping away under it'.[45] Had Espy or Redfield considered this, it might have changed the way they thought about storms. As it was, meteorology had to wait until 1856 for the solution to the storm controversy. It came from a quiet but brilliant mathematical scholar, William Ferrel of Nashville in Tennessee. In 'Winds and Currents of the Ocean', Ferrel explained why the wind should twist around the centre of a storm. He demonstrated that this was caused by the rotary motion of the earth, which deflected air moving towards a centre of low pressure. Once the earth's rotation was added to Espy's theory, a true picture of atmospheric circulation arose. Ferrel based his calculations on the work of a French mathematician Gustave Coreolis, who had calculated the deflective effect of the earth's rotation. Ferrel's achievement was to apply this to winds for the first time. It was a joining of the dots that linked observation and theory in an enticing new way. Puzzlingly, it meant both Espy and Redfield had been half right. The winds did rush upwards as Espy had argued and they did revolve around a central point as Redfield claimed. Neither Espy nor Redfield, who had spent so much of their working lives in internecine warfare, lived to see their theories joined together. Perhaps it was just as well.

One generation had passed but they had bequeathed hitherto unimaginable progress in theoretical and practical meteorology. In London, FitzRoy's weather network was fast expanding; Glaisher's work was intensifying at the Observatory. In Paris, Le Verrier had established a network of twenty-four weather stations, thirteen of which transmitted reports to the Observatory by telegraph three times a day. These reports were published in newspapers like *La Patrie*. In private Le Verrier had gone further, supplying weather probabilities to Napoleon III from a special office at the Élysée Palace. In America Joseph Henry was progressing just as fast. Since 1856 he had been displaying telegraphic weather data on a map in the hall of the Smithsonian. This map was a marvel. It showed, almost in real time, weather at every telegraph station in the USA. Coloured cards – white for fair weather, blue for snow, black for rain, brown for cloudiness – were slung on pegs over specific locations, and arrows showed the wind direction. After its initial refresh at ten in the morning, the Smithsonian's map was updated throughout the day, and, in a shining symbol of this new meteorological science, weather flags were hoisted above the entrance on days of special lectures.

As far as meteorology was concerned, America was no longer in thrall to anyone. In fact, by the late 1850s it was leading the world – firing news of storms here and there with Morse's telegraph. The pace of change was too much for some. In 1858 the *New York Times* in a leader article claimed there was 'no rational doubt the telegraph has caused vast injury' on the mind and morals of the people. It sniped:

> Superficial, sudden, unsifted, too fast for the truth, must be all telegraphic intelligence. Does it not render the popular mind too fast for the truth? Ten days brings us the mails from Europe. What need is there for the scraps of news in ten minutes? How trivial and paltry is the telegraphic column? It snowed here, it rained there, one man killed, another hanged.[46]

There was no such grumbling in Parliament Street where FitzRoy was already plotting his next move. There had been a boom in merchant shipping, fuelled by Peel's repeal of the Navigation Acts in the 1840s and the growth of a wealthy Victorian nation, eager to enjoy the fruits of global trade. By the 1850s the coasts of Britain were thronged with ships. Whalers heading for New Zealand and the South Seas, traders bound for India or the West Indies, emigrants leaving Liverpool for California or New South Wales. Such was the demand for ships, many were of poor quality and at all British ports stories were circulating about lax conditions, inexperienced owners, businessmen out for a quick dollar. The result was inevitable. In 1857 and 1858, a governmental report concluded, an average of 850 lives 'and property to the value of £1,500,000 to £2,000,000 were lost annually by shipwreck on the British coasts'.

By 1859 FitzRoy was certain that science had progressed to combat this evil. He thought the solution lay in turning his growing stockpile of weather data to a different use:

It is obvious that the mere aspect of the sky often deceives the most skilful mariner, but this is not applicable to very bad weather. Fortunately, when nature frowns in earnest, and intends to put forth a stronger storm than usual, she makes it known beforehand. So much so, is this a fact, that it is now known, that no tempest or wind violent enough to involve great danger can approach our shores, or our ships at sea, without advertising its appearance on the barometer. Warning has always been given by nature, and if man is heedless, he must take the consequences.[47]

It was a passage that would have rung in the ears of Captain Thomas Taylor.

CHAPTER 9

Dangerous Paths

An Indian summer settled over England in the autumn of 1859 and there was talk of a changing climate. For two years temperatures had been rising and the last summer had been as hot as the well-remembered July of 1846. Thirteen years on, temperatures had reached an extraordinary 92°F, FitzRoy noted in his records. The harvest had been brought in under a sweltering sun that continued to blaze as Parliament reconvened in September. The newspapers made great play of the politicians, dressed in their inflexible uniforms of chimney-pot hats and midnight-black suits, each being slow-cooked as they worked.

At FitzRoy's office, a minute from the Houses of Parliament, the biggest problem was the state of the Thames. Two years of drought had left it well short of its picturesque best. Noxious vapours rose from its surface, the river beneath loaded with sewage. 'Few Londoners have yet forgotten the state of the Thames in 1859,' FitzRoy recalled. 'Deficiency of water supply during 1858 and 1859, and great evaporation caused a condition of its liquid excessively disagreeable to eye and nose, if not actually pestiferous.'[1]

The hot weather continued until Thursday 20 October when temperatures tumbled overnight. It was not the gentle ebb of summer into autumn. A sleeting blizzard whistled around the Houses of Parliament. The sudden shift of hot to cold was too much for the newly cast great bell 'Big Ben' which at just a month old developed a crack and was left sounding strangely muffled.

At his home in Onslow Square FitzRoy was taking his weather readings. He was alarmed by the drop in temperature. It had plummeted to 22°F, 'a degree of cold not often exceeded in the whole winter'. But it was not only the temperature that had dropped: the pressure had, too, alarmingly so. On 22 October, Maria's family arrived from Yorkshire. They told FitzRoy that they had travelled by train through a fierce blizzard. Soon another friend contacted him, explaining that his barometer had dropped to a disturbingly low level. 'What could it be for?' he asked. 'We shall hear of much wind and snow in the north, the thermometer is so low,' FitzRoy replied.[2]

One man who had been enjoying the recent sunshine was Captain Thomas Taylor of the *Royal Charter* steam clipper. He had spent the last month guiding his ship on her way home from Melbourne in Australia, sailing through warm tropical seas under skies scratched with cirrus. Sleek and graceful, the *Royal Charter* was the latest in an experimental line of steam clippers. Iron-hulled and narrow-beamed, she cleaved the waves like a sharp knife. Just shy of 2,800 tons, equipped with a pair of 200 horsepower trunk engines, and fitted with fireproof and watertight cabins, the *Royal Charter* was a star of her class. She drew her lineage from Brunel's SS *Great Britain* and FitzRoy's *Arrogant*, ships that combined the power of wind with the modern might of steam. When the wind failed Taylor could order the coal-driven engines to be fired up. They were not powerful engines: they produced only 200 horsepower each, but they pulsed away deep inside the iron hull, tiny beating hearts, helping the *Royal Charter* to find wind when it died away.

To her owners the Gibbs Bright Shipping Company the *Royal Charter* was a source of pride. On her first crossing from Liverpool to Melbourne she had cut a third off the usual duration of the voyage and ever since then advertising posters had announced 'Steam from Liverpool to Australia in Under Sixty Days'[3] to a prospective clientele. These were chiefly the ambitious middle classes, aspiring to a wealthy lifestyle and confident they would find it in the famous Ballarat goldfields near Melbourne. Over the last decade thousands from all over Europe had descended on the southern Australian shore. In August 1859 many returning travellers had clambered aboard the *Royal Charter* to sail home to Britain, many with their dreams realised. All told, she carried five hundred or so passengers and a vast cargo of gold. The exact amount has never been established. But locked below in the strongroom, crammed in pockets, hidden away in money belts and stowed at the bottom of portmanteaus was at least £300,000 in gold bullion and specie. The real quantity was probably twice that. For many the *Royal Charter*'s return leg to Liverpool marked the end of a long personal journey. They had left their home parishes behind and travelled an unimaginable distance in pursuit of their dreams. Ninety years after the *Endeavour* had set off, each one of them was returning with stories to tell. Among them was a fifteen-year-old sailor Charles Thomas, who had sent a message home. 'Mother, pray for a fair breeze, and I shall whistle for the wind.'[4]

The *Royal Charter* sailed on 26 August. She had left Melbourne with the wind ballooning in her sails as she glided east across the Pacific. She had cut around the Cape, skimmed up the South American coast and crossed the equator. Further north, in the Atlantic, she sailed under the old, familiar northern stars. By mid-October she had reached the Western Approaches of the Channel. As the weather shifted in London the *Royal Charter* was stealing through the waves off the south Irish coast. On Monday 24 October she reached Queenstown. Fifteen passengers disembarked and shortly after she was at full sail again. It was fifty-nine days since she had left Melbourne and on board there was mounting anticipation as the voyage neared its end. A subscription was got up among the Saloon passengers for Captain Taylor who, at dinner, was heard to say that they would be in Liverpool within twenty-four hours. A good, sociable captain, Taylor had struck up an amiable relationship with the Saloon passengers over the months of the voyage. Playfully he told them that he hoped to be back on the lee side of Mrs Taylor soon.

This flutter of repartee was no more than was expected on the *Royal Charter*, which epitomised the trend for luxurious travel. The ship glimmered on the seas like a floating town house. In return for their 75-guinea ticket Saloon passengers could sip gin and tonic on deck, dine on roasted beef, mutton or pork and, in the evenings, waltz in the tropical moonlight. With one member of crew to every four passengers, the ship ran – half a century before *Titanic* – like a graceful floating hotel. The captain was not a Cook or a FitzRoy but an evolution of these archetypes: a gallant blend of sea captain and manager of a fashionable boutique. This was something new. With the commercialisation of sea travel, the passage from Melbourne to Liverpool had become a show to be savoured.

Captain Taylor had an unblemished record over two decades at sea.

He had sailed on the *Royal Charter* from her maiden voyage and knew her well. The only issue to trouble him on Tuesday 25 October as he navigated the Irish Sea towards the Welsh coast was the murky weather. Just fifty miles away in Dublin a dense fog had gathered over the coast, so thick that pilot boats could not find their vessels, and could navigate only by the knell of the fog bell. FitzRoy later noted this fact, recalling the old sailor's doggerel:

> When morning mists come *from* the hills,
> And the huntsman's horn is free,
> Fine weather reigns: but, woe the time,
> When the mists are *from the sea*.[5]

The haze spread from Dublin to the Welsh coast where the atmosphere was a little clearer. Taylor picked up a pilot boat in the afternoon and by 1.30 p.m. he was off Holyhead. A tremor of excitement fanned down from the deck to the saloon. There, moored at the entrance to Holyhead harbour, was Brunel's colossal ship SS *Great Eastern*, a behemoth of Victorian engineering: 18,914 tons of iron and wood. Straining over the rails the Saloon passengers gazed at the harbour where the ship was tightly moored. Behind them the autumn sun was sinking into a misty atmosphere, gaudy with smudged colour.

One of the Holyhead harbour workers had kept an eye on the weather. The day had started bright with sunshine. So transparent was the atmosphere that he had been able to look across the island to the glorious snow-capped Cambrian mountains rising from the mainland like the knuckles of a clenched fist. In the afternoon the wind had freshened and the sky had grown overcast. At 6 p.m. when the sun had set the sky was obscured by 'a uniform dull mass of vapour'.[6]

Aboard the *Royal Charter* a sailor was heard to say that it would be a 'dirty night'. Whether Taylor looked at any of his three barometers is uncertain. It is also unknown whether he carried or had read a copy of FitzRoy's *Barometer and Weather Manual* – as he might well have done. In any case his years at sea would have told him to be wary of telltale signs: the plummeting pressure, freshening breeze, sea mist, gulls, puffins and razorbills heading for shore. 'So much, at such a time, depends on individual judgment,'[7] FitzRoy would later write.

Confronted with a decision – fifty-nine days out from Melbourne on a sixty-day voyage, passengers toasting him at the dining table – Taylor chose to sail on.

Leaving Holyhead and the *Great Eastern* behind he rounded the north-western tip of Anglesey island. FitzRoy would later point out, 'The simple rule of seamanship is when facing the wind the centre of the storm will be to the right or on the right hand side, therefore you should go to the left, supposing that sea room and circumstance enable you to choose.' Taylor had no such insight. With each wave he crested – and the water was swelling now – he was sailing closer to disaster. A reporter for *The Times* at Holyhead, monitoring events on the *Great Eastern*, watched the weather shift.

Over the mountain came a thin black haze, which rose into the air with ominous rapidity and overspread the sky. The sea and wind kept rising as the glass fell, and before eight it blew a heavy gale from the eastward, with fierce squalls and storms of rain. As night wore on, the wind increased and came in fearful gusts, tearing away among the spars and rigging with a hoarse sustained roar that was awful to listen to, especially when one bore in mind that the glass was still falling, and that what we saw was only the commencement of the gale.

As the wind freshened the *Royal Charter* was navigating the rocky north lip of the island. Taylor responded by trimming the sails and stoking the engine. The ship darted through the swell, sticking doggedly to her path, wheezing and fighting past the skerries, an outcrop of rocks, at about five or six knots. Soon the wind had strengthened and it was whistling through the rigging. By half past six a gale was beating down on the *Royal Charter* from the south-east. Not knowing any better, Captain Taylor simply maintained his course, his chance of reaching Liverpool diminishing by the minute. By now it was dark and the *Royal Charter* had been swallowed up by the night and the spray. Not far away on shore one man noticed the hint of a blue glow emanating from the sea. Later it was thought to have been the first of the ship's distress signals, or perhaps a call for assistance from one of the coastal pilots.

The firing of the distress flares marked the beginning of an awful twelve-hour battle between the *Royal Charter* and the storm. It was a match that pitched the engineering prowess of the nineteenth century against the raw and ancient power of weather. By 8 p.m. the wind was screaming at around a hundred miles an hour. Its force jolted the ship, churning up waves that reared like mountains and then fell away into cavernous gullies. Those sheltering on land had not seen a sea like it. At 10 p.m. the winds which had blown from the south-east twisted around to east-north-east. Having gained no ground over the past few hours, frantically struggling to stay where she was, the *Royal Charter* was now sailing into the teeth of the storm. It bore down on her, an avalanche in the atmosphere: 100 mph winds that not only barred her path but now started to jolt and push her back. Behind lay the rocky beaches of Anglesey's eastern coast.

At eleven o'clock Captain Taylor ordered the port anchor to be let out, a grappling hook that he hoped would fasten the ship to the seabed. It came rattling through the hawseholes into the night, links of iron chain that could barely be heard in the tumult of wind. It tumbled into the depths but, terrifyingly, the flukes did not bite. The ship was still drifting astern. Soundings showed the depth was narrowing, from twenty to sixteen fathoms of water. The implications were clear: the ship had not been shackled to the seabed. She was dragging her anchor, being driven further back.

To remedy the situation Taylor ordered the starboard anchor to be let go. The *Royal Charter* now gripped the seabed like a climber clinging to a cliff face. Taylor's hope was that the chains would hold until the storm passed over. It was logical thinking as by now the ship had been battling the elements for around six hours. In that time a captain might suppose the worst had passed. But the ordinary wind was now blowing at a constant speed of around eighty or ninety miles an hour, and in some desperate moments gusts of more than a hundred would bowl right across deck. It tore the water up into terrifying waves that flung the ship from one acute angle to another, making the simplest of jobs a challenge. For two hours the *Royal Charter* hung on her two anchors. The equation was clear: the resistance of these anchors and the power of the coal engine pitted against the brute force of nature. Caught between, the ship strained under the pressure. For a while it seemed she might hold but, at half past one in the morning, the port anchor snapped at the hawsehole. Suddenly freed from its enormous burden, the chain flew into the air over the ship where it appeared suspended for a second as the ship jolted backwards. Then it plummeted, spiralling beneath the waves.

It was the first telling blow. Inside, Captain Taylor was endeavouring to reassure the passengers, taking coffee with them in the Saloon. But the reality was that the ship was slipping backwards. After an hour of tremendous strain the starboard anchor was also wrenched away from the ship. It left the *Royal Charter* floundering astern, driven straight backwards by north-east winds that were blowing as furious as ever. With her engine beating pathetically and her rudder rendered useless the *Royal Charter* had lost all control. Behind the ship lay the rocky bays of the coast, a notorious site of shipwrecks. At 3.30 a.m., it hit.

At first it seemed Taylor had experienced his first stroke of luck. They had not struck rocks, but had come to rest on a sandbank. The iron hull had not been breached and although visibility was so poor that he could not say for certain where they were, he had confidence enough to tell the female passengers clustered in the Saloon, 'Ladies, we are on shore, and I hope on a sandy beach, and I hope to God we shall all go on shore when it becomes light.'[8] Indeed, it was not long until dawn. Two hours passed. At half past five as the first chinks of smudgy light illuminated the scene the crew were astonished to discover they were just yards from the headland. They had not run aground on a sandbank in the open sea, or been hurled against one of the craggy coastal islands; they had come to rest just yards off the coast, a mile to the north of the fishing village of Moelfre.

So close to safety – and there was a horrific realisation. The gulf between deck and clifftop seemed impossible to bridge. And even now, after all the hours of chaos, the storm was intensifying. It was like one of Turner's nightmarish visions. Flares went flying up into the howling sky. Taylor had ordered the masts to be cut and they were crashing down across the deck. The sea was now filled with debris. Taylor's great hope was that the tide was on the ebb and that they would be left firmly grounded as it withdrew. But, as on so many occasions in the last twelve hours, he was to be proved wrong. The tide was flooding in. Soon the hull would be lifted clear by the rising water.

Yards away villagers from Moelfre were gathering on the clifftop.

But there was nothing they could do. In what would later be celebrated as the swim of the century a Maltese seaman, Joseph Rogers, volunteered to carry a line ashore. He tied it around his waist and threw himself into the darkness. Few had any hope for him. If he wasn't drowned his body was bound to be broken on the rocks. But an outstanding blend of luck, tenacity and courage brought Rogers to the shore, where he clung on. Eventually he was hauled to safety by a string of villagers, linked hand to hand. From the rope a bosun's chair was fashioned and in minutes a perilous relay between clifftop and foredeck was in operation.

The *Royal Charter*'s time was wearing thin. By now the tide was rushing in. With every wave she was wrenched a little higher from her sandy bed. At last a wave lifted her, listing and creaking for a last split second into the open water. Then she slewed astern, shooting wildly backwards on the crest and plummeting down the backslope towards a shelf of submerged limestone rocks. What had been a long and valiant struggle was to become a swift execution. At her last the story of the *Royal Charter* was reduced to a set of tragic vignettes. A clergyman led prayers in the Saloon. A passenger, born and raised in Moelfre, clambered on deck to discover that he had travelled the length of the globe to be shipwrecked in sight of home. In the forecastle Taylor was bereft, in denial. A last sighting of him described him as 'shedding tears, much excited'. Another had him gasping, 'There's hope yet!'[9] Thrown on to the rocks, the *Royal Charter* was split open like a walnut with a sledgehammer.

The end was a scene of complete terror. Passengers were forced to jump into the crashing sea to fight against the deadly undertow, many of them weighed down by the gold in their pockets. Dressed for a wintry night in greatcoats, flannel, duck trousers, scarves, fustian trousers and boots, the men had as poor a chance as women in their stays, stockings and embroidered evening dresses. Not one woman or child would survive. All told, only forty-one passengers lived to see the end of the storm later that day – most escaped on the bosun's chair, a fortunate few were thrown alive on to shore and hauled to safety. Of the rest, some drowned but most were battered against the rocks. Their bodies were ripped and broken or disfigured, in many cases beyond the realms of identification. The most haunting tide Anglesey ever brought in left more than 450 dead.

The aftermath drew visitors from across the kingdom. A Mr W. F. Peacock found the entire coast from Moelfre to Amlwch strewn with wreckage. He threaded his way through the destruction, 'leaping over this cleft and that hole', picking up fragments of mahogany, a foot-length of old rope, a bit of plate glass, 'broken and shapeless', 'the lace-trimmed frock of a little child; a fragment of a fine flannel vest, once purple, but now almost colourless; the spotted and flabby sou'wester of a drowned seaman'.[10] He was shaken by these mangled, yet commonplace objects. Their ordinariness brought home the everyday proximity of disaster. Alongside the twisted, bloated remains of bodies a different type of cargo was thrown up on the beaches of western Anglesey too. Gold coins and gold bars, sometimes in the pockets of the dead, sometimes on the beach in tragic isolation. Many gold watches were salvaged. Almost every single one registered a time between seven and eight o'clock. Of the iron ship itself – graceful, sleek, 3,300 tons, a floating Britain – there was barely anything to be seen.

The British press reacted to the *Royal Charter*'s demise with typical horror. The *Illustrated London News* commissioned a woodcut of the scene, cut by Frederick James Smyth – the artist who had portrayed the London hailstorm in 1846. It was published a fortnight after the disaster, and showed the ship pitched on its side in the blackness of night – the atmosphere a picture of fury. Another to visit the scene was Charles Dickens. The *Royal Charter* was the kind of disaster – the horrid quirk of fate, the drama, the stories of individual spirit and heroism such as that of Joseph Rogers – to catch his imagination. He produced a long and tender piece for his *Uncommercial Traveller* collection of essays. He focused on the reaction of the local community, specifically its clergyman Reverend Stephen Roose Hughes and his efforts in burying the dead and corresponding with distraught families.

As many seemed stunned to death as drowned, Dickens noted, in shock:

They stood in the leaden morning, stricken with pity, leaning hard against the wind, their breath and vision often failing as the sleet and spray rushed at them from the ever forming and dissolving mountains of sea, and as the wool which was a part of the vessel's cargo blew in with the salt foam and remained upon the land when the foam melted. They saw the ship's lifeboat put off from one of the heaps of wreck; and, first, there

were three men in her, and in a moment she capsized, and there were but two; and again, she was struck by a vast mass of water and there was but one; and again, she was thrown bottom upward, and that one, with his arm struck through the broken planks and waving as if for the help that could never reach him, went down into the deep.[11]

It is thought that Captain Taylor was one of the men in the lifeboat.

The storm's effects were not confined to the north Welsh coast. Throughout the weeks that followed, reports of damage surfaced in the press: stories of coasters thrown on to harbour walls, sails blown to ribbons and crews hanging on for dear life. At Minehead in Somerset all the fishing boats were blown in one mass against the harbour walls. At the opposite side of the country, two hundred miles away at Great Yarmouth in Norfolk, a herring sloop, *James and Jessie*, was sent hurtling through the prim new Britannia Pier. Perhaps the most spectacular story of all was that of the destruction of Brunel's iconic railway line between Dawlish and Teignmouth on the Devon coast. There the sheer force of wind and waves had breached a double protective wall crammed with sand and shingle. 'It appears . . . the late lamented Mr Brunel had much under calculated the effect of the waves during spring tides, augmented by strong easterly winds,'[12] pointed out the *Illustrated London News*. 'Such was the terrible force of the impelled water during the late storm that the coping-stones, probably averaging a ton each, were tossed about like corks, and huge fragments of the disjoined wall were rolled upon the metals. The breaking up of the structure is described as having been appalling, surf, foam, and fragments of the debris rising up into the air with a terrific roar.'

It was Victorian engineering might against raw nature and nature had won in every case. The Britannia Pier, the Dawlish railway, the *Royal Charter*, all of these had been held up as paragons of scientific, civilised, Britannia. All had been undone in a single night.

In London reports of the tragedy shocked FitzRoy. Shipwrecks had been on his mind throughout the summer. At a meeting of the British Association he had managed to get a resolution passed calling for a government-funded storm-warning system. The idea had been embraced, but before anything could be put in place the *Royal Charter* had sunk.

It was the kind of national event that jolts political agendas; FitzRoy seized the opportunity. He threw himself into an investigation of the gale. By the time Dickens was clambering over the rocks at Moelfre at the end of 1859, FitzRoy had already revealed his preliminary findings to the Royal Society. He told the Fellows he had collected information from 'lighthouses, observatories and numerous private individuals' and that the storm had proven to be, as expected "a complete horizontal cyclone" that had travelled northwards over the country'.[13] He made sure to push his bigger point, too. 'Another valuable result,' he concluded, 'is that telegraphic communication can give notice of a storm's approach to places then some hundred miles distant, and *not otherwise forewarned*.'[14]

In private he was lobbying hard for his storm-warning idea. On 5 December he had submitted a proposal – based on the recommendation of the British Association – to the Board of Trade. On the 17th he had been granted their approval. After a five-year period of book-keeping, FitzRoy's agenda was changing rapidly. He spent spring hatching plans for the practical implementation of the system, and in analysing what was already being called the 'Royal Charter Gale'.

FitzRoy approached the challenge as Redfield and Reid had done twenty years before. He set out to track the course of the gale. Within weeks he had assembled an arsenal of data. It had hit the

south coast at around noon and progressed north to the Welsh
coast by about half past six. Throughout it had acted predictably.
He had his collegue Babington draw hourly charts of its progression, indicating the changing points of pressure and temperature.
Babington had been drawing such maps under FitzRoy's direction
since 1857 and by now they had gained the name 'synoptic charts'.
The term synoptic was FitzRoy's choice, borrowed from the name
of the Synoptic Gospels – Matthew, Mark and Luke – each of which
told the life of Jesus from a different perspective. Only by studying
them together could a Christian fully understand Jesus' life, and
FitzRoy, tellingly, had integrated this religious detail into his
meteorological work. The synoptic charts showed pressure, temperature and geography all at once, and over the years Babington's
experiments with them had progressed. By the time of the Royal
Charter Gale he had amassed a stock of 'many hundreds', 'as if an
eye in space looked down on the *whole* North Atlantic *at one time*',
FitzRoy wrote.[15]

The synoptic charts of the Royal Charter Gale tracked it at hourly
intervals. They demonstrated how the winds had spun in an anticlockwise direction about a 'lull' in the centre. FitzRoy decided that
its perilous zone had encompassed a space from twenty to fifty miles
on either side of its centre. Within this zone the winds had blown
from sixty to a hundred miles an hour 'in successive eddyings' like
ripples cascading outwards in a pond. It meant that while the *Royal
Charter* was floundering in north-east winds off Anglesey, a north wind
had been ripping down the Irish Sea and a north-westerly gale was
blowing tiles off roofs in Dublin.

FitzRoy was particularly interested in the sharp contrast between
the warm and cold air that he had noticed himself. The shift
between the two had been a key feature of *The Law of Storms*, in
which Dove argued that storms were especially prevalent on fault lines
between air masses; later studies of earthquakes would similarly focus
on the movement of tectonic plates. FitzRoy was fascinated by the
idea. Everywhere he looked he found evidence of the conflicting air
types. One correspondent from Ireland, Captain Boyd, spoke for
many. 'On the 19th I was at Belfast, oppressed with heat, in close
weather, with small rain. It was like a muggy May day. The next three
days I was travelling along the east coast, cut to the vitals by a piercing
north wind, snow and hail squalls.'[16]

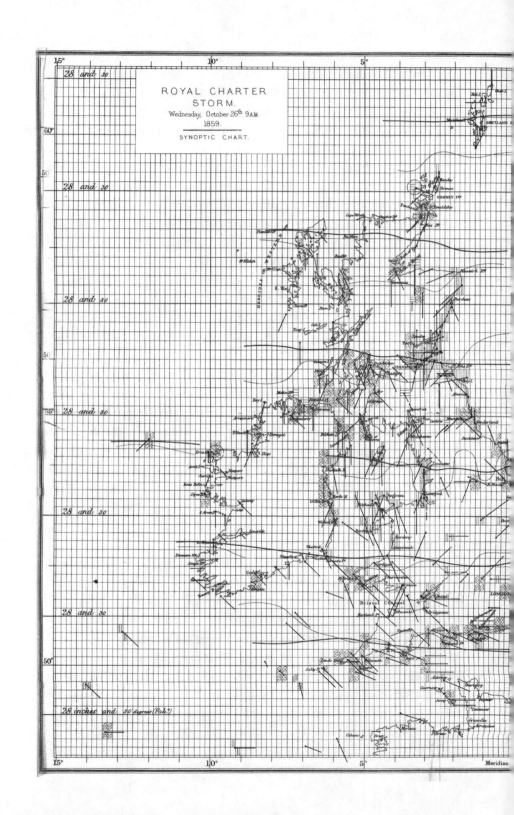

ROYAL CHARTER
STORM.
Wednesday, October 26ᵗʰ 9 A.M.
1859.
SYNOPTIC CHART.

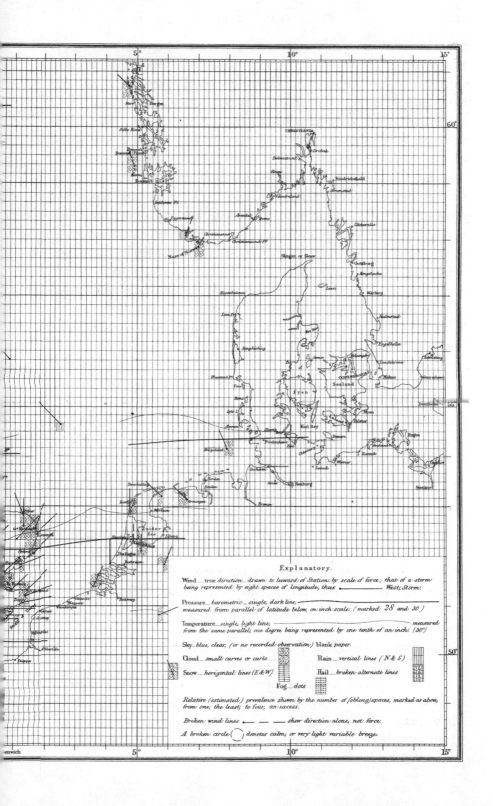

Explanatory.

Wind — true direction — drawn: to leeward of Station; by scale of force; that of a storm being represented by eight spaces of longitude, thus ━━━━━━ West, Storm.

Pressure — barometric — single, dark line; ━━━━━━ measured from parallel of latitude below, on inch scale: (marked 28 and 30)

Temperature — single, light line; ━━━━━━ measured from the same parallel; one degree being represented by one tenth of an inch. (30°)

Sky — blue, clear; (or no recorded observation) blank paper.

Cloud — small curves or curls   Rain — vertical lines (N & S)

Snow — horizontal lines (E & W)   Hail — broken alternate lines

Fog — dots

Relative (estimated) prevalence shewn by the number of (oblong) spaces, marked as above, from one, the least; to four, an excess.

Broken wind lines ━ ━ ━ shew direction alone, not force.

A broken circle ◯ denotes calm, or very light variable breeze.

The temperature drop produced the perfect conditions for one of Dove's storms. The warm and cold air masses were imagined as armies in the sky. The steeper the rise or fall in temperature, the more powerful the competing forces and consequent commotion.

FitzRoy hoped to present his findings to the British Association meeting in June 1860. Just a few weeks before, he received a letter from an American commander, William Johns, of the sailing ship *William Cumming*. Johns' account presented FitzRoy with an alternative narrative. According to his evidence, he had been sailing within ten miles of the *Royal Charter* on the night of the gale. Unlike Taylor, Johns had on board a copy of Maury's wind charts and, apparently, FitzRoy's own *Barometer and Weather Guide*. This proved crucial. Realising he was being hemmed in, Johns had shortened sail and tacked away into the open sea. He had then watched as 'a complete West India hurricane' blew up from behind the mountains. He finished his letter to FitzRoy with the laconic verdict, 'I did not suffer *any damage whatever* more than usual in ordinary blows only a little chafe and some spray.' The contrast with the *Royal Charter* was stark. Over the *William Cummings* the *Royal Charter* had enjoyed a double advantage of a steam engine and an iron hull, yet the *William Cummings* had survived.[17]

Though he did not criticise Taylor openly, FitzRoy was not impressed by his captaincy. With no way of communicating with the *Royal Charter* as she rounded Anglesey, FitzRoy knew that everything had rested on Taylor's judgement.

With her power of steam, in addition to that of sails in perfect order, a few hours on the starboard tack, with but little way, would have saved her. So much, at such a time, depends on individual judgment. Another ship but a few miles off, a wooden *sailing ship*, not a *steamer*, the *Cumming*, and several smaller vessels, acted thus – stood to westwards – and not one was wrecked; nor even injured materially . . .

That ship [the *Royal Charter*] had excellent instruments on board when she left her last port – they *should* have given sufficient notice – but had they not been there, or had their indications been unheeded, those of the heavens should not have been disregarded – overlooked they *could* not have been from any ship – and *were* not by the Cumming, or by numerous coasters.[18]

FitzRoy used Johns' letter as the focal point of his paper 'On British Storms' for the British Association meeting at Oxford. It was powerful testimony but really he was preaching to the converted. At the start of June the Board of Trade had approved his storm-warning plan, four weeks before his paper was delivered on Friday 29 June. The Board of Trade had signed up to a network that would run along the same lines as Glaisher's network of the 1840s. Thirteen telegraph stations – Aberdeen, Berwick, Hull, Yarmouth, Dover, Portsmouth, Jersey, Plymouth, Penzance, Cork, Galway, Londonderry and Greenock – selected by FitzRoy were to be supplied with meteorological equipment. The telegraph companies had agreed to transmit completed meteorological slips – including readings of temperature, pressure, the state of the weather and direction of the wind – to London at nine o'clock each morning. These would be tabulated, reduced and then plotted on a map at Parliament Street. If a storm was noticed a warning message would be sent back down the line.

To signal the storm at the ports FitzRoy had devised a series of cones and drums. Depending on the expected strength and direction of the wind, different combinations would be hoisted in the harbours. Like the telegraph signals of old, devised by Chappe and Edgeworth, FitzRoy's storm cones were designed to be seen at a distance. Three feet six inches high and three feet in diameter, they were raised up a

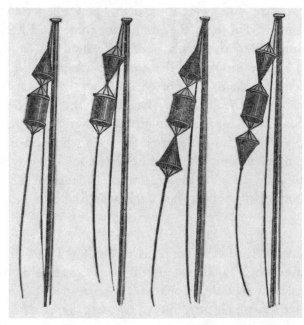

mast at the telegraph or coastguards' stations where they could be seen by ships many miles away. It was a straightforward, practical system, FitzRoy told the British Association, based on simple signals:

A cone, with the point upward, shows that a gale is *probable* from the northward

A cone, with the point downward, shows that a gale is *probable* from the southward

A drum, alone, shows that dangerous winds may be expected from nearly opposite quarters successively

A cone and drum give warning of dangerous wind, its *probable* first direction being shown by the position of the cone – point up, and above the drum, for polar or northerly winds – down, and below, for southerly

Whenever such a signal is shown (In consequence of a telegram from London) it will be kept up (shown distinctly) till dusk of that day *only*, unless otherwise instructed afterwards[19]

The system was set to be launched on 1 September. 'The plan proposed is simple and the machinery is ready,' FitzRoy announced.[20]

FitzRoy's storm cones were to be a vital new weapon in the battle against shipwreck. For many they could not come quickly enough. In Britain, even as FitzRoy's analysis of the *Royal Charter* wreck was drawing to a close in Oxford, the country was absorbing the latest tragedy. 'A terrible calamity,' reported *The Times*. In a chain of summer gales 186 men – mostly fishermen – had died in the North Sea. 'Besides this awful loss of life,' the newspaper noted, 'there has been a destruction of property which has deprived the men who have been spared of the means of earning their future livelihood.' It was an all too familiar story. At Lloyd's, the insurance brokers of London, the infamous Lutine bell – tolled when a ship was lost at sea – was clanging again.

Friday 29 June 1860 had shown one face of FitzRoy: a pioneering meteorologist, advocating a noble plan. But there was to be a dramatic coda to his Oxford lecture. Looking back on the history of nineteenth-century science we see the striking proximity of two defining, catalytic moments.

The first was the wreck of the *Royal Charter* on the night of 25 October 1859. The second was an event that would have a much more profound effect – the publication of Charles Darwin's *On the Origin of Species*.

On 1 October 1859 Darwin had jotted in his journal, 'Finished proofs (thirteen months and ten days) of Abstract on *Origin of Species*.' After decades of denial and anguish, Darwin had written up his theory of evolution by natural selection – an idea that tugged the carpet from beneath the Christian interpretation of the past. Darwin later wrote that to unveil his idea was like confessing to a murder. If he was a murderer then FitzRoy, his captain on the *Beagle*, however unwittingly, was an accessory. The fracture lines that had formed in their friendship would now break open. FitzRoy, the devout Christian, was to have his spiritual core shaken by one of his oldest friends.

As the *Royal Charter* entered British waters on 23 October Darwin was writing to his closest friend Joseph Dalton Hooker, speculating about prospects for his book. As the ship was wrecked the type was being fixed for the first edition of 1,250 copies. By early November, as FitzRoy's inquiry into the disaster was gaining momentum, finished copies of *Origin* were being distributed to British and European scientists. On 24 November the book was published. On the first day, every copy was sold. FitzRoy was among those to receive a copy beforehand. He had long suspected Darwin of nursing radical theories in private, but this was worse than he had ever imagined. He read it in disgust. 'My dear old friend,' he wrote to Darwin, 'I, at least, *cannot* find anything "ennobling" in the thought of being the descendent of even the *most ancient* Ape.'

To FitzRoy, whose sense of personal identity was drawn from his illustrious line of dukes and kings, the idea of evolution was anathema. Over the weeks that followed he seized every opportunity to deride *Origin* – a book that was proving dangerously fashionable. One of his outlets was the letters page of *The Times*. Adopting the gruff pseudonym Senex (Old Man) he railed against Darwin's ideas. Spotting FitzRoy's distinctive style, Darwin realised Senex' true identity. He dashed off a letter to the geologist Charles Lyell:

I forget whether you take in Times: for the chance of not doing so I send the enclosed rich letter – It is, I am sure, by FitzRoy; for he wrote to me the other day on population of world not having increased & in his Voyages there is the pebble theory.

It is a pity he did not add his theory of the extinction of Mastodon &c from the door of ark being made too small. What a mixture of conceit & folly, & the greatest newspaper in the world, inserts it![21]

The day after his talk 'On British Storms' at the British Association in Oxford on 29 June there was timetabled a debate 'On the Intellectual Development of Europe Considered with Reference to the Views of Mr Darwin', to be held at the University's Museum of Natural History. It was high noon for Darwin's theory. For the first time since *Origin's* publication the scientific community gathered as one, and to their number were added professors, tutors and scholars from Oxford University. Even Reverend Charles Dodgson (who would later become known as the author of *Alice in Wonderland*, Lewis Carroll) was present. Surrounded by the Museum's artefacts of antiquity, it was the perfect setting for a debate that threatened to reshape the way people saw the world.

The lecture would be remembered as the great showdown between science and religion. Although Darwin was not present and the initial lecture, delivered by a Mr Draper, was by all accounts a dull affair, the drama soon began. Samuel Wilberforce, the Bishop of Oxford, and a mighty, domineering and ferociously eloquent advocate of the Church, had been rumoured to have chosen the debate as his opportunity to denounce Darwin's ideas openly. Ready to confront him was the biologist Thomas Huxley, who had already earned himself the nickname 'Darwin's bulldog' for his trenchant advocacy of *Origin* in the press.

After Draper's talk Wilberforce rose as anticipated. He spoke passionately for half an hour, attacking Darwin's theory at its root. Have any instances been discovered of a species evolved from natural selection? 'We fearlessly assert not one.' Is it credible that a turnip strives to be a man? Turning to Huxley he rounded off with a last, savage blow. Was it on his grandfather's or grandmother's side, Wilberforce demanded, that he was descended from an ape? Huxley's reply would go down in history for its audacity.

[Whether] I would rather have a miserable ape for a grandfather or a man highly endowed by nature and possessed of great means of influence, and yet who employs these faculties and that influence for the mere purpose of introducing ridicule into a grave scientific discussion – I unhesitatingly affirm my preference for the ape.[22]

This was scintillating combat from two men at the top of their profession. Thereafter the lecture descended into feverish excitement. One of those who stood to record his disgust was FitzRoy. No one knows quite what he said. But reports reached Darwin a fortnight later of FitzRoy standing in the midst of the crowd, waving a bible above his head, shouting of his 'regret' that the book was not a logical arrangement of facts. In a letter to his old mentor Henslow, Darwin seems not to have been in the least surprised – 'I think his mind is often on the Verge of insanity,' he sighed.

It was a crushing verdict from an old friend. Yet FitzRoy was being thrust into an impossible situation. Since his return from the *Beagle*'s voyage his Christian faith had sustained him. Moreover, his faith was bound closely with his memories of Mary. What seemed on the face of it to be an intellectual betrayal by Darwin was actually something more: it was an emotional attack on the memory of FitzRoy's dead wife. Countless individuals would find their own interpretations of *Origin* over the decades to come. Herschel, Lyell, Huxley and others would all take a different stance. For many the challenge to their religious conviction would become the central question of their lives. But hardly anyone could have faced a more complicated moral dilemma than Robert FitzRoy.

At Oxford a schism in FitzRoy's personality was exposed. By nature he was both a devoted progressive and a hardened conservative. He had championed innovation throughout his career – lightning conductors on the *Beagle*, the weather logs, steam power on the *Arrogant*, the development of storm warnings. All these were the mark of a pioneer. Yet FitzRoy was wedded to the past, too: through his relationship with Mary, his upbringing, his faith in hierarchy – the inflexibility of rank and station.

These competing forces tore at FitzRoy. At the Oxford meeting of 1860 he was caught in a position that was increasingly hard to sustain. He was happy to decry Darwin's audacity. Yet if *Origin* was scientific trespass on the past, on ground that had been held for a millennium by the Church, then logically the forecasts he had in mind would mean trespassing into the divine future. Forecast and evolution were scientific twins. One wrenched away the past, one wrenched away the future. One came from Darwin, the other from FitzRoy. United and divided in friendship and in science in increasingly complex ways, theirs was a quarrel that could never be resolved. The two old friends would be divided by belief for the rest of their days.

Now thirty years old the British Association continued to be a testing ground for scientific ideas. Its support or condemnation could make or break a theory, as Espy had discovered. There were now several active land networks, all harvesting useful data. But there was one place that remained obscure: the upper atmosphere.

To address this shortcoming a committee had been formed at the 1858 Association meeting in Leeds and had been given the job of organising hot-air balloon ascents. The idea was to extend the scope of atmospheric observations by putting scientists in balloons, supplying them with instruments and getting them to record data about the air at altitude. An august list of scientists were appointed: Lord Wrottesley, outgoing president of the Royal Society, Michael Faraday, Charles Wheatstone and an old friend of Glaisher's, Dr Lee. Together they took on the buoyant title of the Balloon Committee. The committee had received letters of support from Herschel and Airy and in May 1859 at a meeting at Burlington House in Piccadilly, Glaisher and FitzRoy were added to their number. For meteorology this was a bold new step. It combined the excitement of exploration with the cool rationalism of science. Just like Beaufort's surveying vessels, the basket of a balloon was to be transformed into a centre of philosophic endeavour. What it might reveal no one was quite sure.

Ballooning had been fantastically fashionable throughout the first half of the nineteenth century. A balloon launch was a spectacle not to be missed and tales of aerial voyages had been written up as modern legends, symbolising the ultimate expression of freedom. But while hydrogen balloons had harnessed the power of science for aesthetic pleasure and show, surprisingly little use had been made of them for scientific experimentation. The last significant ascent had happened in 1804 when the young French physicist Joseph Louis Gay-Lussac had risen to an altitude of five miles in a hydrogen balloon called *Coutelle* from the Conservatoire des Artes in Paris. He had carried barometers, thermometers, hydroscopes, compasses to test the earth's magnetic field and flasks to collect samples of air. He wanted to verify stories of teeth-chattering colds, and (hopefully) disprove claims that altitude made heads swell like pumpkins – Luke Howard had calculated that at seven miles high the air would occupy seven times the space it did on earth – and noses stream with blood.[23] Happily he had returned triumphant. By now few who had witnessed that spectacle of bravado were alive (Gay-Lussac was not: he had died in 1850, a celebrated French man of science).

One of the enduring scientific legacies of Gay-Lussac's ascent was a law that he seemed to have set for the decrease of temperature with height. According to his calculations, for every 300 feet of elevation the temperature dropped by 1°F. For half a century this formula had been accepted, although there had been very little meteorological ballooning to test it. One who had attempted a scientific ascent was the indomitable Thomas Forster. Forster had marshalled the funds for an aerial voyage in April 1831, which, as ever, he wrote up in a sprightly book-length narrative, *Annals of Some Remarkable Aerial and Alpine Voyages*. Forster's description of his own ascent was the centrepiece of a book that skimmed merrily through the brief history of ballooning from the craze of the 1780s to the blossoming of the art form in the early nineteenth century. After years of gazing into the skies, the prospect of an aerial voyage for Forster was a delicious treat. He had already experienced clouds from Alpine mountaintops but this was to be something quite different. On Saturday 30 April 1831 Forster climbed into the basket and ascended slowly into a 'soft and serene atmosphere' at Moulsham near Chelmsford in Essex.

In the direction of the Maldon river, and hovering over its marshy lands, we saw what had evidently been a cumulus now subsiding into a stratus or white evening mist, stretching in such a manner over the ground in its descent, that we at first took it for smoke. Higher up there were cumuli in the air, much nimbiform haze still more elevated and some waneclouds. The beauty and the extent of the prospect now increased, and the fields, here and there coloured with the bright yellow of the flowering colewort, green with the young wheat, or richly brown from fallows, chequered with rows of trees whose light foliage and blossoms enlivened the darker hue of their boughs, and the whole country intersected with rivers, roads and villages, had the most enchanting effect.[24]

Had Forster turned his eyes to the east he would have been gazing right over Constable country, the scattering of villages and farmland around Dedham Vale. The balloon rose to 5,000 feet and skimmed gracefully along with the currents. The colour of the horizon was a 'delightful blue', and Forster hung his head over the edge of the car and stared straight down to the depths below. He proudly reported back that he had not felt the least bit giddy, though he admitted that his ears had popped with a 'snapping' sound.

Forster came back to earth with some firm conclusions. He declared the hot-air balloon the marvel of the age, far better than the steamship – 'This new-born Leviathan of the deep is nothing compared to this Pegasus of the air.'[25] Forster was also very severe on the habit of carrying up refreshments – champagne, meats, cheeses and jams had all become common air fare – but this, to Forster, acted only to distract from the *philosophic* experience. The problem was that Forster's expedition had not been scientific at all. Firstly he had forgotten to take up his instruments. The few he had remembered were completely neglected as he spent the voyage delighted by the views, dangling his head over the side and gazing in a trance into the distance. Forster's voyage had generated a few pages of breathless prose, but from a scientific point of view it was useless. He managed just two numeric readings in all: the air temperature and pressure before take-off.

Little more in the way of scientific ballooning had taken place until the 1850s, when the new superintendent of the Kew Observatory, John Welsh, had executed four ascents from London to various altitudes of around 20,000 feet under the command of Charles Green. Green was the most famous aeronaut of his day. He had been ballooning for forty years and had conducted more than five hundred ascents in his iconic *Royal Vauxhall* and *Great Nassau* balloons. His moment of personal triumph had come in 1836 when he had set the world distance record during a voyage from Vauxhall Gardens to the middle of Germany. Since then he had continued to fly, innovating the use of coal gas to fill the balloon instead of hydrogen, which was expensive and precarious to handle. For Welsh, Green had been a perfect pilot. He had steered the balloon while Welsh concentrated on the observations, splitting the responsibilities between navigation and science much as Cook and Banks had done on the *Endeavour*. Their three ascents had been a modest success. The feeling lingered, though, that there remained much more to be done. Now the cream of the capital's scientific community were ranged behind the Committee of the British Association. The time seemed ripe for a fresh attempt. Naturally they turned to Charles Green once again.

In July 1859 FitzRoy had been part of a three-man panel that had approved a payment of £90 to Green to conduct four meteorological ascents from the Wolverhampton Gas Works in Staffordshire. Wolverhampton was a prosaic place for such bold scientific experiments, but it was conveniently located in the centre of the country and the committee had struck a bargain with the owner of the local gasworks to supply fuel for the balloon. The day set for the first ascent was 15 August and Wrottesley, FitzRoy and Glaisher had travelled up to watch the launch. It turned out a fiasco. Attempts to inflate Green's old balloon, the *Nassau*, were scuppered by irregular gusts of wind that left it straining at its moorings, tugging like an anxious horse. By the time it was sufficiently inflated it was too dark to proceed. The following day they had tried again and successfully inflated the *Nassau* by half past one. But again its canopy was caught by the wind. It flapped and tore, and the gas poured out in seconds. Unable to describe the chaos he experienced, Glaisher fell back on tried and tested imagery, calling the disaster an 'aerial shipwreck'.[26] It had brought an end to attempts that summer.

People still lived in general ignorance about the state of the upper atmosphere. No one knew for sure just how far the atmosphere extended. In the 1840s Edgar Allan Poe had written a story about a balloonist who had ascended to the moon, and although few thought this possible no one had actually demonstrated it was not. Over the century various estimates had been made of the height of the atmosphere. John Dalton had suggested about fifty miles, while FitzRoy thought the limit was more likely ten. The highest a balloon had reached was Gay-Lussac's world record of five miles, and breaking this was one objective in the British attempts. But there were plenty of other mysteries too. How many different currents blew over one another as the air extended upwards? Would the temperature continue to decrease with altitude as Gay-Lussac had stated? Would the weather in the high atmosphere correspond with weather below? What would happen if a balloon came into contact with an electrified cirrus cloud? This was something that had seriously worried Forster in the 1830s: he thought the balloon might explode and plummet to earth in a fireball.

Two more summers passed until the British Association meeting of 1862. There a new balloon committee was formed. FitzRoy was too busy this time, but Airy, Herschel, Sir David Brewster and John Tyndall of the Royal Institution were enlisted. Determined to make headway, they set out a list of thirteen objectives, with the principal goals 'to determine the temperature of the air and its hygrometrical states, at different elevations', and to verify whether Gay-Lussac's formula for the decrease of temperature with height was sound. To undertake these experiments they turned to a stalwart of the scientific community to take on the role of meteorological aeronaut: James Glaisher.

Now fifty-three, Glaisher had no practical experience of ballooning and thus far he had shunned the limelight, thriving instead in out-of-the-way offices as secretary of the Meteorological Society or superintendent of the Observatory. Perhaps the decisive voice when it came to the appointment of Glaisher was Airy's. Glaisher had now worked by Airy's side for three decades. He was meticulous, trusted, cool-headed, rational and scientific. For a man with a mania for detail and a track record of getting work done, they could hardly have made a more shrewd choice.

It transpired that Glaisher had a latent interest in ballooning. He had followed John Welsh's ascents of a decade before, watching the arc of one aerial voyage through a telescope from the roof of the Greenwich Observatory. Now Glaisher would get his own chance. He was paired with an experienced pilot, forty-three-year-old Henry Tracy Coxwell. A specially commissioned balloon, the *Mammoth*, was made from best-quality American cloth with a capacity of 90,000 cubic feet. When inflated it would be one of the biggest balloons ever to launch. Reading his account of the voyage in his later book, *Travels in the Air* (1871), you can detect, for the very first time, a note of trepidation creeping into Glaisher's voice. Gone is the cool certainty of his dew or snow-crystal papers. As his 'aerial début' nears he frets over the observations, the speed with which he will have to work and the danger of losing data in a heavy landing. To settle his nerves Glaisher devised a sort of flight test simulator at Greenwich, where he could practise. 'I found myself condemned by the fear of not having been ready when the moment came to observe a phenomenon which perhaps no human eye had contemplated before,' he wrote.

Glaisher did all he could to prepare. To protect his growing arsenal of instruments he invented a simple quick-release mechanism that allowed him to cut away the thermometers, hygrometers and barometers in just a few moments if need be, and stow them in a padded crash box. He planned to take temperature readings, the dew point with Daniell's Hygrometer and a second condensing one. He had wet and dry bulb thermometers. Mercury and aneroid barometers. Vials to collect air specimens. Ozone paper to test oxygen levels. In addition he planned to keep an eye on the clouds and the currents and test for atmospheric electricity. The list ran on and on. It was going to be a bizarre feat of instrument juggling, as much a challenge as Constable undertook with his cloud studies half a century before.

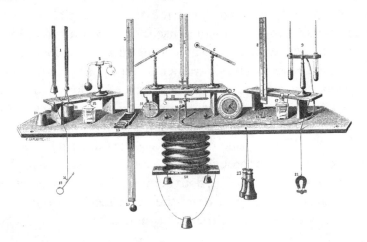

On 17 July 1862, after several aborted launches, his time finally came. It was destined to be a difficult beginning. A terrible west-south-west wind was scowling over Wolverhampton. 'It was a state of affairs by no means cheering to a novice,'[27] Glaisher pointed out with characteristic understatement. By half past nine in the morning the *Mammoth* was fully inflated. Glaisher and Coxwell clambered into the car and at 9.42 a.m. they parted from the ground. What followed was a terrifying minute, full of dramatic suspense as the balloon lurched sideways over the launch site. The basket was pitched at a sharp angle as it gathered speed, causing commotion for Glaisher, who was frantically fumbling with his instruments. '[It] would have been fatal had there been any chimney or lofty buildings in the way,' he wrote.[28]

But there was not. The drama of the lift-off is reflected in the fact that Glaisher errs into approximation for one of the few times in his career ('about 9h 43m AM') – but once the *Mammoth* gathered height, his precision returns. Within five minutes the balloon was in the clouds and Glaisher was busy taking measurements with speed and finesse. 'The meteorologist, like a spirit of a higher order than any, rejoices in the kingdoms of the air,' Ruskin had written twenty years before.[29] Glaisher's account bristles with numeric delight.

> At 9h. 49m, reached the clouds at an elevation of 4,467 feet. Rising still higher, at 9h 51m, with an elevation of 5,802 feet, we passed out of this stratum of cloud, but again became enveloped in cumulo-stratus, at the height of 7,980 feet. The sun shone brightly upon us at 9h. 55m, and caused the gas to expand and the balloon itself to assume the shape of a perfect globe. A most magnificent view now presented itself, but, unfortunately, I was not able to devote any time to note its peculiarities and its beauty, as I was still arranging my instruments in the positions they were to occupy, and we had reached a height exceeding 10,000 feet before all the instruments were in working order. The clouds at this time 10h 2m were very beautiful, and at 10h 3m., at an elevation of 12,709 feet, a band of music was heard. At 10h 4m., the earth became visible through a break in the clouds. At 16,914 feet the clouds were far below us, both cumulus and stratus, however, at a distance appearing to be at the same height as ourselves, the sky above us being perfectly cloudless and of an intense Prussian blue.[30]

As they ascended Glaisher switched from one instrument to another, taking readings, occasionally looking up to enjoy the clouds for an instant before moving on. At last the mysteries were within his grasp. Meteorology was soaring to new, exciting heights.

Afternoon

As the afternoon wears on the cloud base begins to thicken. Strong convective currents propel parcels of moist air higher into the atmosphere where they start to sublime into ice crystals. The cumulus clouds with their cauliflower tops have coalesced and a cumulonimbus is growing fast. In the right conditions a cumulonimbus capillatus can tower for many miles over the landscape, its anvil top scraping along the boundary between the troposphere and the stratosphere. As the tallest structure on earth, the cumulonimbus gives meaning to the expression 'to be on cloud nine', – the nine being the number assigned to the cloud by the *International Cloud Atlas* in 1896.

In supercooled air inside the cumulonimbus, ice crystals like those found in cirrus clouds extend outwards in an enormous canopy. On days like this raindrops can be sucked inside the cloud by powerful updraughts from below and sent see-sawing up and down, creating concentric rings of ice that tumble to earth as hail. But today the ice crystals fall straight. They melt as they descend into the warmer air below. It is the start of a rain shower.

The largest raindrops fall fastest, colliding and splintering as they go. We sometimes notice these, the outriders, splashing on the pavement, acting as a warning sign before the bulk of a shower arrives. The size of a raindrop varies. The smallest can be just a fraction of a millimetre in diameter; the biggest are juggernauts of five millimetres that fall at speeds of nine metres per second.

As the raindrops fall they pass through a shaft of sunlight, refracting it, and on the ground half a mile away someone standing with their back to the sun looks up at an angle of 42°. They see a rainbow.

No two people see the same rainbow. Each one is dynamic and unique to the viewer, while its colours, like the blue of the sky, are an ephemeral blend governed by the size of the raindrops that the light passes through. No pure rainbow has ever comprised the seven celebrated colours – red, orange, yellow, green, blue, indigo and violet; instead blends of colour are seen and rainbows can change as you watch them.

Large raindrops of 1–2mm make rainbows with very bright and vivid greens, vibrant reds but hardly any blue. Average-sized raindrops of 0.5mm produce bows with less red but greater pinks, while tiny raindrops of 0.08–0.10mm create broad bows with hardly any coloration. These are called White Rainbows and are rarely seen.

A second bow is almost always visible over the primary rainbow, at an angle of 51°. Much fainter and not always visible, the colours in a secondary bow are in reverse sequence to the first, beginning with violet on the outside and progressing to red. Once, looking at a rainbow during a storm, a man saw that the boundaries of the colours disappeared every time it thundered, as if the rainbow was being rattled. Whether the vibrations caused the raindrops to coalesce for a split second, destroying the atmospheric palette, no one has been able to say.

PART FOUR

Believing

CHAPTER 10

Dazzling Bright

On 18 August 1862, James Glaisher and Henry Coxwell undertook their second balloon ascent. In comparison to their first all went smoothly. A little after one o'clock in the afternoon Coxwell pulled the spring catch and they rose gently upwards. Glaisher watched as Wolverhampton slipped beneath them until it resembled a model town. Within ten minutes they were passing through a towering cumulus cloud. Emerging from its cauliflower top Glaisher was enthralled by the view. The dome of the sky was a 'beautifully deep blue, dotted with cirri'. Other clouds 'dazzled and charmed the eye with alterations and brilliant effects of light and shade'. For a moment he hung his head over the brim of the car, Forster-like, and gazed about. He saw the outline of *Mammoth* silhouetted against a nearby cloud, the shape of the balloon emblazoned like a 'kind of corona tinted by prismatic colours'. Descending two and a half hours later, Glaisher spotted what he thought was the same cumulus they had risen through over Wolverhampton. It was due north of them and Glaisher fancied that it followed them on their way. It may, 'from its grandeur', he reasoned, 'have been called the monarch of clouds'.[1]

A fortnight later, on 3 September, Glaisher and Coxwell launched from Wolverhampton again. It was a damp, cold day, not perfectly suited to ballooning. But with the blustery autumnal months ahead they were determined to squeeze in another high-altitude ascent if possible. The previous attempt had carried them to about 24,000 feet, just a little higher than Gay-Lussac's pioneering and long-celebrated 1804 voyage. Neither of them had suffered any ill-effects, and this time they aimed to travel even higher.

As before, they left Wolverhampton at one o'clock in the afternoon and within ten minutes they were climbing fast. Breaking through the clouds, 'A flood of strong sunlight burst upon us with a beautiful sky without a cloud and beneath us lay a magnificent sea of clouds, its surface varied with endless hills, hillocks, and mountain chains, and with many snow-white tufts rising from it,' Glaisher wrote. Glaisher kept up his rapid round of readings, moving from barometer to thermometer, from thermometer to hygrometer. The ground temperature was 59.5°F, a mile into the cloud it was 36.5°F, at two miles above the ground it had dropped to around freezing, and by the time they were four miles high it was just 7°F. Glaisher paused briefly, to expose a photographic plate. So dazzling was the sunshine that just a quick exposure would have captured the view. The attempt failed, perhaps due to the swinging of the car. Rising much faster than before, *Mammoth* had begun to lurch in the wind. Soon the balloon was corkscrewing upwards rather than taking a steady vertical path.

By now they were climbing at an accelerated pace. The low-level clouds were shrinking beneath them. The sky was gradually becoming a darker blue. At five miles the temperature was 2°F, colder than the Scottish hills on a frozen winter's day. Dressed in heavy coats, wrapped in scarfs and gloves, Glaisher and Coxwell seemed more prepared for a Sunday afternoon hike than a trip into the upper atmosphere. As they passed 20,000 feet, they felt the cold keenly. Soon Glaisher's equipment began to malfunction. His Daniell Hygrometer stopped registering, he could not get dew to form on the wet bulb thermometer while other instruments were clogged with frost. Determined to climb further they threw out more of the sand ballast they used to regulate height. *Mammoth* lurched upwards, now climbing at a rate of about a thousand feet every two or three minutes.

'Up until this time I had taken observations with comfort,'[2] Glaisher recalled. But what had once been second nature suddenly started to become a wearing task. The last set of readings he made came at either 1.51 or 1.52 p.m. His use of approximation is significant. The barometer was registering 10.8in and the dry bulb thermometer -5°.[3] During the ascent Glaisher and Coxwell had kept to their own tasks working back to back: Glaisher taking his readings, Coxwell piloting the balloon. At 1.54 p.m. Glaisher suddenly called out to Coxwell for help. The pilot, though, was far too distracted. Glancing up from his observation deck, Glaisher saw a terrifying sight. Coxwell had clambered out of the basket and was hoisting himself through the guy ropes towards the ring of the balloon – five feet above the basket.

The *Mammoth*'s corkscrew motion had sent the valve line – crucial for venting gas – swinging from side to side and it had become entangled in the ring.

As Coxwell climbed up out of the car Glaisher was overcome by a strange sensation. One of his arms had gone completely limp. He tried his other, but that was just as powerless. As Coxwell hoisted himself up through the maze of ropes, Glaisher's body began to crumple. His head flopped, jelly-like, on to his shoulder. He tried to lift it but it only fell to the other side. A moment later his legs buckled. He fell backwards. 'In this position my eyes were directed to Mr Coxwell in the ring. I dimly saw Mr Coxwell and endeavoured to speak, but could not.'

> In an instant intense darkness overcame me, so that the optic nerve lost power suddenly, but I was still conscious, with as active a brain as at the present moment whilst writing this. I thought I had been seized with asphyxia, and believed I should experience nothing more as death would come unless we speedily descended; other thoughts were entering my mind, when I suddenly became unconscious as on going to sleep.[4]

Glaisher was in a perilous position. *Mammoth* had reached what is today called the 'death zone', a realm of atmosphere where the concentration of oxygen is not sufficient to support animal life. On the edge of the stratosphere, Glaisher's body was shutting down. At 1.56 p.m., at an altitude of about 29,000 feet, the height of the newly triangulated tallest mountain in the world, Deodunga, or as it was later to be called, Everest – Glaisher's mind went blank.

Minutes slipped by before Glaisher stirred. He was awoken by Coxwell's voice speaking from a distance. He heard the words: 'temperature' and 'observation'. Then he heard Coxwell again. 'Do try; now do.' At length Glaisher realised that Coxwell was trying to rouse him. Information seeped in. The instruments became 'dimly visible'. Then beyond his observation table came the comforting sight of Coxwell himself.

'I have been insensible,' Glaisher said.

'You have, and I too, very nearly,' Coxwell replied.[5]

Glaisher noticed Coxwell's hands were black. He pulled out a bottle of brandy and doused them. By now his instincts were returning. In one of the most triumphant and understated sentences in all Victorian science, he turned back to his desk: 'I resumed my observations at 2h 7m, recording the barometer reading at 11.53 inches, and temperature minus 2°.'[6]

Around ten minutes had passed since Glaisher had slumped back in the basket. As his readings proved, *Mammoth* was descending fast. It was only now that Coxwell was able to tell Glaisher what had happened during those ten minutes. It was a story more dramatic than he could have imagined. Perhaps it was a blessing that Glaisher had not been awake to experience it. Climbing into the ring of the balloon – by Glaisher's calculations they were now as much as seven miles or 35,000 feet high – Coxwell had managed to disentangle the valve cord. Looking down he had seen Glaisher on his back, 'my legs projected and my arms hung down by my side,' Glaisher recalled. He seemed 'serene and placid', Coxwell reasoned. Perhaps he was already dead. A celebrated lithograph later froze this dramatic moment, *Mr Glaisher Insensible at the Height of Seven Miles*. It is a terrifying image of the two men perilously adrift. Glaisher is collapsed in the car an inch from overbalancing. Had he fallen it would have been the death of them both: his body tumbling seven miles to the Staffordshire countryside and Coxwell's fate even more terrible. With the *Mammoth*'s heaviest load discharged Coxwell would have been propelled further and further upwards into oblivion. In the lithograph he is even more perilously placed than Glaisher. He has climbed into the ring of the balloon. His left leg is coiled snake-like around the ropes and he clings on with his right arm. The valve line swings desperately, just out of reach.

In an unprecedented feat of aerial acrobatics, unimaginably courageous, Coxwell grabbed hold of the valve and untangled it. But still their problems were not at an end. Dropping into the safety of the basket, Coxwell found he had exhausted himself. Having shown such dexterity seconds before he now felt paralysed. He could not muster enough power to lift either of his arms to pull the valve cord to vent the air. With a last burst of effort Coxwell tried to clamp the cord between his teeth. It worked. He dipped his head 'two or three times until the balloon took a decided turn downward'.[7]

In ten minutes *Mammoth* had plummeted to around 10,000 feet – a leisurely ballooning height. Had Coxwell and Glaisher known the dangers of subjecting the human body to such fluctuations in pressure they would have realised they were still in grave danger. But Glaisher was oblivious. After his brief spell of 'insensibility', he had resumed his observations as before. It was as if they had suffered nothing more than an inconvenience. At 2.40 p.m. they touched down in a field in the Shropshire countryside. Having tidied away their balloon and belongings they set out for the nearest village. They walked for seven or eight miles until they found a country inn in the village of Cold Weston, near Ludlow, where they promptly drank a pint of beer.

This third ascent in *Mammoth* would make Glaisher famous. In the weeks to follow his account of the aerial voyage would be published in *The Times* where it caught the national imagination. It was not the scientific details that fascinated but the story of human courage and the frightful prospect of what might have happened to them had Coxwell succumbed. The whole account was underpinned by stirring paradoxes. First of all was the cold temperature. It seemed strange that while everyone felt the warmth of the sun, the closer you travelled towards it the colder it became. Then there was an aesthetic paradox. The world Glaisher described was characterised by beauty: the cirrus clouds, the deep blue of the sky, the transparency of the air. Yet it was a place of grave danger. The upper atmosphere did not kill quickly or violently. It lulled senses into a suffocating torpor. It made minds fuzzy, limbs limp. Glaisher's description of the creeping 'insensibility' that overcame him was alarming, unsettling and exciting all at once. He wrote:

Asphyxia steals away the life of the human being as he moves above, suspended in mid-air, as stealthily as cold does that of the mountain traveller, who, benumbed and insensible to suffering, yields to the lethargy of approaching sleep, and reposes to wake no more.[8]

Coxwell's tremendous courage was celebrated above all else. It was clear that had he not succeeded in pulling the valve then they would have died, the two of them drifting away into space. From start to finish the story encapsulated British reserve. Perhaps the best tribute to science was Coxwell's method of rousing Glaisher, tempting him back to consciousness with the catalytic words 'temperature', 'observation'. 'The courage of men of science deserves to have a chapter of history devoted to it,' declared *The Times* in a leader article:

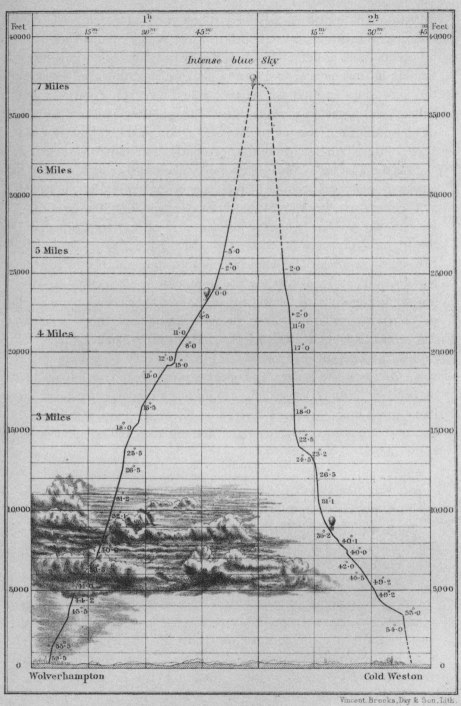

Path of the Balloon in its ascent from Wolverhampton to
Cold Weston near Ludlow.

5th September 1862.

They are solitary, deliberate, calm and passive . . . The aerial
voyage performed by Mr Coxwell and Mr Glaisher deserves to
rank with the greatest feats of our experimentalizers, discoverers
and travellers . . . They have shown what enthusiasm science can
inspire and what courage it can give.[9]

This, of course, was just the personal side of the story. Beyond the
celebrations of derring-do lay a true scientific discovery. For centuries
people had speculated about the height of the atmosphere. The best
suggestion had come from Johannes Kepler in the seventeenth century.
Based on the duration of twilight, Kepler had calculated, the atmosphere must be between forty and fifty miles high. For two centuries
this theory had held, until the early nineteenth century when John
Dalton had calculated a more precise forty-four miles. Now it seemed
that Glaisher and Coxwell had solved the mystery. In doing so they
had reshaped the limits of the inhabitable world, like Columbus or
Marco Polo before them.

After the third aerial voyage Glaisher's ballooning exploits became
national events. Each ascent was anticipated, tracked and written up
in the newspapers with the enthusiasm of reports on sporting
spectacles. They brought Glaisher celebrity, making him one of the
country's most recognisable scientists and giving meteorological
experimentation the kind of public attention it had not enjoyed since
Franklin's lightning exploits of the 1750s. Many scientists would
complain about Glaisher's public profile in the years to come,
muttering that just as *Mammoth* had been filled with coal gas, Glaisher's
ego had been swelled with hot air.

Among the most interested was Admiral FitzRoy in London. He
was galvanised by what he considered a 'truly heroic ascent'[10] and
immediately wrote to Glaisher, asking for a duplicate of the observations. FitzRoy copied out the data, almost as if by doing so he could
experience the thrill of the ascent himself. Although he would not
declare it publicly, Glaisher told FitzRoy that Coxwell had seen the
barometer at its minimum height. It had read between seven and
eight inches, a fact that, if true, meant they had risen above seven
miles to perhaps 36,000 feet. 'There was no moisture – there were
no clouds – but they were far above both: they were nearer to heat-
less, airless, and mysterious space than ever moral man had previously
penetrated; – and into which their daring venture will probably deter

any others from making so desperate, however meritorious an excursion, even for the interests of science.' Glaisher agreed. He admitted to FitzRoy he was glad the high ascents were now over. Understated as ever, he wrote, 'The limit is plainly five miles. Whenever eleven inches is reached – prudence says – open the valve.'[11]

That Coxwell had not opened the valve at eleven inches had, as well as generating a memorable story, produced some exciting science. Glaisher had gathered enough data to cast doubt on Gay-Lussac's theory that air temperature dropped 1°F for every 300 feet of elevation. Instead he had glimpsed a much more complex picture of alternating colder and warmer pockets of air. It was the beginning of a realisation that various strata exist in the atmosphere. This discovery would open up a line of enquiry that would occupy meteorologists for decades and would eventually lead to the delineation of the various stratum. Even in 1862 Glaisher sensed the significance of the voyage. 'This result is most important,' he told FitzRoy. 'It affects refractions, Wollaston's boiling point theory, thermo-electricity, &c &c. More experiments must be made. I have been up 8 times, and if I ascend 8 times in the next year, and be spared to tell my new story, I think others may take the matter up.'[12]

The relationship between Glaisher and FitzRoy was cordial but it never blossomed into friendship. They were remarkably similar yet completely different. Both were experienced scientists in their fifties. Both were fellows of the Royal Society. Both had sat on the British Association Balloon Committees. Each had a reputation in meteorology and a position at the top of the profession. But before the summer and autumn of 1862 it was FitzRoy, the wily old seadog, who was noted for adventure and exploration. Nothing on Glaisher's CV could match him. Had FitzRoy not been so busy in 1862 might he have filled the meteorologist's berth alongside Coxwell? Perhaps the idea is far-fetched. In any case it is worth entertaining for a moment the picture of the celebrated captain of the *Beagle*, now an admiral and bedecked in his austere uniform of state, aloft in the Midland skies. It never happened, of course, for by 1862 as Glaisher was soaring the heights FitzRoy was embroiled in a quite different experiment of his own.

FitzRoy was not the first European to establish a storm-warning system. He was beaten to that milepost by the Dutch meteorologist, Buys Ballot, who began his signal service on 1 June 1860. Appointed to his position as head of the Royal Dutch Meteorological Institute at about the same time as FitzRoy started at the Board of Trade in 1854, Buys Ballot had proved to be an agile thinker. The short Dutch coastline and much smaller country enabled him to establish meteorological stations at four places: Den Helder, Groningen, Flushing and Maastricht. Drawing on a background in mathematics Buys Ballot had spent most of his early years analysing the data recorded at these stations, and by 1857 he had developed a simple rule to determine wind direction. It was based solely on what he called the 'departure of barometric pressure' from a calculated average. Simply put, if the pressure at Helder was higher on a given day than the calculated average of the years before, there was a positive departure of pressure. If lower, there was a negative departure of pressure.

Buys Ballot's theory was the type of rational shorthand that scientists loved. He had worked out that if the departure from the average pressure was higher in the southern stations, Maastricht or Flushing, than the northern stations, Groningen or Den Helder, wind would come from the west. If the situation was reversed, the wind would come out of the east. He would clarify this equation at a meeting of the British Association in 1863:

> More accurately, you may say, the wind will be nearly at right angles with the direction of the greatest difference of pressures. When you place yourself in the direction of the wind (or in the direction of the electric current), you will have at your left the least atmospheric pressure.[13]

Today Buys Ballot's law is still remembered by many sailors. 'If you stand with your back to the wind in the Northern Hemisphere, air pressure will be lower to your left.'

Buys Ballot had relied on a similar analysis of barometric data to calculate the expected force of the coming wind. Using these equations and the electric telegraph he had been able to issue the world's first governmental storm warning, on 1 June 1860, as FitzRoy was preparing for his talk at the British Association in Oxford. Meanwhile, across the border in France, Le Verrier was considering a similar service. He had written to Airy – revealingly, not FitzRoy – to see whether the British would be interested in sharing information about active storms. This was a delicate area. Old scientific rivals, the British and French had long been used to working independently, hoping to claim inventions or discoveries as objects of national pride: most recently the French had triumphed with photographic innovation and the discovery of Neptune, both of which could well have gone the British way. But weather was different. It did not respect international borders and often storms that blew through the Paris boulevards had first toppled trees in Shrewsbury, Dublin or Galway. For the British it was just the same and the reality of the situation meant the old enemies were driven into a suspicious alliance.

Neither Le Verrier nor FitzRoy were designed for diplomacy. While FitzRoy's stealthy work at the Meteorological Department was attracting suspicion from Westminster politicians, in Paris Le Verrier was becoming infamous for his autocratic, bullying regime at the Observatory. Scores of employees would quit through exhaustion or exasperation or be forced out by Le Verrier during the 1860s, all contributing to his reputation as the 'magnificent and detestable astronomer'. The alliance between Le Verrier, domineering and vain, and FitzRoy, proud and secretive, was never destined to be an easy one. Over the next five years, from 1865, the men were forced into a relationship that could best be described as tense and tolerant. Their letters were polite and elusive. They kept a watchful eye on each other.

For the time being Le Verrier and the French government were happy to restrict their meteorological ambitions to publishing the daily weather report in the newspapers and developing their work with the *Bulletin météorologique*, an international daily digest of temperature, wind and pressure, co-ordinated by Le Verrier and circulated to observatories in Europe. FitzRoy, though, had grander plans. On 6 February 1861 he issued his first weather warning for storms on the north-east coast. It involved information sourced by telegraph, interpreted at the Meteorological Department in Parliament Street and relayed back to the coastal stations in time for the cones to be hoisted. From an operational point of view everything worked smoothly. FitzRoy had rightly spotted that a chain of gales was brewing in the Atlantic and would cross the British coast. The biggest problem had come at South Shields on the north-east coast where some fishermen had ignored the cones and had been wrecked as a result.

An even more powerful story came from Whitby, a small, isolated fishing town without access to telegraphed warnings. In the same storm much of the Whitby fleet had been caught in the rocky shallows around the port, and the lifeboat, launched only a few months earlier and manned by the 'finest picked seamen' in the town, had struck out to rescue the coasters. Five times the boat had returned to port with fishermen and she was on her sixth and final rescue, with her crew all exhausted, when she was capsized by a freak wave. Mr Keane, perpetual curate of Whitby, wrote in a letter to *The Times*: 'Then was beheld by several thousand persons – within almost a stone's throw, but unable to assist – the fearful agonies of those powerful men, buffeting with the fury of the breakers, till one by one 12 out of 13 sank, and only one survived.'

The courage of Whitby's lifeboatmen was justly celebrated in the newspapers. There was room too for comment about 'Admiral Fitz-Roy's Storm Warnings'. It had escaped no one's notice that FitzRoy had correctly predicted the gale. In a leader column, *The Times* congratulated FitzRoy on his work, and in the same issue published a letter it had received from him. Here FitzRoy, full of ambition for his system, declared 'that every frequented part of our coasts might have received information of the coming gale three days before it burst. The event was predicted with as much certainty as an eclipse and could have been announced by signals as conspicuous as fiery beacons.'

This was strong stuff. His storm warnings had operated well, but to suggest that each time a gale approached the country could be guaranteed three days' notice was an enthusiastic, if not foolish claim. *The Times*, though, lapped it up. If there was any difficulty at all, it contended, it was in convincing fishermen to heed FitzRoy's warnings. 'It might be vain to tell a boat's crew that in three days' time there would be a tornado,' the paper noted: 'they would, perhaps, disbelieve the information, or still more probably disregard it.' It was a problem of faith more than anything else. People had to believe the science. And as far as the journalist was concerned the scientific question was settled. He finished off in vigorous style: 'We cannot yet forecast the general character of the season but it seems that we can really foretell a gale before it comes and even ascertain the quarter from which the wind will blow. If we have indeed got to this point – and there appears no reason to doubt it – the rest ought to be easy.' The *United Service Magazine* had much the same idea:

Such are a few of the principles of our scientific weather wisdom and such is the system of our Meteorological Office; and the

immense advantages of the institution must be evident when
we consider the number of ships sailing the seas of the globe,
is about two hundred thousands, with more than a million sailors
– all exposed to storms which, for ages before, had come upon
them without a chance of avoidance. It was a pitiful case, and
Providence seems to have taken pity on us in our latter days.
Things are altered now. The Arabs may acknowledge the lion
of their mountains as their master, and concede to him the right
of decimating their flocks without resistance, but, thanks to
science, our navigators may now fairly struggle with the chances
of the winds and the changes of the skies. Science and observation tells us where the wind will blow, 'as it listeth', and the
electric telegraph has vanquished the tempest.[14]

FitzRoy's methods for calculating his storm warnings relied on
Dove's ideas of conflicting fronts of cold and hot air, the collision of
which would spark the 'atmospheric gyration' Dove thought was the
source of all storms. In 1862 FitzRoy had seen into print a second
edition of Dove's *Law of Storms*, a work that analysed the trade winds,
the monsoon and storms of the temperate zone in formidable detail.
By now FitzRoy and Dove had become close friends, and Dove had
dedicated this second edition to FitzRoy for his long-standing support.
For FitzRoy, Dove was the authority behind his warning process, and
through his ideas he had devised his own formula:

Great changes or storms are *usually* shown by falls of barometer
exceeding half an inch, and by differences in temperature
exceeding about 15°. Nearly a 10th of an inch an hour is a full
procedure in the storm of very heavy rain. The more rapidly
such changes occur, the more risk there is a dangerous atmospheric commotion.[15]

FitzRoy had made an auspicious start with his storm warnings of
February 1861 and thereafter his work changed. Until now the
Meteorological Department had been mainly an administrative hub,
but in the course of one year it had transformed into a dynamic body.
FitzRoy knew the key to successful warnings lay in a close and careful
interpretation of the daily meteorological data sourced from the telegraphic network. This shift from the historical analysis of records and

logs to the daily churn of instant data was significant. Everything now became time-sensitive. Should FitzRoy miss any of the twenty or so gales and storms each year there was bound to be a human cost. And as the British weather did not keep regular office hours it was to become an all-encompassing preoccupation.

Since his days at sea FitzRoy had felt a keen sense of personal responsibility. Shortly after the *Beagle* had sailed from Plymouth in 1831 he had sent for the ship's officers and had told them 'he had never known accidents, as they were called, happen in any ship, except when they could be traced to the fault of the officer carrying on the duty; that he was convinced this was almost invariably the case'. He had cautioned them, 'if ever in the *Beagle* a sail was split, a spar was carried away, a man knocked off a mast or yard, or a sea shipped on board, he should consider the officer in charge at that time to blame'.[16] The lecture had unnerved FitzRoy's officers but in retrospect they had come to see it as a vital factor behind the ship's immaculate safety record. Three decades on, he was extolling the same standards of constant vigilance, not this time in relation to oceans of water, but to oceans of air.

In August 1861 FitzRoy decided to expand his storm-warning system to 130 different locations instead of the original fifty. He also took another step. Not wanting to discard his day-to-day interpretations of weather data, he decided he might as well send them to the news-papers too – as predictions for the following day. The first were issued on 1 August 1861. In an understated notice in *The Times* beneath the usual weather chart, there came a new section. It read:

General weather probably for the next two days in the –
North – Moderate westerly wind; fine.
West – Moderate south-westerly; fine.
South – Fresh westerly; fine.

For FitzRoy this was a logical step. He later pointed out that including these predictions cost no additional money and as he thought the information might be useful it might as well be included. It was logical thinking but it was also potentially controversial. At no point had FitzRoy received authorisation to do this. On the face of it this was a trivial matter, but in reality it was not. For the first time in history, FitzRoy was providing scientific weather predictions on behalf of the government. This was scientific experimentation in public, and with no settled theoretical framework for devising the predictions, no predecessor as a benchmark for success rates and no administrative support, it was a bold move. Downriver at the Royal Observatory in Greenwich, Airy had made it clear that any kind of prediction was forbidden. In Parliament Street, FitzRoy was going fast in the other direction.

FitzRoy would soon attempt to tidy up the lexical and philosophical confusion of these announcements by coining a new term, 'Forecasts'. 'Prophecies and predictions they are not – the term forecast is strictly applicable to such an *opinion* as is the result of scientific combination and calculation.' He emphasised this whenever he could. 'Like storm signals, such notices of expected bad weather should be merely *cautionary*, to denote anticipated disturbances *somewhere* over these islands, without being in the least degree compulsory, or interfering arbitrarily with the movements of vessels or individuals.'[17]

Originally storm warnings were intended for the benefit of mariners and those living beside the sea. But with his forecasts FitzRoy opened up the science of weather for everyone. As the storm warnings continued throughout the blustery winter of 1862–3, FitzRoy's forecasts became more and more of a curiosity for readers of any of the six newspapers that carried them. They appealed as much to businessmen making a day trip to London as to a family preparing for their seaside holiday. For the horse-racing establishment they were a particular novelty, as they gave the fashionable classes advance warning for picking their outfit. With the Derby looming at Epsom in June 1862, *The Era* noted: 'With what eagerness in every quarter was the meteorological column consulted in the newspapers where Admiral FitzRoy records the forecast of the weather, and with what satisfaction did the experienced interpreters of the prediction see . . . it would be a remarkably fine day, and that umbrellas might be left behind.'

Reflecting FitzRoy's success was an article published in the journal *Once a Week*. 'In these days of new theories we must not be astonished at the rise of new sciences,' it began. 'One or two of them at least are yearly born of hypothesis and experience. Social sciences is yet a babe in arms; Ethnology, Comparative Philology, and several more are hardly released from the nursery. Meteorology though is growing like a young giant.' The article quoted FitzRoy's claim that a total of 5,500 months of observations had been collected from the logs of 800 merchant vessels. All this was augmented by a growing mountain of data from other sources:

> The amount of fog is measured by the lighthouse-keepers around our shore. The direction of the wind is registered in many places by self-acting anemometers, and a few easy calculations show its force. The amount of dew is daily collected at our great observatories; the degree of heat and cold carefully ascertained; the quantity of ozone in the atmosphere determined; electricity and magnetism taken into account. These statistics are the food for our youthful science.[18]

But more than this, *Once a Week* continued, was meteorology's value as an applied science. It was quick to praise FitzRoy for his forecasts. Not only were they vital for mariners – the article reported that out of fifty-six harbour masters surveyed, forty-six supported the storm warnings, seven expressed 'no decided opinion' and only three were unhappy with them – but forecasts were becoming a feature of everyday life. 'Even plain men by glancing at the public weather reports, consulting their thermometers and barometers, and keeping an eye on the aspect of the heavens, may now with a little practice easily foretell the weather at least a day in advance.'

> So, as meteorology grows, we may speculate on hail and lightning risks being so much at a discount with farmers that they will not care to insure. Nay, those familiar articles, umbrellas and galoshes, may thus gradually lose their occupation, and even macintoshes have to retire into private life.[19]

It was about the time of the *Once a Week* article, one Sunday morning, that FitzRoy's daughter Laura remembered being called to the front door of their home in Onslow Square. Her parents were returning from Communion service and she thought it must be them. Throwing open the door, she was embarrassed to find on their doorstep the Queen's Messengers, 'come to enquire at the Admiral's house about the weather reports and storm warnings for the following day, when Her Majesty was intending to cross to Osborne in the Isle of Wight'.

It became a treasured memory for FitzRoy's daughter. Her father was now deriving an element of fame from his weather work. In

March 1862 he had given a lecture on his forecasts at the Royal
Institution which was so well received that the *Morning Post* had
complained he was destroying the old notions of science. The old
adage, 'As uncertain as the winds has ceased, we are told, to be a
correct comparison, for the winds are said to be governed by fixed
laws . . . To call a man "a weathercock" is no longer a term of reproach;
and to suspect a minister of vacillating policy because he has an eye
to the weather is now so unfounded that the President of the Board
of Trade has a staff of officers paid by the nation for the avowed
purpose of telling which way the wind blows.'[20] On 8 November 1862,
with his forecasts and his storm warnings in full swing, FitzRoy made
his début in the satirical weekly *Punch*. Under the headline, 'The
Cumming of Storms', it announced, 'Since Admiral FitzRoy, that new
Clerk of the Weather, prophesised with such signal success that advent
of the late gale, it is but fair that he should now be popularly recognised as the First Admiral of the Blew.'

This was FitzRoy as he wanted to be seen. Helping others, throwing
his talent into humanitarian concerns: the trailblazer. In line with the
growing scope of his department more employees had been taken on
and the Meteorological Department budget reinforced. By 1862 ten
civil servants now worked for him, including his seasoned deputies
Babington and Pattrickson. There was now little time left to worry
about wind charts and circulating the instruments. Instead most of
the department's resources were directed into analysing data. The
Meteorological Department had become a place of great purpose. A
practical man, a life on the seas behind him, FitzRoy revelled in projects,
goals, destinations. 'Stones may be shaped, bricks may be accumulated,
but without an object in view – within an edifice to be constructed
– how wearily unrewarding to the mind would be such toil, however
animated by true scientific faith in future results.' They could hardly
have expected him to remain content in the role of 'statist'.[21]

The key to the whole operation was the electric telegraph.
Throughout the 1850s it had spread the length and breadth of Britain
and Ireland, funded by ambitious private companies. It meant that in
a flash the station observer could codify a breezy day on Loch Lomond,
blustery showers at Scarborough, bright sunshine at Lowestoft or blue
skies over Penzance. All the variety of British weather could be distilled
into numbers and sent down the wires to 2 Parliament Street. Readings
were taken at nine o'clock in the morning and by ten o'clock thirty

or forty telegrams would be ticking out of the exchange, every day of the week but Sunday. The data was read, reduced, corrected, errors obliterated, and written up into prepared forms for analysis by FitzRoy. By eleven in the morning reports were ready and were dispatched to *The Times* to make the second edition. Others were sent to Lloyd's, the *Shipping Gazette*, the Board of Trade and Horse Guards. In the afternoon a further raft of reports and forecasts were sent out for the evening papers and then, with still more data streaming in, the last circulars were written up for the morning's editions and dispatched before the office closed. FitzRoy prided himself on the smooth, nimble operation. 'These cautionary signals are transmitted so rapidly,' he wrote, 'being shown around the coasts in half an hour from their leaving London.'[22]

Started from scratch, this was a tremendous feat of organisation. To simplify the system FitzRoy had divided the United Kingdom into six geographical areas: Scotland, Ireland, West Central, South West, South East and East Coast. Reports and forecasts for each area were prepared, to make the storm warnings more accurate. He had faced linguistic challenges too. 'As newspaper space is very limited,' he explained, 'and that some words are said in *different senses* by various persons, extreme care is taken in selecting those for such brief, general, and yet *sufficiently definite sentences*, as will suit the purposes satisfactorily.' It is a shame Beaufort was not on hand to help. It was a problem he would have relished.

FitzRoy was ambitious but he was not blind to the difficulties. He knew that his storm warnings were divisive. When storm drums were hoisted, should all ships stay in port? 'Should fishermen and coasters wait idle and miss opportunities?' he wondered. 'By no means. All the cautionary signals imply is, "Look out", Be on your guard.' It was a tricky balancing act. FitzRoy had a duty to warn but he could not claim full responsibility for what he said. To clarify matters he tried to establish himself as a sort of benevolent bystander. He explained that storm warnings and forecasts were principally a matter of faith. Rather than dictating terms he sought to inform, to devolve responsibility, in the hope that people would come to trust his judgement. At the back of his mind he kept one of Aesop's famous fables: The Boy who Cried Wolf. 'Against signalling too frequently, as well as too extensively, caution should be urged, lest "Wolf" should come while unprepared; but on the other hand, it is better to risk occasional error in excess, and the last danger arrive without warning, by which mistake *lives* may be sacrificed.'

The storm warnings worked on some occasions. Other times he was caught out. The daily forecasts were becoming more problematic. On 11 April 1862 *The Times* reported:

> The public has not failed to notice, with interest, and, as we much fear, with some wicked amusement, that we now under-take every morning to prophesy the weather for the two days next to come. While disclaiming all credit for the occasional success, we must however demand to be held free of any responsibility for the too common failures which attend these prognostications. During the last week Nature seems to have taken special pleasure in confounding the conjectures of Science.[23]

Here was a linguistic shift. When forecasts were successful they were championed as science at its best. But when they were not, they were recast as prophecies or prognostications. Depending on the day, depending on the person, FitzRoy might be seen as bold or brash, daring or reckless, visionary or mad, a humanitarian or a fool, a fore-caster or a prophet.

To win over his critics FitzRoy worked harder than ever. 'Certain it is, but although our conclusions may be incorrect, our judgments erroneous, the laws of nature and the signs afforded to man are invari-ably true,' he argued. 'Accurate interpretation is the real deficiency.'[24] It was a bold statement. Science for show had thrived since the turn of the century in Humphry Davy's experiments at the Royal Institution, and Faraday's Christmas lectures had carried on the trend. But this was different. Davy and Faraday had been able to master audiences, performing experiments that had already been tested and were known to succeed. FitzRoy had none of this certainty. Like Morse's telegraphic demonstrations of the 1840s this was science of the moment. Each day the public was presented with a new forecast, a new experiment. Whether it succeeded was a completely different matter. For some it was starting to look as if FitzRoy was on some delirious visionary trip. A trip that would test the public's faith in science.

By now FitzRoy was working with a speed and purpose that surprised even his peers. While on holiday in Brighton in August 1862 he began to distil his vast weather knowledge into a book. In two and a half months the project was finished and by December a first edition of *The Weather Book: a manual of practical meteorology* was on sale. It was a staggering outpouring of intellectual energy. Although he had drawn on previously published pieces for about half of the 450 pages, the rest was new material. It was in complete contrast to twenty-five years before when he had laboured over his *Narrative* of the *Beagle* voyage. Friends were amazed. One wrote to Darwin telling him he had seen FitzRoy 'who was finishing his book as he told me, and looked worn as if he had been closely confined to his work'.[25] The effort seemed worthwhile, though, and he was able to forward finished copies to the Royal Society and to Herschel, explaining that it had been written in his 'so called holiday'.[26]

The Weather Book was an accessible piece of science writing. Taking his lead from Reid and Piddington, FitzRoy wanted it to be useful for everyone. He drew together his two principal themes: keen observation and scientific theory. Again and again he stressed that it was a simple combination of these two ingredients that allowed people to foretell weather. His style was neat, precise and practical:

Perhaps a laudable anxiety to be correct and systematic in making and recording meteorological observations has induced the prevailing idea that *extreme precision* is all-important, and that observations should be very numerous. In Observatories, unquestionably, such should be the case; but to treat all localities, all observers, all circumstances of time, climate, and opportunity alike, and to require a similar registration from each, would indeed be Procrustean, while their application of very refined instruments might be like cutting wool with razors.[27]

You can never imagine Glaisher, who presumably did cut wool with razors, writing such a thing. *The Weather Book* takes on a succession of subjects: instruments, weather wisdom, meteorological progress, a sketch of different climates around the world, the composition of the atmosphere and his own methods of forecasting.

One of the most revealing chapters in *The Weather Book* comes towards the end when he allows himself to reflect on his own experiences. This lets FitzRoy shift into autobiography. The impact of his days as a midshipman are apparent. He writes vividly of a lightning

strike at Corfu, notable because the lightning seemed to come up out
of the water instead of down from the sky. There is also his recollection of the pampero in 1829 when the two sailors were drowned, and
the story of Captain Cable in the Strait of Magellan as they sailed
home from New Zealand. None of his recollections, though, outdo
his description of an electrical storm in La Plata in the 1820s:

> In no part of the world, perhaps, is there more lightning at times.
> In HMS *Thetis*, at sea off the Plata, the whole heavens seemed
> (on one occasion) like an immense metal foundry, so incessant
> and diversified were the lightning flashes in every direction, even
> *from the sea upward*. Repeatedly lightning struck the *water* between
> that ship and another vessel about a mile distant. Indeed the
> whole vault was illuminated, in every direction, though black
> clouds shut out every star. So grand a sight the writer never
> witnessed. Much rain, at times, poured down. Neither ship was
> struck, though forked lightning was seen to strike the water in
> every direction, during about three hours of illuminated dark-
> ness, from 9 o'clock till midnight.[28]

FitzRoy's description is intense and vivid: the flashings of the sky
reflected in his imagination as he wrote at his desk in docile Brighton
half a century later. It captures the drama of weather: close enough
to enthral; not quite enough to hurt.

FitzRoy clung to these experiences of long ago. Like Beaufort, like
Glaisher, like Constable, he was stirred deeply by these youthful memories of nature. A love for weather had stayed with him his whole life.
On 10 March 1863, shortly after *The Weather Book* was published, he
walked through London streets shrouded in fog. When he got to his
office he seized a pen and composed a breathless note to his friend,
the chemist John Hall Gladstone:

> Dear Dr Gladstone,
>   I wish you had been here this morning. So rich and rare a fog
> I was never obscured by in the day. It was a glorious colour – red
> yellow – and its consistency that of pea soup with scattered mint.
> As carbon is so wholesome I expect to feel all the better for my
> morning's supply.[29]

The Weather Book received mostly good reviews. The *London Intellectual Observer* was perhaps the most enthusiastic of all. 'Few men have done so much practical good with so little pretence as Rear Admiral FitzRoy, whose forecasts of the weather are looked for with eagerness all round our shores,' the review began. It continued with a paragraph that FitzRoy, 'the Scientific Sailor', would have relished:

> Among the unintelligent class Admiral FitzRoy may have met with the treatment they habitually accord to benefactors who disturb the equanimity of indolence, or torment the laziness of repose; but the scientific contemporary we have quoted speaks truly when it says: 'The science of which this distinguished man, in spite of sarcasm, is the apostle, is in its infancy. No one is more ready than he to admit this. It may at times deceive us – of this there is no doubt, but we may hope the time will come when it will speak with perfect surety. Even now, when Admiral FitzRoy hoists his alarm signal at the ports a storm is probable, and prudence commands small vessels or weak ones not to tempt the danger of the seas.'[30]

For the *London Intellectual Observer*, FitzRoy's monograph was a welcome addition to any scientific library. It urged his publishers, Longmans, to issue 'a cheap *Weather Book* without delay'. A less rosy tone was taken in another review, by the *Athenaeum*. After praising his 'zeal and energy', it went on the attack. The reviewer (FitzRoy might have guessed at his identity) thought his style 'rather rambling' and, more importantly, omitted 'to supply the fact which meteorologists most need. It is a fault in a book intended to lay the foundations of a new experimental science, that it should be mainly occupied with deductions from unproven hypothesis, instead of careful establishment of axioms by rigorous inductions from observed fact.'[31]

> There may be many among those who have not examined the weather tables published day by day in the journals, who may credit Admiral FitzRoy's statements, under the persuasion that his forecasts are generally just, and therefore give reliable testimony to the correctness of his theories. We do not share their persuasion, but advisedly take the exactly opposite opinion, that

his speculations are *prima facie* open to distrust, because we find
his weather-prophesies to be particularly unhappy.[32]

This was the latest in a series of critical pieces featured in the
Athenaeum, a publication which often carried articles written by the
London scientist and writer Francis Galton. Although the review was
unsigned, Galton's style was hard to miss.

From the Paris Observatory Le Verrier had been monitoring FitzRoy
closely. For three years the British had been issuing storm warnings
and forecasts while the French had remained happy to limit themselves
to reports and private forecasts, like one issued to the celebrated
photographer Felix Nadar before a balloon ascent. Over in America
Joseph Henry's meteorological work at the Smithsonian, once
progressing so well, had been halted by the Civil War, which was
raging just miles from Washington. By 1863 Britain under FitzRoy and
Holland under Buys Ballot were the leading nations in applied weather
science. It was a situation that Le Verrier intended to address.

A French storm-warning system had long been mooted, and it
finally came about in 1863 through the work of E. H. Marié Davy, a
professor from the Lycée Buonaparte who Le Verrier had recruited
in 1862. Clever, hard-working and somehow able to withstand working
at close quarters with Le Verrier, Marié Davy had begun a storm-
warning service in August 1863 like that operational in Britain. 'Night
and day', Marié Davy had worked alone on the telegraphed data. And
from the autumn of 1863 he started to issue storm warnings of his
own.

By the end of the year it seemed that Europe was following FitzRoy's
lead. With Le Verrier influential in Europe as the head of the *Bulletin
météorologique* there was every chance that others – Portugal, Spain,
Italy, Germany and Russia – would follow his lead. If so, it would be
an enormous vindication for FitzRoy, for his vision and the years of
sustained effort. Yet at home opposition was growing. Once again it
would be one of Erasmus Darwin's grandchildren who would bring
him trouble.

CHAPTER II

Endings

The October 1863 issue of the *Westminster Review* featured an appraisal of a recently published book by Francis Galton. Galton's book, *Meteorographica*, had caught the attention of the *Review* for its novel approach, using innovative printing techniques to chart meteorological statistics on a map. 'Mr Galton's work is as interesting as laborious and to a considerable extent a successful endeavour to show in what manner meteorological charts may be prepared for publication,' the *Review* noted. 'It is to be hoped that its production may fulfil the author's wishes by inducing the Meteorological Department of the Board of Trade to take this important matter into consideration.'[1]

At Parliament Street FitzRoy could hardly have failed to notice Galton's latest production. He and Galton moved in similar circles around London's clubland, the Royal Society, Royal Geographical Society and the Athenaeum, and for some time now Galton had been peddling his own meteorological theories – ideas which did not always chime with FitzRoy's. Forty-one years old, introverted, clever and with a healthy disregard for authority, Galton was just the kind of critic FitzRoy didn't need. To make matters worse, he was another of Erasmus' grandchildren and Charles Darwin's half-cousin. Not long before, FitzRoy and Galton had been on amicable terms. But by October 1863 the split between them was clear. A clash seemed imminent.

Of all Erasmus Darwin's grandchildren, Francis Galton had the sharpest intellect. Born in 1822 in Birmingham, as a child he had been a prodigy. As a one-year-old he had mastered capital letters, six months later he was comfortable with the Greek alphabet and by the age of two and a half he had finished reading his first book. At four he was already quite something:

I am four years old and can read any English book. I can say all
the Latin Substantives and Adjectives and active verbs besides 52
lines of Latin poetry. I can cast up any sum in addition and can
multiply by 2, 3, 4, 5, 6, 7, 8, 10. I can also say the pence table. I
read French a little and I know the Clock.[2]

At the age of eighteen Galton wrote, 'a passion for travel seized
me as if I had been a migratory bird' and he then spent years in the
wild lands of southern Africa, dodging hungry lions, shooting
elephants, white rhinos and giraffes. Galton returned to Britain as a
celebrated explorer in 1852, determined to make his mark as a scientist,
and perhaps felling a few more old beasts on the way. The Royal
Geographical Society, which awarded him its Gold Medal in 1853 for
his southern African expeditions, became his power base. In 1854 he
was added to the Geographical Society's council, alongside the ageing
Francis Beaufort, and it would have been there that he met FitzRoy
for the first time. Initially relations between them were friendly
enough, and in 1859 FitzRoy even signed Galton's nomination form
for his Royal Society election.[3]

By then Galton had established himself as one of the rising talents
of British science, albeit an idiosyncratic one. He generally worked
alone, on a wide range of inventions or problems that he tackled with
numbers and statistics. Often his investigations were whimsical. In
1859 he attempted to define a formula for the ultimate cup of tea.
Fashioning a research teapot fitted with a thermometer he tested the
following variables, n – Number of ounces of water used, e – excess
of its temperature above that of the teapot, t – additional temperature
attained by the pot after the water has been poured in, C – required
capacity. From careful research he devised the formulaic expression
for the perfect cup, $C + ne = (C + n)t$.[4] It was a commitment to
scientific truth that would have impressed Glaisher immensely. At
other times he calculated the sum total of all the known gold in the
world, and worked out that he could cram it into his living room
with ninety-four cubic feet to spare. Later in his career he would
produce an incisive essay entitled, 'Cutting a Round Cake on Scientific
Principles'. If Galton's philosophy were boiled down to a single motto,
it might be: 'Count, whenever you can.'

Galton thrived in the outdoors and perhaps his meteorological
awakening can be traced to one of his walking tours around Luchon

in the French Pyrenees in 1860. Keen to try out a new sheepskin sleeping bag, Galton abandoned the comforts of his hotel and the company of his wife and walked into the foothills, where he settled down for the night.

A heavy storm was gathering, but before the evening closed and before the storm broke, I had time to find a good place on a hill some 1,000 feet or more above Luchon, and there to await it inside my bag. Nothing could have been more theatrically grand. The thunder-clouds and the vivid lightning were just above me, accompanied by deluges of rain. Then they descended to my level, and the lightning crackled and crashed about, then all the turmoil sank below, leaving a starlit sky above.[5]

The immediacy of Galton's experience is reminiscent of Beaufort or Glaisher in the Irish hills, or FitzRoy in Tierra del Fuego. It was an aesthetic seduction that in each case foreran the desire to understand. And so it was with Galton who, in the year following his Royal Society election, with FitzRoy's pioneering weather warnings as a backdrop, plunged with his usual ferocity into meteorology.

Galton's initial complaint was that the data that appeared in neat columns in the meteorological tables of the newspaper presented 'no picture to the reader's mind'. He announced in the *Philosophical Magazine* in 1861 that he was working on a new system, 'not lists of dry figures, but actual charts which should record meteorological observations pictorially and graphically, without sacrificing detail'. What Galton had in mind was a weather map. Ever since Elias Loomis' first maps of the 1840s meteorologists had been drawing up similar charts. Glaisher had done so in private in 1849, Espy had included maps in his annual reports, the Smithsonian had projected the weather map on the wall and FitzRoy and Le Verrier had both made good use of charts in private. Galton's idea, then, was not an innovation in itself. Rather he wanted to bring this art of chart making out of the private sphere and into the newspapers, where the map could replace the data table. In his *Philosophical Magazine* piece, Galton sketched out his plans and included a specimen map, printed with movable type. The clouds were shaded, a concoction of horseshoe shapes denoted wind strength and direction, and temperatures were shown in figures.

Taken with his idea, Galton decided to model it with real data and he proposed, like Loomis before him, to study a narrow period of history in great detail. He settled on December 1861 – a month of extremes – and he would plot the month's meteorological activity over the whole of Europe. It was here that he first ran into difficulties with FitzRoy, who was less than supportive when Galton asked for the retrospective data. FitzRoy may well have felt territorial, taking Galton's new idea as criticism of his methods. Equally, having learnt from past experience, he might have been indisposed to aid another of the Darwin clan. Whatever the case, Galton found little encouragement from Parliament Street. Elsewhere, Galton also had mixed success. He had drawn up a letter to the leading observatories and meteorologists in Europe – written in English, French and German – asking for help. Buys Ballot in Holland had been useful, as had meteorologists in Belgium, but the French, Swiss, Danish, Swedish, Italians and Irish had mostly ignored him, meaning that he was forced to plough months of effort into sourcing weather data from individuals and newspapers. At length, though, Galton built up a stock of data for December 1861, which he plotted on his prototype charts. What he saw surprised him.

By now Redfield's conception of storm whirlwinds, refined over the years by Reid and Piddington and given the title cyclones, was an established part of weather theory. Now that Ferrel had accounted for the Coreolis Effect – the spinning force of the earth – it was clear why wind should spiral around a centre of low pressure, as Redfield had first observed. Indeed, others had shown that wind did not just circle round, but was drawn towards a central point where it was blown up an aerial chimney, as Espy had argued, like water swirling around a massive inverted basin. Galton was familiar with these theories, and as more and more weather data was collected he was able to view these cyclones sweeping over Europe.

He also saw something else that nobody had noticed before. On the map for 2 December 1861 Galton had plotted a weather system that was operating in completely the opposite manner to a cyclone. Instead of winds spiralling towards a centre of low pressure, they were fanning outward around a point of high pressure, forming, as Galton wrote, 'a sort of anticyclone'. Completely the reverse.

Galton wrote up his observation, 'A Development of the Theory of Cyclones',[6] at Christmas 1862 and sent it to the Royal Society, where it was studied warily. It was a stroke of inspired observation. Galton had defined the second great macro power at work in the atmosphere. Today cyclones and anticyclones are understood as complementary opposing forces that help to transfer air around a global circulatory system. While cyclones suck air inwards and blow it out of the top, anticyclones push air outwards and down towards the ground, where it can once again be recycled by cyclones in an endless toing and froing, passing the air back and forth. Galton was the first to see this grand vision of the atmosphere at work. In his paper to the Royal Society he explained his discovery with a mechanical simile: they were like two gears, he wrote, 'and make the movements of the entire system correlative and harmonious'. For all the years Redfield, Reid, Piddington and FitzRoy had struggled to establish the cyclonic theory, Galton had done the same for anticyclones in less than a year, completely independently.

Satisfied with his discovery Galton continued with his December weather charts. It was a long, tedious exercise of sourcing data, reducing it and drawing it on the map. But by October 1863 his efforts had borne fruit. He called his book *Meteorographica* – a fusion of meteorological data and graphs – and it included ninety-three charts showing the weather across the European continent on each day of December 1861 in the morning, afternoon and evening. Galton used symbols to show rain, snow, clear blue sky or the degree of cloudiness at a specific point as well as pressure, temperature and the direction of the wind. Two years in the making, it showed that weather maps could be brought to the masses in a lively, comprehensible manner. In an introduction to the charts Galton took the opportunity to swipe at FitzRoy, whose daily reports, he thought, were 'insufficiently numerous, extended, or frequent to afford a just knowledge of the winds even of the general kind'.[7]

This was the least that Galton was thinking. Over the past two years he had become deeply frustrated with FitzRoy. He suspected him of stockpiling weather data and not making good use of it. For Galton, looking at the charts confirmed to him how little was known. When he looked at one of Le Verrier's meteorological maps – he had started to issue them in 1863 – he was struck by the massive size of weather systems. Europe, north to south, east to west, was too small to contain the whole of one weather phase. He saw bits of cyclones, bits of anticyclones, clumps of cloud, bands of rain, but he constantly found himself wanting to look beyond the page – to see what was happening out in the Azores, or in the Baltic. Galton went on to contrast the 'narrow limits of what is really known [with] the audacious dogmatism too common among meteorologists'.[8] Few would have missed the point. Galton was writing about FitzRoy and his forecasts. How, with such a paucity of theory and such limited data, could FitzRoy possibly be predicting weather?

In January 1863 a succession of grumbling notes appeared in the letters page of *The Times* complaining about the accuracy of the forecasts. FitzRoy chose to respond in person and, far from being abrasive, he drafted a conciliatory, warm-hearted reply, penned in the persona of 'The Clerk of the Weather'. 'I need hardly repeat, Sir, what has been so often explained, that the "forecasts" are expressions of probabilities – and not dogmatic predictions.' The forecasts, he went on, were the only 'reliable and satisfactory' way of warning of storms and as time had gone on their accuracy had improved. 'It is "a race", as I said, to warn our outposts before the gale reaches them; yet this is now often done. It could not have been effected a year ago, as we had not then the *savoir faire*.' He finished off playfully:

> In conclusion of this enforced letter may I add that, notwith-standing the fair or unfair jokes or criticisms of those whose hats have been spoilt from umbrellas having been omitted, or whose views are reasonably opposed, 'as at present advertised', the Weather Clerk himself, after a life of practice, as well as study, values the 'forecasts' more and more as a scientific ground on which to hoist the drum.[9]

FitzRoy's letter was well received, and even sparked a defensive riposte in *Punch*, not the typical ally of a governmental employee under attack. But the larger picture was more complex. The winter of 1862 to 1863 had been difficult for the Meteorological Department. The storm warnings and forecasts could no longer survive on novelty alone. Most troubling to scientists was FitzRoy's method, or lack of it. Although he had published his *Weather Book*, many people were still not exactly sure how each forecast was formulated. Were they done on an individual basis? Or had he managed to devise a standardised, repeatable process?

There was still no consensus among meteorologists about theory.

The country remained full of weather theorists of every stamp: lunarists, Espians, astro-meteorologists and cyclonists. With no fail-safe way of proving any particular theory, people were free to cherish their old pet ideas, safe in the knowledge that they would be very difficult to overturn. FitzRoy was left to navigate a route, as tricky and rock strewn as the Strait of Magellan, between the different schools of thought. As his *Weather Book* showed, he subscribed to Dove's views and believed in Redfield and Reid's theories but beyond that he was elusive in his methods and, as he had in *The Times*, fell back on his somewhat nebulous 'life of practice' when cornered.

A typical attack came from Mr W. H. White, the London correspondent of the *Park Lane Express*. To White, FitzRoy was the high king of weather prophets, surpassing the lunarists and the planetary meteorologists because of his position at the top of the hierarchy, 'backed up by Government support and high patronage throughout the length of the land'. Though White conceded that FitzRoy had saved the life of many 'brave seamen' and the loss of much property, his achievements were diluted by the fact that he had not 'the remotest hint' of a transparent method. 'Whenever it shall please the God of storms to remove the Admiral from among men,' he wrote, 'the theory of storms will die with him, and therefore man will still be as ignorant as before the Admiral commenced his career.'[10]

> This is a grievous state of things, but it is nonetheless true. Many of the storms which were predicted by the Admiral did not come at all, while others came unpredicted; facts which clearly prove that no *theory* can be attached to such a course of prediction, notwithstanding that his system has produced many important results – results that arose from an unknown, and, consequentially, an unforeseen cause.[11]

White wasn't alone. One of FitzRoy's most outspoken and longest-standing critics was a pamphleteer, 'B'. 'B' was one of a number of committed Espians who refused to believe in the idea of cyclonic storms. He accused FitzRoy of succumbing to 'cyclonic mania'. Paper after paper flowed from 'B's' pen, scorning FitzRoy. In particular he made good play of one of Espy's old quotes, 'I fear Captain FitzRoy will not do much for the science of meteorology, he is deficient in the elements. On the subject of vapour in the air, he is entirely *befogged.'*[12] 'B' often found an excuse to hash up this old sentence when complaining about FitzRoy. Those who had followed the development of meteorology over the years knew that Espy had written scathingly about almost everyone who did not share his views, from Herschel to Reid, but the general public saw the line as further evidence that FitzRoy was a hapless and blundering amateur, in contrast with the suave theoretical champion. The attacks, from 'B' and from White, neatly portray FitzRoy's dilemma. If he eluded theory, he was criticised. If he championed a theory, others took issue.

FitzRoy had long understood this. In the early 1860s he had attempted to mollify his critics by launching into theory himself. His one major foray into atmospheric physics came in the eighteenth chapter of *The Weather Book*, and involved a concept he called lunisolar action. FitzRoy's idea was unmistakably that of a seaman who had spent his working life in contact with oceanic tides. His idea was to apply the same principle to the atmosphere. 'The moon also acts on every particle of air, as on water and earth, by universal gravitation,'[13] he wrote in *The Weather Book*. 'Tidal effects must therefore be caused by the moon and the sun, in earth's atmosphere, and their scales may be *large* in proportion to its depth and extreme mobility.' Searching for his universal law, FitzRoy had travelled back to the seventeenth century and found inspiration in the precise mathematics of Newton. It was an ambitious idea that left him with a vivid picture of the invisible atmosphere:

> As the world turns on its axis, successive waves are raised, and drawn on; so that before the effect of one is lost another advances, and the result is *continual* motion: constant *elevation* of atmosphere being avoided by overflow towards each side, in *augmentation* of the solar heat action.[14]

To win support, he turned to Sir John Herschel. FitzRoy had now known Herschel for twenty-five years. Still the pre-eminent scientist of the day, Herschel had remained a cheerful correspondent throughout FitzRoy's time in the Meteorological Department. His ideas on lunisolar action coming together, FitzRoy had bombarded Herschel with letters on the subject through late 1862 and early 1863. On Christmas Eve 1862 he wrote, 'May I beg you to pass over all the earlier chapters of my weather book and dip into the 18th? I cannot dismiss the idea that there is something more than nonsense in the new part till you put down your foot upon it (as Pres Lincoln is said to do on the trouble) and entirely condemn such notions.'[15] Then on 10 March 1863: 'Having added a short paper – in reference partly to my 18th chapter I ventured to turn a spare proof of it – and to say that there is the revised second edition shall be submitted to you in a few days.' And six days later: 'Were you not yourself alone – I should hesitate to trouble you again with my conglomeration of ideas about lunisolar action. But as I feel that my manuscript (sent on the 10th to 12th) expresses my view very ill – I now beg you to burn it and allow the enclosed paper to take its place.'[16]

All FitzRoy's earnestness would be to no avail. His ideas were too tainted by lunarism for Herschel to have anything to do with them; indeed from the start Herschel saw them as a heap of clotted nonsense. He wrote back, telling FitzRoy as much. FitzRoy was many things: industrious, clever, pioneering, honest, practical, innovative. A talented administrator in his own fashion with powerful humanitarian instincts. But he was certainly not a theoretical scientist. FitzRoy took Herschel's verdict badly but at least he accepted it. 'Thank you kindly for your opinion,' he replied. 'It will save me from exposure by a paper which I thought of writing for the Royal Society. It is too late to alter my book. – Having appealed to Caesar – I am content. To no other authority am I disposed to go.'[17]

His hopes for a new theory dashed, FitzRoy was left to concentrate on the practical operation of the forecasts and storm warnings. Over the last years the parliamentary vote for the department had jumped sharply to cover the cost of telegraphy, from £3,107 in 1861 to £5,325 in 1862, and up to £7,104 in 1864 – more than half a million pounds in today's money. For FitzRoy the country was amply repaid in the lives saved, but others were becoming uncomfortable with the arrangement. Back in 1854 the Meteorological Department had been founded on the premise that its winds charts would help to speed British trade ships around the oceans, bringing a net profit to the country. With less and less time being spent on the charts, and more time put on the warnings and forecasts – which had no obvious financial return – the government was left in the curious position of losing money from what had been planned as a simple, profit-led initiative. Among those to spot this discrepancy was the Liberal MP for Truro, Augustus

Smith. A free-trade enthusiast, Smith had a businessman's eye for a balance sheet that FitzRoy never possessed. After the budget increase of 1864 Smith became a belligerent foe on the green benches. In May 1864 he argued that FitzRoy's forecasts were getting poorer by the day, that meteorology would be better conducted by Airy and Glaisher at Greenwich, and the French were operating a 'much more scientific establishment'.

Replying to Smith in *The Times*, FitzRoy was not as civil as he had once been. In a provocative letter to the editor, John Delane, he pointed out that there had been a statistical improvement in recent forecasts, that the Meteorological Department had been founded only because Airy was unprepared to deal with the work, and that the French service was actually modelled on the British one, making Smith's claim that the French one was better seem rather odd. Obviously nettled and having noticed that Smith held the lease to the harbours in the Scilly Isles, FitzRoy rounded off in waspish style:

> I am told that since our notices of weather have been published the number of disabled vessels has diminished about the west of England, and that the excellent harbours of the Scilly Islands have been much less frequented by vessels in distress.[18]

The inference was clear. Smith struck back furiously the following day, refuting FitzRoy's 'insinuation', something he thought 'congenial to the Admiral's habits of thought and action, judging others naturally by his own gauge in such matters'.[19] There the spat ended, for now.

Outside the Westminster bubble, FitzRoy's reputation had become that of a mischievous celebrity, as if he was involved in some devilish game. It was the start of our attitude to weather forecasters today that lies somewhere between familial fondness and deep suspicion. Back in the 1860s a sense of novelty could be added to this mix. In October 1863 FitzRoy appeared as the central character in a poem by W.G. Herdman of Liverpool. Herdman's poem, '*Aeolus Redivivus*', opened in dreamy style:

> I wish I was Admiral FitzRoy,
> Who up in the clouds calmly sits – high

Ordering here, and ordering there
The wind and the weather – foul or fair[20]

Herdman's poem wound playfully on, transporting FitzRoy out of Parliament Street into the sky, where he toyed with the atmosphere below like a child with his lead soldiers. For the many outside the scientific and political community, this was how he was perceived. A dabbler, a dreamer, a schemer, FitzRoy was an old Jack Tar who had washed up in Whitehall to begin his meteorological game, playing God.

Back in Onslow Square this inference could have only perplexed FitzRoy. Into the 1860s he had sustained his ultra-conservative, evangelical interpretation of the Bible. Whether this was born of conviction or whether, after Mary's death, he had felt an emotional duty to persevere in the faith they had shared is uncertain. But in the decade since her death in 1852, and particularly since Darwin's *Origin* was published in 1859, the world they had known together, a world regulated by Christian certainty, had started to fracture.

For years FitzRoy had relied on the old orthodoxy that through scientific investigation a person could experience spiritual fulfilment. It was a pact that allowed them to combine religious piety and scientific curiosity without letting one overshadow the other. It had worked for Edgeworth, for Beaufort, for Reid, and for a long time it worked for FitzRoy. Writing of Glaisher's balloon ascent in *The Weather Book*, FitzRoy's mind had wandered off track. He penned a short religious meditation:

All-pervading, ubiquitous, incomprehensible, *now* to man, almost infinite in power, rapidity, and extent, though but an agent of the Almighty. Marvellous, indeed, are the effects of this subordinate influence as studied in these forms, and in other combinations, such as magnetism and gravitation. They indicate the power of Divine will *looming* through the mist of man's materialistic philosophy.[21]

This glimpsing 'through the mist' epitomised the old attitude: nature was the prism through which God's majesty and mystery could be experienced. Yet since the publication of *Origin* in 1859, this perspective had been called into question. Darwin's sensational demonstration that nature was largely Godless heralded for each scientist a personal dilemma. In private Darwin wrote often to his friends Lyell and Hooker, wondering how far people were prepared to go with him. Everyone differed. John Herschel, John Stuart Mill and William Whewell, three of the major philosophers of the day, were so perplexed by Darwin's argument that they wrote hardly anything in public about it – though Herschel dubbed the theory of natural selection the 'law of higgledy-piggledy'[22] in private. Charles Kingsley, professor of modern history at Cambridge and Queen Victoria's chaplain, was more transparent about the effect of the book on him. He wrote to Darwin telling him, 'that if you be right, I must give up much of what I have believed & written'.[23] Bolder still was Francis Galton, who found liberation in his cousin's argument. He would later write in his autobiography that *Origin* began 'a marked epoch in my own mental development, as it did in that of human thought generally. Its effect was to demolish a multitude of dogmatic barriers by a single stroke, and to arouse a spirit of rebellion against all ancient authorities whose positive and unauthenticated statements were contradicted by modern science.'[24]

In contrast FitzRoy seemed anchored in the past. His religious philosophy was immovable: conservative and literal. Yet his meteorological work was just the opposite: innovative, probing, experimental. On one hand he was advocating not just the study, but the prediction of weather; on the other he was aligning himself with the belief that weather was part of Providence.

God's control over the weather remained a live issue. In 1860 Bishop Wilberforce had ordered his clergy to read a prayer for dry and sunny weather before the harvest. Prayers against cholera, the Crimean War and the Indian Mutiny had been called for in the 1840s and 1850s. To call for divine intervention in weather, though, was a step further. It may have been a defensive ploy by Wilberforce. Whatever the case, it was controversial. Still mulling the implications of *Origin*, Charles Kingsley responded sharply. He composed a sermon accusing Wilberforce of wilfully misinterpreting the Bible and God's influence over natural events:

> Shall I presume, because it has been raining here too long, to ask God to alter the tides of the ocean, the form of the continents, the pace at which the earth turns around, the force,

the light, the speed of the sun and the moon? For all this, and
no less, shall I ask, if I ask Him to alter the skies, even for a
single day.[25]

The spat was noticed by Galton who, in private, was already
contemplating another controversial project. Never one to quake in
the shadows of old institutions, Galton had turned his mind to the
practice of prayer. The question was troubling him. Did prayers
work? In his usual methodical way he began to tackle the issue
statistically. He started his experiment with a hypothesis: 'We are
encouraged to ask special blessings, both spiritual and temporal, in
hopes that thus, and thus only, we may obtain them,'[26] taken from
Smith's Dictionary of the Bible. He then devised a way of testing the
claim. He found a copy of the *Journal of the Statistical Society*, which
listed the mean life expectancy of kings and queens with those of
other classes of people. Galton pointed out that at every church
service, Protestant or Catholic, it was customary to pray for the
sovereign: 'Grant him/her in health long to live'. If prayer worked,
Galton argued, specifically such targeted and constant prayer as this
should result in longer lives for kings or queens. But according to
the *Journal of the Statistical Society*, this was not the case. A member
of the royal house lived an average of 64.04 years, while clergy,
lawyers, medical doctors and the aristocracy lived much closer to 70
years. 'The sovereigns are literally the shortest lived of all who have
the advantage of affluence,' Galton summed up. 'The prayer has
therefore no efficacy.'[27]

Galton's paper would not see the light of day until it appeared with
predictable controversy in the *Fortnightly Review* in 1872. But its very
existence was significant. If he had written such a paper three hundred
years before he would have been burnt; two hundred years before he
would have been thrown into prison, or a hundred years before into
a lunatic asylum. Yet by the 1860s such questions about the power
and integrity of religion had found their place in contemporary debate.
Galton was only writing what many were already thinking.

In this same climate, FitzRoy was left with the question: who
governed the weather? Was it God or was it science? Once again
FitzRoy was to find himself caught in the vice. On one side he was
the pioneering scientist looking into the future; on the other the
righteous Christian, anchored where Mary had left him in 1852, shifting

through the data that streamed into the office day after day, coining theories and then, on a Sunday, going to church to pray for clement weather. It was FitzRoy's peculiar paradox. He was like a growing boy in a suit of armour.

Throughout 1863 Glaisher continued with his balloon ascents. They came with increasing regularity and were seized upon by the press as vibrant copy. Mr Glaisher's Ninth Balloon Ascent, Mr Glaisher's Thirteenth Balloon Ascent, Mr Glaisher's Fifteenth Balloon Ascent. There seemed no end to his aerial ambitions, but for some the novelty was wearing off. In June 1863 *Punch* published the first of its parodies. The craze for ballooning, it pointed out, was no better than the newfangled craze for underground railways. Anticipating a union of these two novelties, *Punch* revealed, 'The proposed plan for an Underground Balloon Railway is still under consideration.' On 17 October an entire feature, *The Sitific Count of MESSERGLAISHERAROXWELL*, appeared. It was a terrific put-down. The imaginary ascent was launched from 'under an arch in the gas works of the British Association', but soon the basket had risen high enough for Glaisher to commence his measurements:

> The Balloon was 200005 feet above HM 59. 3000, 1 h ground at 2p.m. The temperature of the air was 0000000000000000, &c, and Mr Coxwell decreased to two and a half when he varied and declined to snow.
>
> The view at this point was like huge swans harmoniously grouped. On the plain the trees moved with great rapidity, and after feeling MR COXWELL's bumps, we avoided a farm house and bounded on the light earth. It was most painful on opening up my packages to see the *débris* of MR COXWELL quite uninjured. As for ourselves we had several bruises about the size of the equinox.
>
> We descended at Temple Bar, six miles NW of —— Blackburn Esq, and our best thanks are due to the Balloon, who, in the kindest hospitality, sent his carriage to meet us at the station.[28]

As with all the best satire, *Punch* was making a serious point and the undertone of the feature was clear enough: what on earth was the point of all this ballooning? Now Glaisher had reached the roof of the world, why go on? Why bombard the public with such regular volleys of indecipherable statistics?

In fact Glaisher had sound scientific motivations for carrying on with his ascents. On one ascent in the south of England he had been able to study 'the curious' formation of clouds in the sky above the River Thames. 'The clouds followed the river in its course through all its windings, not departing from it', suggesting the power of convection. He had also gathered much more data on the changing temperature at altitude, as well as the number of conflicting air currents in a single place. Furthermore, Glaisher's ascents had yielded some interesting biological research into the effect of 'rarefied air' (with low concentrations of oxygen) on the human body. His resting heart rate of 76 beats per minute would jump to 90 beats per minute at 10,000 feet, and 110 beats per minute at higher elevations. At 17,000 feet his lips would turn blue, while at 19,000 feet both his hands and feet were dark blue. 'At four miles high,' he wrote, 'the pulsations of my heart were audible, and my breathing very much affected.'[29]

This information had real scientific value, providing some of the earliest data into the study of altitude sickness. But making the public understand this was a different matter. For the cynical Glaisher's exploits seemed to be frivolity at best and vanity at worst. It was as if he had become a society bore, charmed by his new hobby and desperate to share his experiences with anyone who would listen. When the news leaked out that Glaisher was to take a dog and a pair of rabbits on his next ascent, *Punch* went to town. It published a letter, supposedly from Glaisher's 'aeronautical dog', an erudite canine called 'The *Sky* Terrier'. Told from the dog's perspective the ascent was a chaotic affair. From the off the Sky Terrier was consumed by a desire to eat the rabbits, 'a pair of dullards', and Glaisher and Coxwell had to struggle frantically to keep them apart. Soon the Sky Terrier began his own observations:

At 2h. 45m. by GLAISHER's watch, I growled.

At 2h. 46m. by the same COXWELL kicked me.

At 3h. 31m. I tried to bite GLAISHER's calf, but couldn't comfortably.

At 3h. 36m. I thought I saw the moon and howled.

At 3h 37m. 'Kick him!' says COXWELL kindly. GLAISHER's an old man, but he is also a wise one, and he refrained. 'No' says he.

'Let dogs delight to bark and bite. And scratch and tear and Howl. Let bears and lions dance and fight –' 'But don't let that dog growl' says COXWELL, finishing the verse offhand.[30]

Ridicule is a dangerous thing for an idea or an experiment. It can stop it dead: transform support into derision. And if Glaisher had become one staple *Punch* personality along with Gladstone, Disraeli, and the outrageously mustachioed Italian, then FitzRoy, the 'clerk of the weather', had become another:

We have lately heard, at the beginning of June, several people speaking of May as 'the last month;' we applied to ADMIRAL FITZROY who immediately pitched out his barometer, hoisted his cone, beat his drum, blew his own trumpet, and then telegraphed off to us that May was not the last month; that we're in another month now, and there'll be plenty more up to the end of the year.[31]

It was a light-hearted quip that did nothing to hide a depressing truth. FitzRoy was floundering and a recent leader article in *The Times* suggested as much. The piece was a response to FitzRoy's annual report, which had claimed an increasing hit rate for the forecasts, but *The Times* was not so sure. It began by parroting Arago's famous cautionary quote, warning against forecasting, 'whatever may be the progress of science'. Then it cast doubt on the science itself, claiming that even now the best signs of coming weather were not those of Dove or Reid but the ancient signs of nature: the halo round the moon, the braying of asses, the cackling of ducks. That better signs than this exist, the leader continued, 'Admiral FitzRoy has still to convince the public and at this task he labours yearly with much praiseworthy assiduity'.[32]

Reading the piece, detecting the faint praise, FitzRoy must have known what was coming. Often the warnings are late, or the wind blows from the wrong quarter, 'but there can be no doubt that when Admiral FitzRoy telegraphs something or other is pretty sure to happen'. There were improvements to be made, too, in FitzRoy's written communication. Even for the most educated of readers many of his explanatory sentences were indecipherable. *The Times* leader picked out several offending passages, one – 'It is the prescience of the dynamical consequences arising out of statistical facts that enables a really scientific calculation of probabilities to be made' – it held up as an example, adding that it was 'lucid and grammatical with some that we could quote'. More laughable still, it went on, was FitzRoy's impenetrable simile: 'Facts are as the ground; telegraph wires are roots; a central office is the trunk; forecasts are the branches; and cautionary signals are as fruits of this youngest tree of knowledge.'[33] For the writer this image was nebulous and confused, almost a perfect microcosm of FitzRoy's Meteorological Department itself: grand, ambitious, overstated.

There was no reply from FitzRoy in the letters pages this time. *The Times* with its enormous circulation was among the leading voices in British intellectual life. To have his project, his writing style, his ideas and his ambitions so derided was a humiliation.

The early optimism of his work had faded. Over the autumn of 1864 the Board of Trade took an axe to his budget, stripping away funding for eight of his most vital telegraph stations, on the south coast, Wales and Scotland. Already starved of information, FitzRoy was having to issue forecasts on less data than ever. Writing a feature on the Meteorological Department, the *United Service Magazine* began on a plaintive note, quoting John Bunyan, 'From the very gate of happiness there is a short bye-road to the gulf of perplexity and chaos.'[34]

In late 1864 FitzRoy was selected for inclusion in the third volume of an anthology of biographical sketches – *Portraits of Men of Eminence in Literature, Science and Art*. The book tapped into the growing Victorian taste for biography but its real attraction lay in the accompanying albumen photographs of each of the featured individuals. The prints were by the society photographer Ernest Edwards and were intended as formal portraits. So far in the series Edwards had taken photographs of Faraday, Huxley and Thackeray, and this third edition would feature Sir George Back, the famous Arctic explorer and Richard Burton, the celebrated author, linguist and swashbuckling swordsman who had made his celebrated hike to Mecca disguised as a Muslim.

Drawing such a vibrant cast of characters together was an achievement. For FitzRoy being set alongside the august men of the age must have been satisfying. Edwards' photograph of him, though, is unsettling. Taken either in late 1864 or in early 1865 it is a formal, full-length portrait. FitzRoy sits bolt upright in a chair, arms folded across his chest, a picture of defiance. Only a few years had passed since another photograph of FitzRoy – an albumen print taken by the London Stereoscopic & Photographic Company – depicted him as the dashing captain of old: one arm resting jauntily on a table, in a greatcoat with a necktie, and a chequerboard waistcoat. In the earlier portrait his face was full; he stares dreamily off into the distance as if to show off his side whiskers to best effect. Now, just a few years later, he looks entirely different. His face is drawn and haunted. His eyes are glazed. The hair has completely gone from the top of his head, leaving a cold, white pate. Sitting in a spartan room, FitzRoy's unease is unsettling. He looks torn between loss and anxiety, resignation and panic.

At fifty-nine years old FitzRoy's body was failing him. Once so energetic he could race up and down mountains in Tierra del Fuego, now old age had come to him suddenly. His health had begun to deteriorate when he was working on *The Weather Book*. Written outside office hours, often long into the night, it had begun to test FitzRoy's natural stamina. To his alarm he had started to fall asleep after only a few minutes' reading. 'In vain he struggled against this propensity, trying every possible means to overcome it, but without avail.'[35] In August 1864 he had been forced to take a period of rest. One of his friends noted that he was 'looking much broken and thin'. His hearing was going and, worst of all, his eyesight. For years he had complained to Herschel about 'these days of eye-destroying small print'.[36] Now he was struggling to read at all. The problem was borne out in his handwriting. Once that of the jaunty naval officer, with elegant Georgian loops, dashes and frills, it had become large, almost comic in execution. Some of his letters appear frantic. You can sense his pen pressing hard into the paper.

The *Portraits of Men of Eminence* was not quite as lavishly produced as it first seemed. While all of Edwards' albumen prints were original and carefully executed, the same could not be said of the accompanying prose sketches. In FitzRoy's case, at least, he was asked to supply an account of his own life in the third person. Busy and harassed with work he pulled out an old biographical sketch – the one written a decade before, in March 1852 – and added to it. He scrapped much of the detail of his early career at sea and reshaped it to include his meteorological work. He described the establishment of the Meteorological Department 'as an experiment', and how it grew from 'collecting, tabulating, and deducing results from meteorological observations made at sea' into 'the telegraphic system of Weather-warning or forecasting'. FitzRoy rounded off this short autobiographical sketch:

> He has laboured to make the principal facts and conclusions of this somewhat complicated subject evident to all earnest inquirers. How far he has succeeded is shown by the great increase of interest taken by all classes, but especially the seafaring or maritime, in all that relates to the weather.
>
> It is difficult to estimate the lives saved, and the property, on the occasion of each successive storm on our coasts, of which warnings have been duly given by the signal drum, if not also by the forecasts published, daily, in most newspapers; but we find that a general confidence is expressed around the coasts of England, and those adjacent, in the successful results of these cautionary notices.[37]

Here the sketch ends and FitzRoy's voice quietens. It would be the last public word of his life. His friends were already anxious. They knew that he had no money. They worried that he was working too hard and that he could be easily shaken by criticism. A figure of fun in *Punch* was one thing, quite another was *The Times* rounding on him. It was a cruel about-turn for a paper that had once encouraged him.

On 27 March 1865 FitzRoy attended a meeting of the Royal Geographical Society where he was asked for his opinion on a forthcoming Arctic exploration. He reportedly left in a bad mood, certain he had been slighted by the president, Sir Roderick Murchison. Worn out, FitzRoy took a period of rest, leaving Babington in charge at Parliament Street. To get away from the mill and bustle of central London and his continual round of social and professional obligations, he moved with his wife and their children to a three-storey Georgian house at 140 Church Road in Upper Norwood. Norwood lay just to the south of London, 'a most salubrious spot, surrounded with beautiful views of hill and dale'. FitzRoy's energy, once so formidable, seemed all but gone.

For the next few weeks FitzRoy avoided everything to do with meteorology. He kept to his bed. It was the last week of April before he started to brighten. On Sunday 23 April he walked for two hundred yards down Church Street to All Saints Church, where he took communion. During that weekend he was well enough to take a drive with a friend and sit in the garden while his daughter Laura played croquet. All the while he seemed to be gathering spirit for his return. In the week that followed he made three trips up to London, on the Croydon Railway to Charing Cross, once to visit Babington at Parliament Street. Every time he returned, shattered.

By now this pattern was well established. A period of rest would lead to a brightening of spirits, but as soon as he was exposed to any sort of mental stress he would crumble. Maria FitzRoy was aware of the pattern. By now they had been married ten years, and throughout his breakdown she had nursed him, keeping a watchful eye on his behaviour as the spring days lengthened.

It was nearly May, a quiet time for storms, a perfect time to relax. But relaxation had never come easily to FitzRoy. Some time during that last week of April FitzRoy read about the assassination of Abraham Lincoln at Ford's Theater in Washington, DC. The news seemed 'to absorb his whole being'.[38] Days later he heard that Matthew Maury was visiting London. Whatever his earlier misgivings, FitzRoy had come to regard Maury as a friend. He had followed reports of the Civil War and had wondered how Maury was faring, fighting for the Confederate States. Now they had been defeated Maury's future seemed bleak, and FitzRoy was suddenly overcome with concern. With presidents being shot, what might the future hold for Maury? The terror gripped him. 'Think of poor Maury without a home; his wife and children away, he knows not where.'[39]

Private Journal of Maria FitzRoy

[Thursday 27 April]

Just before going to bed he received a letter from Mr Tremlett inviting him and myself to come and stay with him from Sunday till Monday to see the last of Capt. Maury. This note seemed completely to upset him, between desire to comply with his request and his just expressed wish of remaining quiet. Of course he did not sleep well that night; the only advice I gave him was to do that which would give his mind the greatest ease.[40]

FitzRoy took the train into London that Friday morning, visited the Meteorological Department, and returned, telling Maria that he had written to Tremlett refusing the invitation. Getting a grip on affairs, he decided to spend the afternoon writing. Soon after he began work he called Maria – 'distressed at the quantity of unanswered notes'. She comforted him, and together they replied to the 'two or three most pressing'.

The idea of Maury's visit was still bothering FitzRoy on Saturday morning. He told Maria that he had a 'strong desire to see Maury again'. Maria said he had 'better gratify it if he had' but FitzRoy replied that he was completely 'incapable of exertion, and could only lie down and rest'.

After lunch on Saturday FitzRoy told Maria that he thought a walk would do him good, and he set off out with his two daughters. Maria left separately, for a drive in the carriage. When she returned to Church Road she found out that FitzRoy had left his daughters and caught a train to London to meet Maury.

He did not return home until eight o'clock that night. When he did, Maria saw that he was 'worn out with fatigue and excitement and in a worse state of nervous restlessness than I had seen him since we left London [for Norwood]. He seemed totally unable to collect his ideas or thoughts, or give any coherent answer, or to make any coherent reply.'

Private Journal of Maria FitzRoy

[Saturday 29 April, evening]

He generally went to sleep after dinner for a little time, but this evening he did not even close his eyes for an instant. When the girls had gone to bed he said to me he wished to talk over with me about his idea of going to London on Sunday to see Maury once again. I asked him if he had not wished him good-bye; he said he had. I then said I was very tired, and the best thing we could both do was to go to bed, and talk [it] over the following morning.

He agreed with me, saying how worn and tired I looked. I went away and then went downstairs again as he had not come up to his dressing-room. He thanked me for coming to see after him, and said he was coming up directly; he was standing up by the table, with the newspaper open before him, and was not long in coming to his room.

I was in bed when he came to bed; he came round to the side where I was, asked me if I was comfortable, kissed me, wished me good night, and then got into bed. It was just 12 o'clock. I was soon asleep, and when I woke in the morning I said I hoped he had slept better, as he had been so very quiet. He said he had slept he believed, but not refreshingly; he complained of the light, and I said we must contrive something to keep it out. Just then it struck six. From 6 to 7 neither of us spoke, being both half asleep I believe.

Soon after the clock struck 7 he asked if the maid was not late in calling us. I said it was Sunday, and she generally was later, as there was no hurry for breakfast on account of the train at 10 o'clock as there was on other days. The maid called us at ½ past seven. He got out of bed before I did. I can't tell exactly what time, but it must have been about ¼ to eight.

He got up before I did and went to his dressing room kissing Laura as he passed and did *not* lock the door of his dressing-room at first.[41]

The news broke the next day. *The Times*, *Morning Post*, *Herald* and *Pall Mall Gazette* all got hold of it. It travelled the length of the country on the telegraph wires.

SUICIDE OF ADMIRAL ROBERT FITZROY

A sensation which is almost impossible to describe has been caused through the whole of Norwood district, and, when this notice becomes public, will be shared with great pain by the whole of the scientific world. Admiral Robert FitzRoy, the great prognosticator of the weather has destroyed his life by cutting his throat with a razor.[42]

CHAPTER 12

Truth Telling

The news fanned out in dreadful ripples. The botanist Joseph Dalton Hooker wrote immediately to his great friend: 'My dear Darwin, We feel very much the shock that poor FitzRoy's death must be to you . . . Poor old FitzRoy – I am very sorry – for though I did not know him much I always regarded him in joint association with you & I did admire his Scientific pluck, as a Meteorologist, & his wonderfull kindness & goodness.' Hooker added a rushed postscript. 'I hope to heavens that they will not appoint Glaisher to the post, or Maury, or any of those cattle, who seem to live on self glorification.'[1]

Darwin replied two days later.

May 4th, Down

My dear Hooker,

I was astounded at news about FitzRoy; but I ought not to have been, for I remember once thinking it likely; poor fellow his mind was quite out of balance once during our voyage. I never knew in my life so mixed a character. Always much to love & I once loved him sincerely; but so bad a temper & so given to take offence, that I gradually quite lost my love & wished only to keep out of contact with him. Twice he quarrelled bitterly with me, without any just provocation on my part. But certainly there was much noble & exalted in his character. Poor fellow his career is sadly closed. You know he was nephew to L^d. Castlereagh, & very like him in manners & appearance.

Very much thanks to you for writing to me about FitzRoy. – Poor fellow how kind he was to me at first during the voyage.[2]

Even before Darwin wrote this first posthumous appraisal of FitzRoy's character the official inquest into his death was done and dusted. It had been a grim affair. Those who had seen FitzRoy in his last weeks and days had filed into the White Hart tavern in Upper Norwood. There they told similar stories, tales of an exhausted man desperate to return to work but lacking the stamina to do so. He had seemed panicked by a letter from Le Verrier. He told his doctor that once he could have replied in a quarter of an hour, but now it would have taken twenty-four hours. Struggling to sleep, FitzRoy had turned to medication. A few weeks earlier, he had swallowed an opium pill that had almost killed him. When he asked his doctor, Dr Hatley, whether he might return to work, Hatley had cautioned against it. 'I replied that such a condition of the brain would ensue that paralysis would follow. FitzRoy seemed to appreciate the opinion fully and he said, "I am indeed grateful to you. It may have saved my life."' Just a few days afterwards he was gone.

The verdict was: 'That the deceased Admiral FitzRoy came by his death by his own act and that at the time of the said act he was in an unsound state of mind.'[3]

It was too much for Maria FitzRoy; she stayed away from the inquest and her evidence was never heard. Suicide was difficult to comprehend. For believers there was nothing worse: it was cheating God, snuffing out life before the time had come. It seemed to fulfil a terrible destiny, FitzRoy's bad blood bringing him down just as Castlereagh's had before him. Most tragic of all was the fact that FitzRoy, that most Christian of men, could not be buried in consecrated ground. Having written so movingly and so piously following Mary's death in 1852, 'May God grant that her children and her bereaved husband may so profit by her example that they may all join her after and that not one head may be missing of those she so tenderly loved,'[4] he had destroyed any prospect of a reunion in the afterlife. Was his suicide a final declaration of atheism? It was a shattering thought. The cruelty of his fate held no bounds.

Heartbroken, Maria FitzRoy withdrew from public view. No replies were made to letters from the Board of Trade or Admiralty asking for details about his funeral. His old friend Bartholomew Sulivan, who rushed to London from Cornwall on hearing the news, only found out about the plans when he bumped into one of Maria's brothers at Parliament Street. He was busy collecting FitzRoy's personal papers and he told Sulivan that in the circumstances only family were invited. Sulivan nonetheless turned up, on Saturday 7 May, 'early on' at All Saints Church, Upper Norwood. There was a scattering of others there, including Babington and an admiral of the French Navy. 'They thought so highly of him,' Sulivan wrote in a note to Darwin the next day:

> It was a very quiet and plain funeral, just what I think all funerals should be. Poor Mrs FitzRoy would go, & the two daughters were with her. We all waited outside and walked after her carriage – & the same back the brothers only going into the house. It was a trying scene at the grave. Poor Mrs F and the girls looked dreadfully ill, & Mrs F gave way very much. The coffin was plain black wood with 'Robert FitzRoy, born —— — died —— —' on a brass plate.[5]

To satisfy the church and the family, they had come to an arrangement to bury his body just beyond the churchyard gates.

Significantly, one of the first organisations to pay their respects was the Royal National Lifeboat Institution at a meeting on 4 May 1865. The president, Earl Percy MP, gave a stirring eulogy to FitzRoy for his meteorological work, proclaiming that 'the National Life Boat Institution and science had lost a sincere friend in the gallant Admiral'. He ordered that a message of condolence be forwarded to Maria. A week later she replied in a short, heartfelt note. 'My noble husband sacrificed his life far more than the man who loses it on the field of battle, or the deck of a man-of-war, hotly contending with a foreign foe – more even than those brave men of whom England is so justly proud who man the life-boat to rescue their fellow creature – for he continually periled his life.'

When this correspondence was published in the newspapers it so moved a Miss Hannah Harvey from Cheltenham that she instantly donated £600 to the Royal National Lifeboat Institution for a new boat at Anstruther in Fife. Her only stipulation was that it was to be called the *Admiral FitzRoy*. That Hannah Harvey had been so touched by FitzRoy's story was not surprising. After news of his suicide broke, he became something of a morality story. An obituary in the *Gentleman's Magazine* dwelt on FitzRoy's inability to delegate and the constant strain of his responsibilities. A weatherman's life could never be totally 'quiescent'. 'The whistle of the coming breeze, the rattling of the windows, the pelting of rain, lightning, thunder and sudden change . . . all tend to keep up an excitement not to be understood by others than the worker in Observatories.'

Glaisher echoed this sentiment in his obituary in the *Athenaeum*, and the *United Service Magazine* also picked up on the theme in a tribute to 'Admiral FitzRoy, the skilful sailor, the travelled naturalist, the earnest Christian, and the best friend of the population which fringes our sea girt isle'. It continued:

Would that the advice [to rest] have been taken. While we mourn his untimely loss, let us hope that it will be eloquent to many of our overwrought thinkers and workers, who, in their limitless enthusiasm, forget their limited powers of endurance, and discount the future.

Another who recorded his shock was Sir Roderick Murchison, president of the Royal Geographical Society. FitzRoy was a true loss, Murchison said: a founding member, a winner of the Society's Gold Medal and one of Britain's most esteemed geographers. 'Being of a high-strung nervous temperament, and imbued with the loftiest sense of honour and fidelity to his charge, and agitated with over-work, the strain proved too great for the brain, which had surmounted so many difficulties, and the spirit of this high-souled man fled from this world, to the grief of his many friends and admirers, the anguish of his widow, and the deep regret of all his countrymen.'[6]

These eulogies, full of feeling and candour, at least put the finishing gloss on a life that had seemed to end in disaster. Unable to explain the theory behind his forecasts, FitzRoy had come up short as a scientist, his literal commitment to the Old Testament seemed misguided; his professional career, too, seemed a disappointment. Returning from the *Beagle*'s circumnavigation in 1836 he might, with his background and his intellect, easily have gone on to become the First Lord of the Admiralty or perhaps a member of government, squabbling not with minor clerks at the Board of Trade but with political heavyweights like Disraeli and Gladstone. But he had managed none of this. His political career had stalled. As governor of New Zealand he had turned out a shambles, and he had even managed to whip up controversy in a small, out-of-the-way statistical department in the Board of Trade. Not once after 1854 had he been promoted within the Board of Trade. He had even been overtaken by his old junior Bartholomew Sulivan.

If all this was not tragedy enough the story was rounded off in the summer of 1865 by the discovery that FitzRoy had died heavily in debt. This was an embarrassing stain for a family who, outwardly, were part of London's fashionable, affluent elite. The FitzRoys had lived in style at their town house in South Kensington, waited on by five servants. But as news of his financial woes became known over the summer of 1865 this picture of domestic tranquillity was pulled to pieces. To the public it was hollow wealth, an ostentation. In June the lease to his house in Onslow Square was auctioned off, and later in the month his massive library 'of about 2000 valuable works'. So broke were the family after FitzRoy's death that even his own copy of the *Beagle Narrative* was put up for sale.[7] 'What a melancholy career he has run with all his splendid qualities,' Darwin wrote to Hooker.[8]

The bankruptcy and the auction played into the narrative that FitzRoy was a gentleman fraud – both in his public and private life. With no regard to sentiment, and possibly still bristling from their quarrel the previous summer, FitzRoy's old foe Augustus Smith was soon up on his feet in the Commons, calling for an urgent review of the forecasts and storm warnings. His speech drew a stinging response from Maria, who accused him of 'hunting a good and noble man almost to his grave'. If storm warnings were so useless, she asked, 'then why do other countries eagerly adopt them? Why is it the cry of the wives of fishermen on the Northern coast of Scotland, Who will *now* take care of our husbands?'

> You can no longer hurt him – though I do not suppose you act from personal malice towards one who never injured anyone. Your motive may be to gain a little popularity by being the economical Member of the House – but you can add one drop of bitterness to the cup of misery already overflowing in the heart of his unfortunate Wife who now feels doubly every slur cast on *his* work and most gratefully all the deservedly high terms in which that work is spoken by thousands and will be long after I am gone and can be no more affected by praise or blame than he now is.[9]

Many in the country shared in Maria's loss. For weeks letters continued to be printed in the newspapers, regretting FitzRoy's death, giving examples of timely storm warnings and handy forecasts that had saved lives. The horse-racing community, who had grown particularly fond of the forecasts, decided to honour him in their own special way.

On 8 July the *Sporting Gazette* reported that Mr H. Morris – one of the leading trainers of the day – had named a young colt Admiral FitzRoy. In a delicious detail for the historian, the *Gazette* reported that Admiral FitzRoy had been sired 'by Predictor out of Duchess'.[10]

In cold, unsentimental Westminster, though, Smith's suspicion that FitzRoy had gone rogue was widely shared. Worries about his methods had been making officials in the Board of Trade uncomfortable for some time. Now he was gone there was a chance to put things right.

Speculation about the future of the Meteorological Department began as soon as FitzRoy's death was announced. Hooker's note to Darwin touched on the chances of Glaisher getting the role, a rumour that was already circulating Whitehall. On 12 May *The Times* republished an article from the *Medical Times* advocating Glaisher as the heir apparent. Twenty-five years after his initial appointment as super-intendent of the Magnetical and Meteorological Department at Greenwich, Glaisher was still working under Airy. It was a combi-nation of this experience, his continued work for the Meteorological Society checking instruments and publishing results, and his new-found profile as an aeronaut that, according to the *Medical Times*, made him a perfect fit for the job. *The Times* ran the syndicated piece, suggesting its tacit agreement. But in Whitehall other plans were already under way.

On 26 May Thomas Farrer, an assistant secretary at the Board of Trade, wrote to Beaufort's old friend Major Sabine, now President of the Royal Society. Sabine was a grand old man of science. Born in Dublin into an Anglo-Irish family at the end of the previous century, he had shared a heritage and a friendship with Beaufort and had colluded on meteorological matters with John Frederic Daniell throughout the 1820s. A practical administrator, Sabine had become FitzRoy's ally in a hostile political world, giving him free rein to inno-vate. Now he would try to protect his posthumous reputation. Farrer, however, was a different kind of man. Educated at Eton and Oxford, he had trained as a lawyer before switching to the civil service. A talented administrator, in 1865 his stock was rising fast. Farrer had long been wary of FitzRoy's activities and his increased annual expen-diture. He wrote to Sabine, 'The vacancy in the meteorological depart-ment, occasioned by the death of Admiral FitzRoy, has seemed to My Lords to present a fitting opportunity to review the past proceedings and present state of the department.' Farrer's letter included nine direct questions about FitzRoy's activities.

It took three weeks for Sabine to reply. 'The system of forecasting which Admiral FitzRoy instituted and pursued has been expressly described by himself as "an experimental process",' Sabine pointed out. This experiment, he went on, had been sanctioned because it had been popular. He noted that in a report to the Board of Trade in 1862, FitzRoy had reported back the opinions of shipmasters at fifty-six ports. Forty-six had been favourable to storm warnings, three decidedly unfavourable while seven showed 'no expressly decided opinion'. On that basis, Sabine told Farrer, he felt there was a mandate to continue.[11]

To bolster his point Sabine quoted more statistics. From 1 April 1863 to 31 March 1864, FitzRoy had made a total of 2,288 signals to coastal stations. According to FitzRoy's own reckoning, 1,188 of these, or more than half, had been borne out by subsequent weather. Thereafter the accuracy of the signals had continued to rise. Sabine told Farrer that he anticipated further improvements. He rounded off with recommendations from himself and the Royal Society's Council. It would be best if Babington continued with FitzRoy's storm warnings 'respecting the daily forecasts of weather, however, they decline expressing any opinion'.[12]

This distinction between forecasting and storm warnings was not one that FitzRoy would have recognised. As he had often argued, the warnings arose from the forecast. Removing one was to undermine the other. Whether aware of the argument or not, Sabine's letter created uncertainty. To settle the issue the government commissioned a report on the daily practice of the Meteorological Department. The investigation was set to take place in the autumn and winter of 1865, and the report was to appear in the spring of 1866. Three men were commissioned by the government to conduct the investigation: Farrer, representing the Board of Trade, Staff Commander Evans of the Hydrography Department and – a contentious choice – Francis Galton, put forward by the Royal Society. With the benefit of hindsight, it was like asking Anne Boleyn to sit in judgement on Catherine of Aragon.

In many ways Galton was a perfect choice. He was highly intelligent and hard-working, seemingly immune to flattery and could be relied upon to get to the bottom of an issue. His background, too, was an advantage. For at least five years he had been publishing articles on meteorological matters, he had improved theory with his anticylonic discovery in 1863, and he was firmly abreast of all the latest developments in Paris to the point that he had just written a long, meditative review of Le Verrier's achievements for *The Reader*. This closeness, however, was the very thing that made Galton a divisive choice. Since 1863 he had been critical of FitzRoy in public. He was irritated by FitzRoy's refusal to give him access to data that he coveted for his projects. His antagonism had surfaced several times, notably in his scathing review of *The Weather Book* for the *Athenaeum*. He also brought to the task a deep statistical bias. Clements Markham of the Royal Geographical Society – a man who himself had crossed swords with Galton – argued that, 'His mind was mathematical and statistical with little or no imagination', 'He was essentially a doctrinaire not endowed with much sympathy . . . He could make no allowances for the failing of others and had not tact.'[13] These were hardly the qualities of an impartial judge of an objective report. Yet in the autumn of 1865 the files of the office were opened up for him and his two colleagues.

For eight months work continued as before at 2 Parliament Street as the committee researched, considered and wrote up their report. In April 1866, just shy of a year after FitzRoy's suicide, it was published. It was dense, direct and damning. If there had been doubts over FitzRoy until now, then these were turned into certainties. The report spoke of a department dangerously off course. Long forgotten was the original, statistical foundation of an office set up after Maury's Brussels conference. The early work had been impressive. In the late 1850s the number of registers 'was steadily increasing'. No doubt much would have been achieved if 'the attention of Admiral FitzRoy and of his department had not become gradually diverted from the objects recommended by the Royal Society, to those belonging to a wholly different department of meteorology namely the Prognostications of Weather'. This was the axis on which the report spun. The inclusion of 'prognostication' instead of 'forecast' was telling. The committee explained the lexical shift. 'The word "forecast" seems to have been used for the reason that it expressed a less degree of prediction and certainly more than the usual words "Predict" and "Foretell". Whether the reason is a sound one may be doubted. The use of vague phraseology has a tendency to make those who use it satisfied with uncertain conclusions.'[14]

In the 1860s no scientist could be happy with uncertain conclusions, least of all Galton. Science was to be precise, definite, transparent. Yet from the time FitzRoy abandoned the initial aims of his project – which he had begun with 'scrupulous care and assiduity' – he had forgone the hallowed ground of mathematics and statistics for the madness of prediction. As a result, the committee declared that much of the original work of the department – the aims of which were still sound – remained to be done. For a start FitzRoy's wind stars could be vastly improved upon, reduced from 10° squares of the ocean to 5° squares. A better, more complex skeleton form was required to yield more raw data. The authors thought that about 1.65 million extra observations needed to be processed as a matter of high priority.

Instead of restricting himself to such a goal, the committee continued, FitzRoy had been lured by the prospect of prognostication. They reckoned that 'as early as the year 1856 the late Admiral FitzRoy's attention had been directed to the daily observation of the changes of weather over the British Isles, with a view to such changes.' In their eyes the forecasts were not a noble undertaking but a wild application of science such as had never been seen before. That there was no allowance for this in the 1854 brief was plain, they said, and even when the British Association passed its resolution for storm warnings in 1859 there was nothing to show that it 'intended anything more than that storms already known to exist at one place should be announced by telegraph to another place . . . And that at any rate there is nothing in them upon which to found such an elaborate system of foretelling probable weather.'[15]

The committee portrayed the department lurching forward without specific aims from 1859 onwards. The system was chaotic. None of FitzRoy's rules for prediction were ever reduced to formulas or maxims. 'At any rate no such conclusions and no such rules now exist in the Department.' On questioning Babington, they discovered that 'the grounds on which the Department acts in foretelling weather are not capable of being stated in the form of Rules of Law'. They accepted that FitzRoy had spelt out his ideas in his various Board of Trade reports and *The Weather Book*. But they did not feel this was an adequate basis for such a public experiment.

That many of these conditions and probabilities are capable of being stated in the form of Laws, and that some of them are

Laws that would be accepted by Meteorologists generally we do not doubt; nor do we doubt that the probabilities are in many cases considerable, and especially in the important cases of sudden and violent changes of weather. But we do not find that these conditions or probabilities have been reduced into any definite or intelligible form of expression, or are, as they now exist in the Office, capable of being communicated in the shape of instructions. Were the gentleman now in the department to leave it, no rules would be found in the Office for continuing the duties on their present basis.[16]

It was the same accusation that W. H. White had levelled against FitzRoy two years before. The weather-forecasting system existed only in FitzRoy's head. For four years he had forecasted and for four years he had failed to distil his process into a legible form. This was not able administration. It was megalomania. The only telling historical record that remained, they revealed, was a sort of scrapbook, filled with jumbled accounts of the effects of gales and cuttings from the papers.

But for all its venom, the report had to face up to a fundamental truth – one argued by Maria in her letter to Augustus Smith. Storm warnings were popular. As Sabine had pointed out in his initial letter, the reaction of harbour masters and fishermen, the men who mattered, had always been overwhelmingly positive. The report itself conceded that 'from inquiries we have made through trustworthy persons at most of the principal ports, we find that seafaring men look on them more favourably than they did at first'. Even this, though, was explained away.

But in estimating this at its true value it must not be forgotten how eagerly the world at large is disposed to base an unreasoning belief on the occasional successes of weather predictions, and how easily it forgets the failures. We need not say that we do not wish for a moment to compare the efforts of the Department with the predictions of the ordinary weather prophets who attempt to connect the changes of the weather with the stars or the changes of the moon. It is not, however, irrelevant to refer to these prophecies, and to the belief which has been so often placed on them, when we are estimating the value of popular feeling as evidence of the value of the Storm Warnings.

The report went on to cast doubt on all the statistics that supported the forecasts and warnings, particularly those calculated by FitzRoy and cited by Sabine. The committee had devised a new mathematical framework and had found the predictions much less effective than FitzRoy had claimed in his Board of Trade reports. Their recommendations were clear. FitzRoy's forecasts must be halted at once: 'we cannot say that there is evidence that the Daily Forecasts have been correct in point of fact, or that "we are enabled" . . . *to know what weather will prevail during the next two or three days*, and, as a corollary, when a storm will occur.'

> As regards the utility of the daily Forecasts, we have to observe, in the first place, that if there is no sound basis on which they are founded, and no evidence they have been correct in point of fact, they are wanting in everything which can render them practically useful. But even independently of this, we doubt whether intimations of ordinary coming weather, so vague as these Forecasts must necessarily be, can be of any real value.[17]

On storm warnings the committee were not so trenchant. They thought that about half of these had turned out well in relation to the strength of the wind, with a much lower degree relating to wind direction. As such, they thought warnings should continue, but they should be curtailed in their operation to simply telling of a coming gale. Overall they thought the Meteorological Department should be sliced in two. One half should concentrate on the statistical work that had been badly neglected. The other should be overseen by the Royal Society and be concerned with storm warnings and science. It was a death knell for FitzRoy's meteorological office. Reading the report, Babington was outraged. He resigned immediately.

The following months saw the old system destroyed in several sharp blows. First to go were the forecasts and soon on their tail were the storm warnings. This move exceeded even what Galton and the committee had called for, but with the whole idea of prediction called into question the Board of Trade had decided to distance itself from the controversy. It could, Farrer had argued, have limited the storm warnings to simply notifying different ports of storms that had already hit. But the President of the Board of Trade, Milner Gibson, had dismissed this argument.

Very interesting – but after all, there is not much practical utility in foretelling storms if the warning precedes the storm by so short a time as twelve hours – the people of the locality with the use of a barometer – can generally form a pretty good idea of the weather a few hours before a gale begins.[18]

On 29 November 1866 a circular was issued stating that the storm warnings were to be stopped from 7 December. The watchmen had been called down from the turrets. The experiment was at an end.

December 1866 marked the close of a remarkable period of activity in the Meteorological Department. One of the youngest offices in Whitehall, in its twelve years of existence FitzRoy had established a marine network, conceived and published his wind star charts and his *Barometer and Weather Guide*, founded a telegraphic network, launched a storm-warning system, become the world's first ever governmental forecaster, written his *Weather Book* and pioneered meteorological relationships with other European powers – chiefly Le Verrier at the Paris Observatory. FitzRoy, at least, felt a sense of achievement. As he had noted in his submission to *Portraits of Men of Eminence*, in 1864:

It is difficult to estimate the lives saved, and the property, on the occasion of each successive storm on our coasts, of which warnings have been duly given by the signal drum, if not also by the forecasts published, daily, in most newspapers.[19]

FitzRoy's administrative career had been full of vigour and purpose. Had he not been chosen for the role in 1854 and it had instead gone to Glaisher, the Meteorological Department would have been a very different place. More statistical, certainly better ordered, but also more cautious.

That FitzRoy was an outrider was acknowledged in his lifetime. What distinguished him from the average Whitehall bureaucrat was his affinity with the work of his department. For him the weather was not a trivial matter. It was not a shower in Surrey or a blustery wind at Epsom. Instead it was a lightning strike at Rio, a pampero in La Plata, a hurricane squall in the Strait of Magellan or storm after storm at Cape Horn. For a captain who had seen sailors blown to their death, weather could never be a mere collection of statistics – it was an emotional force to battle against. If FitzRoy's story was to be a tragedy then it was a noble tragedy. He had spent his energy trying to improve conditions not just for sailors or fishermen, but for anyone affected by weather. Two and a half centuries earlier, Francis Bacon – the patriarch of science – had written a meditative passage. Just what draws someone to science?

> For men have entered into a desire of learning and knowledge, sometimes upon a natural curiosity and inquisitive appetite; sometimes to entertain their minds with variety and delight; sometimes for ornament and reputation; and sometimes to enable them to victory of wit and contradiction; and most times for lucre and profession, and seldom sincerely to give a true account of their reason, to the benefit and use of men: . . . [science is] a rich storehouse, for the glory of the Creator and the relief of man's estate.[20]

FitzRoy embodies Francis Bacon's final, rare group: altruistic, inspired by religion, 'for the glory of the Creator and the relief of man's estate'.

He also had a remarkable gift for carrying companions with him – from his crew on the *Beagle*, who followed him down sinewy rivers and up great mountains without question, to Pattrickson, Babington and the rest at Parliament Street, who would all believe in the ambitions of the office long after FitzRoy's death. It was a paradox that lay at the heart of FitzRoy's character. He had a rare talent for making people believe, yet at his core he was racked with uncertainty. These competing forces were always at work, but FitzRoy nonetheless managed to function. So revolutionary was his meteorological work that he propelled a whole government department far ahead of its time, sparking a true scientific crisis. He raised fundamental questions. How fast should practical science progress? Does every solution require a transparent method? How can theoretical science be communicated to the public? How emotionally involved should a scientist become in their work? Should it be, as Charles Darwin wrote, that 'A scientific man ought to have no wishes, no affections – a mere heart of stone'?[21]

Many of these dilemmas still trouble us today. How quickly should a transformative drug – say a new cancer drug – be made available to the public? How important is it for people to understand the complex particle physics at work in the Large Hadron Collider in Geneva? In the 1860s these questions were all the more acute as science was in the ascendance. According to *Once a Week*, the rise of new sciences was 'astonishing': statistics, philology, sociology, ethnology, genetics. These all promised to transform lives, make the world a better, happier, more efficient place. Yet at the same time FitzRoy's forecasts seemed a step too far. Lampooned in *Punch* and gossiped about all around the country, it is little wonder they made many feel uncomfortable. They threatened to rob science of prestige and honour at its moment of triumph. For many, like Airy and Galton, that was too much. FitzRoy, a rebel with a conviction that he was doing good, carried on regardless.

Progress does not come at a uniform, linear pace. Sometimes there is a lurch forwards, as there was in the years after Franklin's kite experiment in the 1750s or Howard's cloud classification in 1802. On other occasions a generation can pass before an idea takes root, as happened with Redfield's twisting winds.

There is a passage in Chapter 4 of *The Pickwick Papers* when Dickens sends Mr Pickwick scurrying after his hat:

> There are very few moments in a man's existence when he experiences so much ludicrous distress, or meets with so little charitable commiseration, as when he is in pursuit of his own hat. A vast deal of coolness, and a peculiar degree of judgment, are requisite in catching a hat. A man must not be precipitate, or he runs over it; he must not rush into the opposite extreme, or he loses it altogether. The best way is, to keep gently up with the object of pursuit, to be wary and cautious, to watch your opportunity well, get gradually before it, then make a rapid dive, seize it by the crown, and stick it firmly on your head: smiling pleasantly all the time, as if you thought it as good a joke as anybody else.[22]

The hat here is an excellent symbol for the idea. Go after it too fast, as Espy did with his rain-making or his upward columns, and you risk the scorn of all. But if you don't pursue it quickly enough then it can fade as Edgeworth's telegraph did. FitzRoy, too, faced this dilemma with his forecasts. They were as much a novelty as Morse's telegraph and it took Morse a decade to put his idea into successful production. For years he struggled to make people believe. It is a challenge that runs right to the heart of philosophy and science: how to make someone believe? In 1867 or 1868 Emily Dickinson wrote a short poem about faith, 'Tell All the Truth But Tell It Slant':

> Tell all the Truth but tell it slant—
> Success in Circuit lies
> Too bright for our infirm Delight
> The Truth's superb surprise
> As Lightning to the Children eased
> With explanation kind
> The Truth must dazzle gradually
> Or every man be blind—[23]

It is a precise exploration of the challenges of telling truths. Each of Dickinson's lines is a remodelling of her first declaration: 'Tell all the Truth but tell it slant'. The image of the frightened children in their bedroom being comforted with an explanation of lightning is powerful and lasting. The trick is not to explain the lightning through physics; instead you have to come at it from an oblique angle. You cannot force an idea upon people, they have to learn with explanation, story and narrative. 'The Truth must dazzle gradually, Or every man be blind'.

FitzRoy used what he could to get his message across. He published his *Barometer and Weather Guide* for fishermen and sailors, he wrote *The Weather Book* for the general public, he lectured at the British Institution, gave a paper at the British Association and interacted with detractors in the letters pages of *The Times*. There was little else available to him.

In 1902 the moral complexities of weather forecasting were encapsulated in fiction for the first time in Joseph Conrad's *Typhoon*. *Typhoon* was the sort of scientific parable that FitzRoy would have loved. Set on a steamship called the *Nan-shan*, on the China Sea under the command of the sonorously christened Captain MacWhirr – a man 'having just enough imagination to carry himself through each successive day, and no more' – the story examines MacWhirr's lack of scientific faith. On a stifling hot day he sees the barometer drop but does nothing in response. 'Omens were as nothing to him, and he was unable to discover the message of a prophecy till the fulfilment had brought it home to his very door.' Unable to anticipate and unwilling to move his ship off course, MacWhirr sails dead straight towards his destination through the lurid sunshine of a dirty sunset. 'The coppery twilight retired slowly, and the darkness brought out overhead a swarm of unsteady, big stars, that, as if blown upon, flickered exceedingly and seemed to hang very near the earth.'

MacWhirr was a perfect evocation of Captain Taylor rounding Anglesey in the *Royal Charter* in 1859. The barometer has been falling for six hours. On the bridge MacWhirr is troubled enough to dig out a book on storms – perhaps one written by Reid or Piddington? Soon the inevitable typhoon arrives.

Conrad's typhoon is a behemoth. 'The gale howled and scuffled about gigantically in the darkness.' Water washed over the deck, bursting into the mouth of the first officer, sometimes fresh water, sometimes salt. The *Nan-shan* climbs up mountains of black water and races down the far side. MacWhirr glares out like a boxer from the bridge. 'He was trying to see, with that watchful manner of a seaman who stares into the wind's eye as if into the eye of an adversary, to penetrate the hidden intention and guess the aim and force of the thrust.' The typhoon, though, is not to be understood, it is to be endured or avoided through sound judgement or scientific understanding – neither of which MacWhirr possesses. Pointing at his storm book several hours earlier, MacWhirr had said to his first officer: 'The truth is that you don't know if this fellow is right, anyhow. How can you tell what a gale is made of till you get it?'[24]

MacWhirr narrowly lives to learn his lesson, limping into port with a ravaged ship, caked in salt, looking for all the world as if she had been fished out 'somewhere from the bottom of the sea and brought in here for salvage'. Conrad's point was stark. By ignoring the science MacWhirr had got the least of what he deserved. He was one of 'the greatest fools that ever sailed the seas'. *Typhoon* was serialised in *Pall Mall Magazine* in 1902 and published in book form a year later. It was a story about anticipation, about believing in science. FitzRoy would have nodded as he turned the pages. It was a vindication of everything he stood for.

It took only a week from the suspension of FitzRoy's storm warnings in 1866 for the first letters of complaint to arrive at the Board of Trade. 'I trust the suspension will not be of long continuance,' wrote a clergyman from Silloth in Cumberland, 'for the *"warnings"* are *invaluable* on this coast, and I think that if the President and Council of the Royal

Society could have witnessed the growing attention paid to the Signals by the sailors and were aware of their general accuracy they would not have recommended even a temporary suspension.'[25]

The MP for Aberdeen, Colonel Sykes, took up the issue in the Commons. On 26 February 1867 he asked Sir Stafford Northcote, President of the Board of Trade, whether he had read any of the memorials from the Meteorological Society of Scotland or the 'mercantile bodies at Leith, Glasgow, Dundee, Aberdeen and Edinburgh' asking that the storm signals be resumed? Northcote referred Sykes back to the Royal Society report, but Sykes was not swayed. At seventy-six he was one of the fathers of the House. Behind him was a glittering career in India where he had served in the Army and as a director of the East India Company. Sykes also had scientific pedigree, having studied Indian wildlife and published several catalogues on the subject. More recently he had sat, along with FitzRoy, Airy and Glaisher, on the British Association's Balloon Committee and after the Royal Society's report he had taken the view that FitzRoy had been badly treated. Over the next three months he took every opportunity to question the government about warnings in the Commons, and he began his own calculations of their effectiveness.

Sykes released his calculations. He reckoned that over a three-year period about 75 per cent of FitzRoy's warnings had been verified by subsequent weather. This was quite different to what had been claimed by Galton and the committee, who had FitzRoy's hit rate at less than half. Colonel Sykes' work encouraged others, including Christopher Cooke of the Astro-Meteorological Society who wrote a pamphlet entitled *Admiral FitzRoy: His Facts and Failures*. Although Cooke disagreed with FitzRoy when it came to theory, like Sykes he believed that the committee had gone too far. Tallying up the parliamentary records, Cooke estimated that FitzRoy's work had cost the government about £45,000 in all.

Surely Britannia, who rules the waves more potently than did Canute, if she desires to preserve her dominion, should not grudge this national mite in favour of those who do her work, who fight her battles, and who pay taxes out of their earnings for her support![26]

Cooke's pamphlet ended with a summary of the evidence currently in favour of storm warnings. None of it had been included in the many appendices of the committee's report. It began with a list of testimonials. The Aberdeen Marine Board, the Marine Board in Dundee, the South Shields Ship Owners' Society, the Sunderland Pilots, the Sunderland Mercantile Marine Office, Collector of Customs at West Hartlepool, the Receiver of the Wreck at Great Yarmouth, the Collector of the Customs at Deal, all expressed the same opinion: they would have preferred the storm warnings to continue. The only dissenting voice came from the Marine Board in Plymouth – and, as Cooke explained, at the south-western end of the country without observations from the Atlantic to protect it Plymouth was the one place where a storm warning was less likely to work. It seemed incredible that such a powerful, useful tool should be done away with because it did not fulfil the demands of practical science.

So often the British Association had been the forum for meteorological debate. In the 1830s it had played host to William Reid's pioneering storm work and two years later Espy had used these meetings as the platform to present his ideas. More recently Dove and then FitzRoy had delivered papers and now, in 1867, the forthcoming meeting presented the scientific community with the perfect chance to pass judgement on the forecasting controversy. At Dundee in September the number of meteorologists and meteorological papers 'was quite up to the average'. A host of contributions had been advertised. One on the behaviour of the aneroid barometer, another on magnetic disturbances, others on the luminosity of phosphorus and rainfall and, most alluring of all, 'Storm Warnings, Their Importance and Practicability, a lecture by Colonel Sykes'.

Sykes' talk began with a retrospective account of the meteorological progress of the past thirty years. He then moved on to FitzRoy's appointment 'and the long and most valuable and assiduous labours of that gentleman in advancing meteorology, and in applying it to the most practical and useful purposes'. Sykes soon arrived at his point. 'In the course of three years he gave 405 storm warnings,' he reckoned, 'and of this number 305 were correct.' There was applause from the audience. Sykes went on to state his disappointment that the Royal Society had refused to continue with FitzRoy's signals. Instead they had decided to wait fifteen years, collecting data, 'and when these were obtained they would then issue these warnings, if the nature of these observations warranted that course'.

In order to do this, however, a large number of new Observatories
were to be established with self-recording instruments, and this
at a very much larger cost to the country than the Meteorological
Department of the Board of Trade had cost. He had no hesita-
tion in saying that the argument employed in the refusal of the
Committee of the Royal Society was a pedantic affectation of
science – literally the coxcombry of science [laughter].[27]

Sykes wondered what other countries must make of this peculiarly
'British' debacle. In France they had persisted with the 'very highly
appreciated' storm warnings. In St Petersburg they had adopted the
FitzRoy method:

> Yet here are we, the most maritime nation in the world – having
> set the example to other countries in this matter of storm warn-
> ings – and yet we are now dropping them [Applause]. We were
> too scientific for the work [Laughter].[28]

Sykes had struck a chord. He ended his witty performance with a
simple plea – that the people of Great Britain should not be deprived
of the warnings 'by the mere crotchet of individuals'. He sat down
to more applause.

John Don, President of the Dundee Chamber of Commerce, was
the first on his feet. Don told the audience that he agreed entirely
with Sykes that the country was being held hostage by the 'pedantic
affectation of science'. He called for a resolution, demanding the
warnings be resumed immediately. D. Milne Home, President of the
Meteorological Society of Scotland, then voiced his support, pointing
out that due to Britain's position as Europe's geographical outpost,
situated at an unstable crossroads of Atlantic gales, Arctic freezes and
warm continental air, she was the country generating the most useful
information – the Continent's lookout. Also in support was the elab-
orately named William Montagu Douglas Scott, better known as the
Duke of Buccleuch. He revealed that he too had been lobbying in
private for a return of the warnings.

Another luminary in the audience that day was Admiral Edward
Belcher, an acquaintance of FitzRoy and one of Beaufort's old
surveyors. Belcher was now sixty-eight years old, a survivor of the
Napoleonic Wars, a veteran of the fabled days of the Royal Navy and

famous for his role in the hunt for Sir John Franklin in the Arctic. He
began a fiery speech by declaring that Admiral FitzRoy had been 'tried
by a civil court-martial, and by a set of men who were not qualified
to judge him'. He went on:

> They should hear the naval side of the question, for it was really
> after all a naval question. It had been said that we could not
> foretell events of the weather, but he could tell them facts to the
> contrary [Applause]. As far back as 1812 he recollected that it was
> the constant habit of the Admiral on the Bordeaux station to
> give signals every hour when the barometer changed, and the
> ships struck masts and sent down yards accordingly. On one
> occasion, when they were chasing the enemy and just as they
> were on the point of coming up with them, their captain looked
> at the barometer, and suddenly gave the order to reef topsails,
> but before that could be done many spars were lost. On another
> occasion, he himself had predicted a storm in a certain quarter,
> and his prediction was fulfilled to the very hour. He knew they
> could continue the storm signals if those in authority would
> only be stirred up to do their duty. It was all nonsense to say
> they could not do so. [Applause]. Why, the cattle, the birds, the
> fish, and the reptiles, and in fact everything gave indications of
> coming storms, and, with the assistance of meteorological
> observations, scientific men should be able to give very precise
> notice of atmospheric changes.[29]

The words could have come from FitzRoy's own mouth. His project
might have been derided by administrative and cautious men in
London, but here the cream of the scientific community were uniting
in his support.

The last contribution came from the new president of the
Meteorological Society, James Glaisher. So far Glaisher had listened
in silence to a subject he understood far better than anyone else. Now
nearing sixty, his name known across the country, his reputation as
one of the founding fathers of British meteorology was assured and
his opinions were treated with the respect they deserved. Glaisher told
the audience he was 'much surprised by the decision of the Royal
Society's Committee'. He could find 'no good reason' why the warnings should not be resumed, a statement that was met with applause.

'He concluded by earnestly stating his opinion that the signals should be restored.'[30] It was vindication, and from the highest authority.

The resolution was passed, leaving the Board of Trade and the Royal Society with no choice but to act. They settled on a compromise. FitzRoy's forecasts and storm warnings were not going to return. They were political dynamite. So they settled on storm intelligence. Plainly put, this meant spreading word of an active gale – one whose existence had been verified – around the coasts by telegraph. This allowed the government to use their expertise and technology without straying into prediction. It seemed an adequate halfway house. But as FitzRoy had envisaged in his *Weather Book*, there turned out to be problems with storm intelligence. For the south-western ports it was useless as they were often only told of the storm after it had happened. In the North Sea intelligence was equally unsatisfactory for the long-haul fishermen and merchantmen who were often out of port longer than the ten-hour warning they were likely to get. The abandonment of FitzRoy's old system left a gap, a desperate, gaping hole. There is an extra chill to the shipwrecks during this lull after 1865. Questions of *what if* abound.

The North Sea storm of 1869, for example, when the full combined herring fleets from Scarborough and Filey had set out on a late-season dash on 25 October. It was a fine sparkling morning, the yawls skimmed over the waves thirty miles out from port. The storm soon came. Pounding waves, dark skies, hopelessly vulnerable ships. It was a familiar story. George Jenkinson, the skipper of *Good Intent* and a Methodist lay minister, prayed for his crew's deliverance. 'In the evening I perceived what I considered to be the unmistakable signs of an increasing storm, and asked the Lord to stretch forth his mighty hand, through the night, on our behalf.' Jenkinson had turned to his crew and told them to prepare to trust their souls to God. The *Good Intent* limped back to Scarborough the next morning, one of the few to make it safely home. It was a narrow escape. 'We cannot sufficiently praise God for his signal deliverance. May we all continue to believe in the Son of God, till we have navigated the sea of sorrow, and all land at length in "the harbour of comfort",' Jenkinson wrote afterwards. Science had retreated. Fishermen were once again left to place their faith in Providence.[31]

Had James Glaisher read of the storm he would have shuddered. Glaisher's life had travelled in a different direction. Instead of growing more cautious with age he had become more daring. He was now a grand figure in British science. Pictures of him from this time suggest an imposing man: penetrating eyes, bushy side whiskers, domed forehead. His reputation was international. In 1866, having read of his ballooning feats, Alexander II, Emperor of Russia, had sent him 'a magnificent diamond ring' as a token of his appreciation. It was more than Glaisher could have foreseen when he had begun his career thirty years before, high on the Irish hills.

Between 1862 and 1866 Glaisher made twenty-eight ascents, seven into the high atmosphere. His retrospective accounts became stellar copy for the newspapers. They blended the familiarity of travel writing

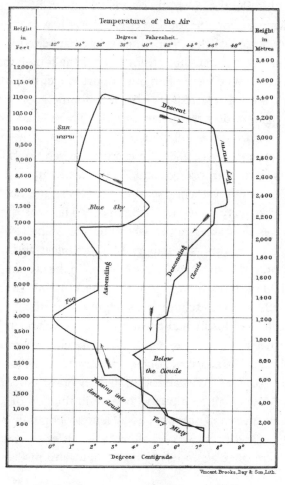

Vincent Brooks,Day & Son,Lith.

Temperature of the Air at different heights
observed in the Ascent and descent.
6th April 1864.

with something different to the warm-blooded Victorian tales of lion hunting, elephant shooting and colonial rabble rousing. There was something attractive about Glaisher's tales from the sky. This was not a place to be tamed. The sky could not be owned or colonised. And through this intangibility it retained a powerful grip.

This grip took hold of no one more than Glaisher himself. While in the balloon, banking through the clouds, he felt like 'a citizen of the sky': his body keener, his brain more active, 'every sense increased in power to meet the necessities of the case'. In *Travels in the Air* he wrote a dreamy projection of what life was like in this other world.

> Above our heads rises a noble roof – a vast dome of the deepest blue. In the east may perhaps be seen the tints of a rainbow on the point of vanishing; in the west the sun silvering the edges of broken clouds. Below these light vapours may rise a chain of mountains, the Alps of the sky, rearing themselves one above the other, mountain above mountain, till the highest peaks are coloured by the setting sun. Some of these compact masses look as if ravaged by avalanches, or rent by the irresistible movements of glaciers. Some clouds seem to be built up of quartz, or even diamonds; some, like immense cones, boldly rise upwards; others resemble pyramids whose sides are in rough outline. These scenes are so varied and so beautiful that we feel like we could remain for ever to wander above these boundless plains.[32]

Although he was charmed by the beauty of his surroundings, Glaisher never quite forgot his obligations to science. In July 1863 he ascended on a boisterous summer's day, determined to watch the formation of rain inside a nimbus cloud. Floating through wet fogs and mists, he studied the force and size of the raindrops at different heights. At 2,000 feet he found they were the same as 'a fourpenny-piece'. On this and other ascents he proved beyond doubt that Gay-Lussac's theory about the decrease of temperature with height was wrong. 'It is necessary to renounce this ideal of regularity,' he decided. Even on bright, still days Glaisher found it impossible to construct a formula for temperature by altitude. Equally complex, he discovered, was the direction of wind.

In January 1864 he launched from London in a south-east breeze. At 1,300 feet he met a strong south-westerly current, which at 8,000 feet veered around to the south-south-west. This series of embroidered winds was far more complex than anyone had thought years ago. The discovery led Glaisher to speculate on the importance of the winds to the British climate. Later during the same winter ascent, Glaisher located a warm south-westerly current of 3,000 feet in thickness. 'Above this the air was dry, and higher still very dry,' he wrote. A peculiar feature of this current, Glaisher noticed, was that it came from the same direction as the Gulf Stream.

> The meeting with this S.W. current is of the highest importance, for it goes far to explain why England possesses a winter temperature so much higher than is due to our northern latitude. Our high winter temperatures have hitherto been mostly referred to the influence of the Gulf Stream. Without doubting the influence of this natural agent, it is necessary to add the effect of a parallel atmospheric current to the oceanic current coming from the same region – a true aerial Gulf Stream.[33]

Decades before it was located or understood, Glaisher's words can be seen as an anticipation of the jet stream, today considered a chief factor in determining British weather. Though his south-westerly current was at a much lower altitude than the jet stream – which operates at heights of about six miles – Glaisher's observation hints at the effect currents of wind can have on our climate.

Aloft in his balloon, Glaisher symbolised the progress of meteorological science. The past seventy years had seen an exploration of the atmosphere like never before. It was no longer a realm of chaos or a citadel of God. It was a fundamental part of the living world: part of us. In his masterful synthesis of scientific knowledge, *Kosmos* (1845), Alexander von Humboldt had called for a wider appreciation of the atmosphere. 'The earth, robbed of its atmosphere, like the moon, presents itself to the imagination, as a desert brooded over by silence.'[34]

Luke Howard had lectured on the same theme, years before:

> Supposing mankind and other animals, and even vegetables, to have been so formed as not to need [the atmosphere] for respiration, for nutrition and for warmth, still, *without the air* what a dull scene, what a blank in nature, in place of our many enjoyments abroad! No refreshing breezes! No blue skies alternating

with kindly showers, no waving branches and rustling leaves: none
of the beauty and variety of summer clouds, no rainbow, no rain![35]

In 1870 the US Army, augmented by the remnants of the Smithsonian
observation network, set up a Weather Bureau. The next year it began
to issue 'probabilities' of coming weather, derived from telegraphic
reports.

In 1875, the first daily weather map appeared in *The Times*, drawn
by Francis Galton. Four years later the British forecasts were reinstated
after a thirteen-year hiatus. Another decade would pass before the jet
stream – the great artery of the atmosphere – was discovered. It was
the end of the century before a Norwegian professor, Vilhelm Bjerknes,
and others finally established forecasting on a sound statistical footing.

Glaisher remained at the Royal Observatory until an argument with
Airy in September 1874. 'It would be convenient and advantageous,'
Airy wrote in a note, 'that Mr Glaisher should not quit the Observatory
till 2pm each day. No other officer of the Observatory leaves before
2pm.' Glaisher did not waste too many words on his resignation letter:

Your fault-finding note to me this morning is so painful that in
consequence I wish to resign.[36]

It was a laconic end to forty years of scientific partnership. His own
master at last, Glaisher continued with his many interests. Having already
spent five years as President of the Photographic Society, he took the
post again in 1875 for another seventeen-year stint. He also continued
with his meteorological work, compiling his quarterly statistics for the
Registrar General until 1901 – fifty-five years after he had begun.

To the end Glaisher recorded air temperature and pressure, wind
speed and direction, with the finest instruments at a weather station at
his home in Croydon. Born in 1809 into a world of sailing ships, muskets
and cannon, Glaisher died at ninety-three in 1903, the year Orville Wright
made the first aircraft flight with a petrol engine at Kitty Hawk Bay in
North Carolina. He would be followed eight years later by Sir Francis
Galton. The lives of them both had spanned an age of meteorological
enlightenment. With their deaths the last links with a generation who
had codified the atmosphere, quantified it, painted it, mapped it,
described it and predicted it as never before were gone.[37]

Dusk

As the afternoon stretches on the strength of the sun diminishes, the thermals fade away. There will be no more cumulus clouds today. The clouds weaken as their energy source disappears. At sunset the sky is clear once again.

The sun has dipped towards the horizon. It blazes from a shallow angle across the atmosphere, softening its rays into the golden hour, a time treasured by landscape photographers when the earth is bathed in rich yellow light and the texture of objects is revealed. The sun is now 5° above the horizon and different colours are starting to appear. The milky-white lip over the horizon of daytime is replaced by a yellowy red. Zones of red and orange and purple appear over it, almost in horizontal bands. As the sun sets in the west, in the east the earth's shadow comes into view, a bluish-grey horizontal band, rising as high as 6° over the horizon.

On a clear day like this, the drama intensifies after sunset. Although the sun has gone, twilight fills the skies for several hours. Eventually the distinct bands of colour even into a purple hue of striking opalescence, 'more pink and salmon coloured than true purple'[1]. This light remains for about an hour after sunset, until it is replaced by a cool blue glow 20° above the horizon. This is the twilight glow.

The further away from the equator, the longer twilight lingers. In the Orkney Islands north of Scotland twilight fills all the night hours from the middle of April to the middle of August and it is quite possible to read from a book or to garden at midnight. For twilight to properly end and night begin, the sun must be 19° beneath the horizon. Until this final tipping point light continues to be scattered, and it glows over the landscape. Soon, though, stars of different magnitudes are visible. High up in the mesosphere, fifty miles above ground, noctilucent clouds made up of ice crystals glimmer and twinkle like chandeliers. These are the highest of all clouds.

By now the heat of the day has faded away. The atmosphere is cold and clear. Down below in a meadow, blades of grass are cooling fast. On the stem of a dandelion, in a space so infinitesimally tiny that even James Glaisher or Francis Beaufort could never glimpse it, a tiny fleck of dew is forming. The start of another day.

West Winds

It had been on the news all day, the storm was on its way. As I shut the door of my west London home it was somewhere out in the Channel, perhaps blowing up great waves like those FitzRoy and Sulivan had encountered two hundred years before on the *Thetis*. It was a quarter to nine in the evening on Sunday 27 October 2013. By Christian custom the next day would mark the ancient feast of St Jude, patron saint of lost causes. Someone had noticed this and begun a trend that had gathered momentum on social media. Even before it crossed the coast it was St Jude's Storm.

In the street the wind was already lively. Clusters of brittle, golden leaves were twirling from the plane tree outside, adding to the mound of several hundred that had accumulated on the pavements over the last few days. There were a few hours left of this uneasy calm. I pulled up the hood of my battered parka and set off for the river.

It was three days since the Met Office had first put out a forecast for potentially damaging winds across the south of England. On Friday they had upgraded this to an amber warning, to be 'prepared for potentially hazardous conditions'. Over the weekend we had watched the satellite images as the clouds began to swirl in their ominous anticlockwise spiral out in the Atlantic. The forecasters' demeanour changed, their easy playfulness gone. They stared gravely into the camera, repeating the maxim that this storm was one 'you would not see every year'. On Sunday, tension had built. Railway companies were anticipating delays during Monday morning's rush hour. Firemen and paramedics had been put on standby. To emphasise the message to anyone still not listening, erstwhile BBC weatherman Michael Fish, famous for missing a violent extra-tropical cyclone in 1987, had been fetched out of retirement to broadcast his own personal warning. 'This is totally amazing,' Fish said in an interview on BBC News. 'The modern computers are literally able to invent these things in thin air. No human being could have done this.'[1]

Down by the Thames at Fulham Reach the atmosphere was thick, wet and cold. I couldn't see far in the gauzy light – I could just make out the green outline of Hammersmith Bridge a hundred metres upstream. I sheltered beneath a white willow and listened to the breeze. Orange tunnels of sodium light illuminated individual, fine drops of rain that were falling at an angle. There was a smell of musk and the river seemed swollen. From somewhere above the iron lid of sky came the metallic whirr of an aeroplane descending towards Heathrow. A jogger bounced by with his headphones in. Then a dog walker, his Labrador tugging impatiently at his hand.

Before St Jude came I had wanted to feel this moment for myself. The unsettling mix of anticipation and suspense. This is the power of weather. It is what Constable experienced in the East Bergholt windmill, a wide-eyed boy of fifteen watching the horizon as scudding clouds flew overhead. It is the sense of anxious tension FitzRoy must have felt during that night in the Strait of Magellan in 1846, on his passage home from New Zealand, with Captain Cable gone below and his barometer dropping fast.

I waited under the willow for half an hour. Then the wind began to rise, so I turned for home.

St Jude swept across the south of England four hours later, and we woke the next morning to news of its effects. Winds of 99mph had been recorded on the Isle of Wight. Six people had died, mostly from the hundreds of fallen trees. At Heathrow 130 flights had been cancelled; 850,000 homes were without electricity. The Port of Dover had been shut for three hours, as had Dungeness nuclear power station.

But that was not the end of it. St Jude turned out to be the opening volley in one of the most unsettled winters in recorded history. As I sat inside reading about Reid and Redfield and Espy and FitzRoy, one storm after another was flung by a jet stream that seemed locked like a missile on the south coast of England. Six major storms struck in January and February 2014, with volumes of rainfall that broke all records. Not since 1766 – two years before Cook set sail for the South Seas in the *Endeavour* – had the recorded rainfall been so high. As it had during the Royal Charter Gale, the Dawlish railway collapsed. At Milford on Sea in Hampshire thirty-two diners had to be rescued from a seafront restaurant when gigantic waves sent shingle from the beach flying through the windows. The floods stretched east from the Somerset Levels to the Thames Valley. By the middle of February Eton's playing fields were underwater.

The weather thrust the Met Office centre stage. With its IBM Power7, one of the most powerful computers in the world – able to calculate 1,000,000,000,000,000 sums a second – it could track and forecast each storm. Almost a hundred and fifty years after he died it was the realisation of FitzRoy's vision. Rather than being cast aside as a costly extravagance, the Met Office was at the heart of the action – briefing politicians, businesspeople, the media and the public.

Today the Met Office has a budget of more than £80 million; it employs around fifteen hundred staff, five hundred of them scientists.* Although the weather, particularly British weather, retains its potential for surprise, we now live in a world where forecasts are predominantly trusted. By the Met Office's latest statistics 94.2 per cent of its maximum temperature forecasts are accurate within 2°C and 85 per cent of minimum temperatures to within 2°C; 73.3 per cent of rain forecasts turned out correct while storms are almost never – apart from the rare Michael Fish case – missed.[2]

The value of the Met Office was illustrated by a 2007 consultancy report. It concluded that it delivered 'an exceptional return on investment', that it saves lives, protects properties and provides wide-ranging social and environmental benefits; all told, it brings £353.2 million of savings to the British economy. Although much has changed since the early fractious days of the 1850s and 1860s, its initial vision has not. FitzRoy's ideal of a public weather service, provided by government for the good of all, has not only survived but has become integral to our way of life.[3]

For his foresight FitzRoy is fondly remembered by those at the Met Office today as their founding father. Its headquarters is located on FitzRoy Road in Exeter and, in a wonderfully apt tribute, back in 2002 one of the BBC's fabled shipping-forecast regions was renamed from Finisterre to FitzRoy. He has been well served by three excellent biographies and in 2005 FitzRoy's life was turned into thrilling fiction in Harry Thompson's Booker-nominated *This Thing of Darkness* – proving

---

* In October 2014 the Met Office announced that it had secured the funding for a new £97m supercomputer. The machine will be capable of performing 16,000 trillion calculations a second, around thirteen times faster than the current IBM Power7 model.

beyond any doubt that history loves a rebel. Far away from Britain's shores soars the awe-inspiring Monte Fitz Roy in southern Patagonia. Named in FitzRoy's honour in 1877 by Francisco Moreno, the Argentine explorer, the summit juts up above a spectacular landscape like a fang. The weather at Monte Fitz Roy is often wild. For climbers it is the ultimate ascent, and few have ever stood on its summit.

Today FitzRoy is chiefly remembered for the part he played in the story of evolution as Darwin's taciturn captain on the *Beagle*. This history has always overshadowed FitzRoy's later life and his meteorological work. To be labelled 'Darwin's captain on the *Beagle*' is a fate FitzRoy would have hated. A much better epitaph is inscribed on his gravestone at All Saints Church, a breezy passage from Ecclesiastes:

> The wind goeth towards the south, and turneth about unto the north; it whirleth about continually, and the wind returneth again according to his circuits. (Eccles. 1:6)

Among those who know the story of FitzRoy's days at the Meteorological Department, there remains a sense of injustice. 'FitzRoy was treated very badly by the scientific community,' Dame Julia Slingo, the Chief Scientist at the Met Office, said when I spoke to her. I asked her, with the benefit of hindsight, whether she thought what he had done was unscientific? 'No,' she replied. 'He was just at the start of a very long journey.'[4]

And if Robert FitzRoy was at the start of one journey, then Dame Julia Slingo is at the beginning of another.

On 7 February 1861, the evening before FitzRoy issued the first ever British storm warning to the north-east ports, John Tyndall, Professor of Natural Philosophy at the Royal Institution, stood to deliver the prestigious Bakerian Lecture at the Royal Society. Forty-one-year-old Tyndall was an Irish scientist and one of the rising stars on the London scene. He was a gifted experimenter, communicator and popular author, well known for his written accounts of his climbing exploits in the Alps, where he had summited many of the hardest peaks. Already he had been at the Royal Institution for about a decade and his reputation as a lecturer was well established. In a year's time he would be invited to serve along-side FitzRoy, Glaisher, Herschel and Airy on the British Association's Balloon Committee, but this night he had other things on his mind.

Tyndall's lecture was titled 'On the Absorption and Radiation of Heat by Gasses and Vapours', the latest update on a scientific enquiry that had been occupying him since 1859. Like Glaisher, Tyndall had developed an interest in the transfer of heat throughout the global system. And just as Glaisher had tracked the flow of radiation through solid bodies, Tyndall had resolved to do the same – but this time with gases. His interest was born of the realisation that for the earth to be hot enough to support life some of the gases had to trap and retain some of the sun's heat. This seemed obvious but, as Tyndall realised, the question had been almost completely ignored by science. It was, he announced, 'perfectly unbroken ground'.[5]

For two years Tyndall had sought to answer the question, testing which gases were the strongest absorbers of radiant heat – what we today call infrared radiation. He had constructed his apparatus at the British Institution, a rig which let him pass heat through tubes of gas and monitor the amount of absorption. The task had been difficult but he had stuck at it and from 9 September 1860 until 29 October he had 'experimented from about eight to ten hours daily'. Now Tyndall was ready to reveal his results. He told his audience it seemed that a negligible amount of heat was soaked up by the typical atmospheric gases: oxygen, hydrogen, nitrogen. Other gases, however, had dramatic absorptive powers, as did water vapour. One of his discoveries related to carbonic acid (carbon dioxide). He was eager to correct a misapprehension:

In the experiments of Dr Franz, carbolic acid appears as a feebler absorber than oxygen. According to my experiments, for small quantities the absorptive power of the former (carbonic acid) is about 150 times that of the latter (oxygen); and for atmospheric tensions, carbonic acid probably absorbs nearly 100 times as much as oxygen.[6]

No one could have guessed that February night at the Royal Society, as FitzRoy prepared himself for his first ever storm warning, that Tyndall was laying the theoretical foundation stone for one of the most contentious scientific disputes in history. The implications of Tyndall's discoveries were clear. The more water vapour, carbon dioxide and other 'greenhouse gases' present in the atmosphere, the warmer the atmosphere would be. In the weeks that followed the lecture Tyndall put out a press release in the London papers. 'All past climate was now understood, and all future climate changes could be predicted simply from a knowledge of the concentrations of these "greenhouse" gases.[7]

For years Tyndall's research remained little more than a neat, if somewhat obscure, foray into the properties of gas, remembered by some, forgotten by most. In the late nineteenth century the Swedish meteorologist Svante Arrhenius dabbled with the riddle, producing calculations on the correlation between carbon dioxide levels in the atmosphere and surface temperature on earth. It was not until 1938 that the subject was revisited again, this time by G.S. Callendar, a British steam engineer, who wondered what the consequences of a high-carbon atmosphere would be. By then Britain was producing about 250 million tonnes of coal a year, the burning of which along with other hydrocarbons was emitting increasing volumes of carbon dioxide into the atmosphere. Callendar calculated the upshot in temperature that should result from a 20 per cent rise in carbon dioxide levels and concluded that it was probably a good thing: rising temperature helping to stave off another ice age.

For a few decades scientists occasionally speculated about this quirk of atmosphere – known to them casually as the Callendar Effect – while all the time the volumes of carbon dioxide continued to rise. The increase in concentrations was dramatic. From about 1805 when FitzRoy was born to the end of the twentieth century the level of carbon dioxide rose from 280 parts per million to 380 parts per million. In the last decades of the twentieth century there was a rejuvenation of interest in the problem. No longer was Tyndall's discovery a scientific curiosity or mathematical puzzle. Politicians realised that the experiments Tyndall had performed in a sealed tube at the Royal Institution were now being played out on a massive scale in the earth's atmosphere. The problem was given a title – global warming – and it became the defining scientific issue of the age.

The issue hit the political mainstream in 1988. That year saw Margaret Thatcher give an anxious address on global warming to the Royal Society, cautioning that humanity had 'unwittingly begun a massive experiment with the system of the planet itself'.[8] The same sentiments were expressed by the NASA scientist James Hanson before a Congressional committee in Washington, DC, and in response the Intergovernmental Panel on Climate Change (IPCC) was founded. In the years since, the IPCC has issued five reports on the state of the atmosphere and the likely impact of increased levels of carbon dioxide and other greenhouse gases. The latest was launched with a press conference in September 2013. It stated that 95 per cent of scientists now agree that global warming is happening. In their notes for policy-makers, the IPCC declared that 'warming of the climate system is unequivocal'.[9]

The fifth IPCC report is loaded with statistics, many of them disconcerting. There had been a rise in global surface temperatures of 0.85°C between 1880 and 2012; 'the atmospheric concentrations of carbon dioxide, methane, and nitrous oxide have increased to levels unprecedented in at least the last 800,000 years'. As evidence of a changing planet it points to the shrinking of the Greenland and Antarctic ice sheets; the disappearance of massive glaciers; Arctic sea ice and northern hemisphere spring snow. Should the atmosphere continue to fill with increasing volumes of greenhouse gases, before the end of the twenty-first century we will find temperatures rising between 3° and 5°C, with sea levels up by half to one metre. Thomas Stocker, the IPCC co-chair who launched the report, made the point starkly. Climate change 'threatens our planet, our only home'.[10]

The climate-change debate has evolved very much like a slow-motion weather forecast, strikingly reminiscent in tone and tenor to the early forecasts of the 1860s. Indeed what FitzRoy and others faced then seems like small beer compared to the passionate arguments that have developed over the past twenty years. The difficulties of the Brussels conference of 1853 are as nothing when set beside the political wrangling of summits at Rio, Kyoto, Bali or Copenhagen. Augustus Smith in the Commons in the 1860s barely registers when we consider the lobbying of powerful free-market capitalists of today. Yet the debate is eerily familiar. Just as before, it lies with human faith in a meteorological prediction. Can we trust scientists to warn us of coming danger? What economic and social costs should we be expected to bear? How can scientists know that they are right?

All these questions are fought over with the same volatility as FitzRoy's forecasting debate. Now, as then, the language is polarised. For both sides there are pejorative and celebratory terms: you can be a warmist or a denier, a believer or a sceptic. In the US where – keeping on the Espy/Redfield tradition – the debate is most febrile, Democrat-leaning news programmes have been found to use the term 'climate change' while their Republican opponents have preferred 'global warming'. Each term calls up a particular set of Pavlovian reactions in the audience. Climate change is valiant, scientific, crucial to the survival of humanity. Global warming is expensive, a pseudoscience, a propaganda trap of the first magnitude.

For the sceptics climate change is the latest in a chain of alarmist phenomena, from Malthus and his population theories to fears over the depletion of fossil fuels and, more recently, the Millennium Bug. Climate change is the latest beating of the drum by a powerful environmental lobby that does not have evidence to support its theory. They point out that there has been no observed warming over the past fifteen years and deride the IPCC's argument that excess heat has been soaked up by the oceans as 'creative science' deployed to patch up a stumbling theory. For sceptics climate change is a hysterical cult that is too big to fail and has become intolerant of criticism. Its political and economic edge give it a dangerous power with governments and businesses overly invested, fumbling away in an advanced state of paranoia, trapped in a cult of *omertà*.

Perhaps the most sustained and potent attacks on climate-change policies have come from libertarian economists, exasperated at the millions of pounds wasted on carbon taxes. They prefer to stick with tried and tested economic theory: the lowest-cost principle. Hydrocarbons, like fossil fuels, are the richest sources of energy and, as a result, provide us with the means for fastest growth. The most lucid explanation of the economic argument is made by Rupert Darwall, an ex-investment banker, in his forensic analysis, *The Age of Global Warming*. He argues, 'Across every dimension, global warming has been a costly fiasco':

> Unsustainable commitments to solar and wind energy in Germany and Spain; the morally abhorrent diversion by rich countries of resources from growing food into making biofuels; the collapse of the EU's carbon market; the transformation of the UK's liberalised energy market producing some of the cheapest electricity to become Europe's most expensive electricity producer; the scandals associated with the Clean Development Mechanism; the destruction of tropical rainforests to make way for palm oil plantations – all provide material for students of policy failure.[11]

Lord Lawson of Blaby, once Chancellor for the Conservatives under
Margaret Thatcher, added his own conclusion in early 2014: 'Global
warming orthodoxy is not merely irrational. It is wicked.'[12]

Dame Julia Slingo, Britain's most visible climate scientist, has worked
on the subject for forty years, becoming the UK's first ever female
professor of meteorology at Reading University in the 1990s and then,
in 2008, returning to the Met Office where she had started her career.
She has been leading the climate-change debate, communicating the
science and the dangers of a warmer world to the British public.

Julia Slingo told me that her curiosity for weather began during
her schooldays in the 1960s. 'I spent a lot of time sat at the desk in
my bedroom,' she said, 'while I was revising for my A-level physics.
The window looked out to the south and I watched the clouds and
wondered why the wind should always blow from the west.'[13]

A few years later, in the early 1970s, as one of the Met Office's
young researchers Slingo started work on some of the earliest ever
climate models – then a completely new area of study. 'When I entered
the Met Office we had barely seen a satellite image,' she says; 'no one
knew what clouds looked like from above, we didn't have any models.'
From the start Slingo worked on the carbon dioxide question. One
of her first publications, *Carbon Dioxide, Climate and Society*, examined
the climate's sensitivity to increased volumes of the greenhouse gas.
'I didn't think this was going to be the biggest problem for humanity
in the twenty-first century, as it is turning out to be,' she said.[14]

Forty years later, in November 2013, Slingo gave the Burntwood
Lecture at the Institution of Environmental Sciences, her subject 'Why
Climate Models are the Greatest Feat of Modern Science'. The 1970s
were a world away. She showed the evolution of climate modelling,
how meteorologists understood how heat is moved around the global
system – something pioneered years before by Glaisher and Tyndall.
To the layman these climate models are indescribably complex. They
combine advanced mathematics, Newtonian physics, thermodynamics,
radiative transfer, particle microphysics, chemistry and biology to
create forecasts that can then be projected on to increasingly tiny
squares of the earth to suggest how the climate might evolve in the
years ahead. All the formulas are processed by the IBM supercomputer
in Exeter, informing IPCC reports. It is a system that for sheer
complexity would have baffled, amazed and thrilled nineteenth-century

scientists, who had little more to go on than their barometers, thermometers and weather maps. 'People end up with this idea that
meteorology is just a soft environmental science, when in fact it is
very difficult,' Slingo tells me. 'It is a postgraduate science. You need
to understand maths, physics and chemistry at the very least.'[15]

But for all the grounded theory, meteorology's lot remains a controversial one. Over the decades sceptics have uncovered bad scientific
practice within the climate-change community. They accuse scientists
of fiddling the numbers, loading the dice so that the machine pumps
out confirmatory data. When the *Daily Telegraph* published a review
of one climate-change book in 2014 it received 9,093 reader comments
in reply: 'sceptics' attacking 'warmists', 'believers' hitting back at
'deniers'.[16]

'There are so many parallels with what is happening now and what
happened with FitzRoy,' Slingo says. 'We are the only science that
really has to predict and that brings trouble. I have had criticism from
FRSs [fellows of the Royal Society], they have this idea that you are
not a very good scientist if you work in this field, which is wrong.
Other scientists are not criticised in the same way. You wouldn't
criticise the Chief Medical Officer. No one would dare to pick apart
medical science the way people do with climate change.'[17]

In a BBC interview in 2014, Slingo was asked why she continued
to work in such a contentious area. She went back to Francis Bacon's
quote in her answer. 'This has become more and more important to
me. It is why I applied for the Met Office Chief Scientist's role after
many years in academia. I needed to see my science working for
society – "the relief of man's estate". To save lives and livelihoods and
to make life better for people who are affected badly by hazardous
weather, climate extremes . . . People don't like what I say about the
science because it doesn't suit them, but it's never going to stop me
saying it.'[18]

I suggested that listening to her words reminded me of FitzRoy,
and how he might have replied a hundred and fifty years before. I
asked if she thought there were any similarities between the two of
them. 'Well, I've no intention of committing suicide,' she replied.[19]

Like no other science, meteorology demands faith. For some, such a
demand is unscientific. In *The Age of Global Warming* Rupert Darwall
writes at length about the 'inscrutability of the future'. What can

really be known? For a benchmark he uses Karl Popper's twentieth-century theory of falsification – Popper's idea that a theory can only be described as scientific if it is capable of being falsified. The best theories are capable of being falsified, he argues. 'The most,' Darwall points out, 'that can be said is global warming *has* happened between two dates in the past.'

He bolsters his philosophical attack on global warming with a quote from Percy W. Bridgman, a Harvard physicist and Nobel Prize winner. Bridgman wrote:

> I personally do not think that one should speak of making statements about the future. For me, a statement implies the possibility of verifying its truth, and the truth of the statement about the future cannot be verified.[20]

And here we come to the crux. For the good and safety of humanity meteorologists are forced to forecast. Although not falsifiable by Popper's method or watertight by Bridgman's standards, the weather forecast is something we generally today have faith in. Few ignored it in October 2013 as St Jude's Storm approached, just at many kept indoors in 2012 when Hurricane Sandy was bearing down on the US's eastern seaboard. Neither Hurricane Sandy nor St Jude could have been falsified, yet they are not unscientific.

Nowadays forecasts are a vital source of protection, a paradigm shift since the days when they were a source of hilarity in the Commons. Who knows how good climate models could become in the future? In the meantime, as the meteorologist Sir Brian Hoskins told the BBC:

> By increasing the greenhouse gas levels in the atmosphere, particularly carbon dioxide, to levels not seen for millions of years on this planet, we are performing a very risky experiment and we're pretty confident that that means that if we keep going like we are that temperatures are going to rise somewhere around 3 to 5° by the end of this century, sea levels up to half to one metre rise.[21]

This is our own experiment, and forging a coherent pact between the science, the politics and the economics will be one of the challenges of our age. We can have faith in the science or we can let nature take its course, just as Captain MacWhirr did in Conrad's *Typhoon*.

There is a moment of prophetic lucidity early in Conrad's story, when MacWhirr surveys the atmosphere from the chart room of the *Nan-shan*.

. . . he stood confronted by the fall of a barometer he had no reason to distrust. The fall – taking into account the excellence of the instrument, the time of the year, and the ship's position on the terrestrial globe – was of a nature ominously prophetic; but the red face of the man betrayed no sort of inward disturbance. Omens were as nothing to him, and he was unable to discover the message of a prophecy till the fulfilment had brought it home to his very door . . . The lurid sunshine cast faint and sickly shadows. The swell ran higher and swifter every moment, and the ship lurched heavily in the smooth, deep hollows of the sea.[22]

Stars in FitzRoy's Meteorological Galaxy

In *The Weather Book*, FitzRoy writes, 'a galaxy of distinguished names occurs to mind as having largely contributed to the meteorological knowledge now generally available'. He goes on to list those he feels have contributed most. Here is a selection, with a few additions, of FitzRoy's Meteorological Stars.

George Airy – 1801–1892

British mathematician and Astronomer Royal from 1835 to 1881. Key figure in Victorian science who pioneered meteorological data collection at Greenwich with James Glaisher in the 1840s and encouraged the first weather reports of the *Daily News*.

François Arago – 1786–1853

Influential French mathematician and astronomer, Director of the Paris Observatory 1843–53. His *Meteorological Essays*, chiefly dealing with lightning was translated into English by Edward Sabine in 1855. Cautioned against forecasting.

Sir Francis Beaufort – 1774–1857

Hydrographer, scientist, sailor and *éminence grise*. Author of the first widely adopted wind and weather scale. Mentor to Robert FitzRoy.

James Capper – 1743–1825

British Army officer and employee of the East India Company. Credited by William Reid as the first to argue that storms had a circular motion, though his ideas were not popularised in his lifetime.

John Dalton – 1766–1844

Quaker, weather diarist and schoolmaster. Started recording his own weather data in 1787 using home-made instruments. Published his *Meteorological Observations and Essays* (1793) in which he began to speculate on the nature of matter and atoms, the beginning of modern atomic theory.

John Frederic Daniell – 1790–1845

Professor of Chemistry at King's College London, his *Meteorological Essays* of 1823 were popular. Invented the Daniell Hygrometer and Daniell cell battery – which became a vital component of Morse's telegraph. Died at a meeting of the Royal Society in March 1845, halfway through a description of the new water barometer.

Heinrich Dove – 1803–79

Prussian meteorologist and winner of the Royal Society's prestigious Copley Medal in 1853 for research on the distribution of heat in the atmosphere. FitzRoy had Dove's *Law of Storms* translated into English in the 1850s, and Dove's ideas on conflicting air masses underpinned much of the early forecasting method.

James P. Espy – 1785–1860

Maverick American meteorologist, classicist and mathematician. Strongly advocated his atmospheric chimney model of air circulation, much to the annoyance of his great rival William C. Redfield. Later became notorious for claims about making rain.

William Ferrel – 1817–91

American teacher and meteorologist who successfully applied the laws of earth's rotation to the standing theories of atmospheric circulation in the 1850s. The Ferrel Cell – one of three great atmospheric cells of air circulation – is named in his honour today.

Thomas Forster – 1789–1860

English naturalist, astronomer, meteorologist and physician. His 1813 book, *Researches About Atmospheric Phaenomena*, was perhaps the most popular meteorological work of the early nineteenth century and attracted a wide readership that included John Constable and François Arago. Forster had a particular interest in natural weather signs and he published many lists of them throughout his career. He made a memorable balloon ascent in 1831.

Benjamin Franklin – 1706–90

Celebrated US philosopher of the eighteenth century. Notable in meteorology for flying a kite in a Philadelphia thunderstorm and drawing sparks out of the sky, proving that lightning was an electrical phenomenon. He also pioneered the study of storms, tracking their progression up the US east coast.

Francis Galton – 1822–1911

Victorian polymath. Notable in meteorology for identifying and coining the term 'anticyclone'. Also pioneered methods of mapping the weather from 1860 onwards, culminating in *The Times'* first ever weather map in 1875. Professional rival of Robert FitzRoy.

James Glaisher – 1809–1903

Astronomer, meteorologist, photographer and balloonist. Long meteorological career began in 1840 when he was appointed Superintendent of the Magnetical and Meteorological Department at the Greenwich Observatory. Wrote many papers, notably on dew and snowflakes. Broke the world altitude record for a balloon flight with Henry Coxwell in 1861. President of the Meteorological Society in 1867–8.

Joseph Henry – 1797–1878

Leading American scientist, first Secretary of the Smithsonian Institution and famously the inventor of the doorbell. Made many contributions to science, particularly in electromagnetism, and led the Smithsonian Meteorological Project from the 1840s onwards.

Sir John Herschel – 1792–1871

Famous British scientist and administrator. Made many contributions to astronomy and mathematics and retained an interest in meteorology. Supported Redfield's theory of circular winds and later corresponded with FitzRoy on the mathematics of rainbows, landscape art and lunisolar theory.

Luke Howard – 1772–1864

British pharmacist and Quaker, widely considered the father of modern meteorology. Achieved fame in the early nineteenth century for his cloud-classification system but went on to publish many other meteorological works. Credited with first describing the urban heat island effect and pioneering climatic studies with his *Climate of London*.

Urbain Le Verrier – 1811–77

French geometer, mathematician and scientific administrator. Became famous for his discovery of Neptune by mathematical calculation in 1846 and, later, for being the autocratic successor to Arago at the Paris Observatory. Steered French meteorological efforts from the mid-1850s.

Elias Loomis – 1811–89

American mathematician, meteorologist and astronomer, for many years a professor at Yale College. In 1834 was one of the first Americans to observe Halley's Comet. Worked extensively on a December 1836 storm and pioneered early, colour-coded weather maps.

Matthew Maury – 1806–73

Lieutenant in the United States Navy, oceanographer and administrator. From the 1840s onwards distilled meteorological data into his sailing charts, earning him the nickname 'Pathfinder of the Seas'. Encouraged European nations to follow America's lead at the Brussels conference of 1853.

Henry Piddington – 1797–1858

Captain in the East India Company and meteorologist who became interested in storm physics in the 1840s. A follower of Redfield and Reid, he published *The Sailor's Horn Book for the Law of Storms* in 1848, a work that brought the theory of circular winds to a mass market. He also suggested the name 'cyclone', which is still used today.

William Redfield – 1789–1857

American businessman and meteorologist. His *Remarks on the Prevailing Storms of the Atlantic Coast, of the North American States*, published in 1831, introduced the US scientific community to the idea of inward-spiralling winds and began a decade of storm analysis. Friend and ally of William Reid, he went on to be the first President of the American Association for the Advancement of Science.

William Reid – 1791–1858

British Army officer and colonial administrator. Became interested in meteorological science following the hurricane of 1831. Spent many years collecting data on West Indian storms and was influenced by Redfield. Published his *Law of Storms* in 1838. Later a governor of various British overseas territories, remembered by Dickens as 'The Good Governor'.

John Tyndall – 1820–93

Popular Victorian physicist who had a long association with the Royal Association of Great Britain. Significant for demonstrating in the early 1860s that some gases were capable of retaining more heat than others, a phenomenon later called the Greenhouse Effect.

Baron Alexander von Humboldt – 1769–1859

Prussian explorer who pioneered the idea of the outdoor scientist. A great influence on Robert FitzRoy and Charles Darwin during their voyages on HMS *Beagle*. Distilled his great scientific knowledge into his book *Kosmos*, which was translated into English in 1845.

Abbreviations

BL	Beineke Rare Books & Manuscript Library, Yale University
FB	Francis Beaufort
GL	Gladstone's Library
HC Deb.	House of Commons Debates
HL	Huntington Library
NA	National Archives
NLI	National Library of Ireland
NMA	National Meteorological Library and Archive
RLE	Richard Lovell Edgeworth
RS	Royal Society Archive

Notes

The Weather Experiment

1 Psalm 19, Bible Gateway: https://www.biblegateway.com/passage/?search=Psalm+19

2 John Frederic Daniell, *Meteorological Essays and Observations* (London: Underwood, 1823), p. 2

3 François Arago, *Meteorological Essays*, translated by Colonel Edward Sabine (London: Longman, Brown, Green & Longmans, 1855) p. 219

4 Jan Golinski, *British Weather and the Climate of Enlightenment* (Chicago: The University of Chicago Press, 2007), p. 18

5 Luke Howard, *Seven Lectures on Meteorology* (1837; Cambridge: Cambridge University Press, 2011). p. 2

6 John Ruskin, *Transactions of the Meteorological Society Instituted in the Year 1823 Vol. One* (London: Smith, Elder & Cornhill, 1839) p. 57

7 Ibid. p. 59

Chapter 1: Writing in the Air

1 NLI. Francis Beaufort to Fanny Edgeworth, MS 13176 (11)

2 Ibid.

3 *The Annual Register, or a View to the History, Politics, and Literature for the Year 1794* (London: Auld, 1799) p. 51. For more on the discovery of Chappe's telegraph see the *Universal Magazine* for October 1794.

4 Charles Dibdin, *The Professional Life of Mr Dibdin, written by himself, together with the words of Six Hundred Songs Vol. III* (London: Dibdin, 1803) p. 315

5 *Daniel Beaufort Journal Entry*, Trinity College Dublin, MS4031, 7 March 1789

6 Alfred Friendly, *Beaufort of the Admiralty: The Life of Sir Francis Beaufort* (London: Hutchinson, 1977) p. 50

7 Howard, *Seven Lectures on Meteorology*, pp. 16–17

8   NMA. Private Weather Diary of Admiral Beaufort, box 1 HMS *Latona, Aquilon* and *Phaeton*, MET/2/1/2/3/539

9   NLA. Francis Beaufort to Charlotte Edgeworth, 24 January 1803

10 Daniel Augustus Beaufort, *Memoir of a Map of Ireland* (London: Faden, 1792) p. ix

11 Richard Lovell Edgeworth, *Memoirs of Richard Lovell Edgeworth, begun by himself and concluded by his daughter Maria Edgeworth Vol. II* (London: Bentley, 1844) p. 260

12 Jenny Uglow, *Lunar Men: The Friends that made the Future* (London, Faber & Faber, 2003) p. 292

13 Edgeworth, *Memoirs Vol. I*, p. 140

14 Ibid. p. 141

15 Ibid. p. 147

16 Ibid. p. 142

17 Desmond King-Hele, *The Collected Letters of Erasmus Darwin* (Cambridge: Cambridge University Press, 2006) p. 74

18 Samuel Johnson, *A Dictionary of the English Language Vol. 3* (1755; London: Longman, Hurst, Rees, Orme & Brown, 1818)

19 Ibid. p. 305

20 James Lequeux, *Le Verrier – Magnificent and Detestable Astronomer*, edited and with an introduction by William Sheehan; translated by Bernard Sheehan (New York: Springer, 2013) p. 271

21 Edgeworth, *Memoirs Vol. II*, p. 159

22 NLI. Maria Edgeworth to Mrs Ruxton, 4 November 1803 (Friday morning)

23 H. F. B. Wheeler and A. M. Broadley, *Napoleon and the Invasion of England: The Story of the Great Terror* (Cirencester: Nonsuch, 2007) pp.272–3

24 NLI. Charlotte Edgeworth to Emmeline King, 11 July 1804

25 NLI. Maria Edgeworth to Sophy Ruxton, 18 December 1803

26 NLI. Francis Beaufort to Charlotte Edgeworth, MS 13176 (11)

27 HL. 24 December 1803 (Dublin) FB to RLE

28 Ibid.

29 HL. 26 March (Athlone) FB to Daniel Beaufort

30 NLI. Charlotte Edgeworth to Emmeline King, 11 July 1804

31 Friendly, *Beaufort of the Admiralty*, p. 120

32 *Freeman's Journal*, 7 July 1804

33 HL. 27 May 1804 (Galway) FB to William Beaufort

34 Friendly, *Beaufort of the Admiralty*, p. 129

35 Daniel Defoe, *The Storm* (1704; London: Penguin, 2005) p. 24

36 Ibid. p. 24

37 NMA. Private Weather Diary of Admiral Beaufort, HMS *Woolwich* 1805–7, MET/2/1/2/3/540

Chapter 2: Nature Caught in the Very Act

1   HL. RLE to FB, 1 June 1810

2   HL. FB to RLE, 9 December 1809

3   Ibid.

4   HL. Joseph Banks to Richard Lovell Edgeworth, 26 December 1813, and 1 December 1813, FB to RLE

5   *The Scots Magazine and Edinburgh Literary Miscellany, Vol.* 76 (Edinburgh: Constable & Company) p. 152

6   Howard, *Seven Lectures on Meteorology,* pp. 115–16

7   *Nicholson's Journal,* January 1814

8   Ronald Brymer Beckett, *John Constable's Correspondence Vol.* 2 (Ipswich: Suffolk Records Society 1970) p. 118

9   C. R. Leslie, *Memoirs of the Life of John Constable, Esq., RA: composed chiefly of his letters. Second Ed.* (London: Longman, Brown, Green, & Longmans, 1845) p. 132

10 Ibid. p. 16

11 Andrew Shirley, *The Rainbow: A Portrait of John Constable* (London: Joseph, 1949) p. 128

12 Ibid. p. 141

13 Beckett, *John Constable's Correspondence Vol.* 4 (Ipswich: Suffolk Records Society, 1970) p. 101

14 George Harvey, *A Treatise on Meteorology* (London, 1834) p. 155

15 Leslie, *Memoirs of the Life of John Constable,* p. 49

16 Beckett, *John Constable's Correspondence Vol.* 10, p. 83

17 Thomas Forster, *Researches About Atmospheric Phaenomena. Second Ed.* (London: Baldwin, 1813) p. 126

18 Ibid. p. viii

19 Sir John Barrow, *Autobiographical Memoir of Sir John Barrow, Bart, Late of the Admiralty* (London: John Murray, 1847) p. 10

20 Golinski, *British Weather and the Climate of Enlightenment,* p. 19

21 Luke Howard, *Essay on the Modifications of Clouds. Third Ed.* (London: Churchill, 1865) p. 1

22 Forster, *Researches About Atmospheric Phaenomena,* p. 56

23 Ibid. p 7

24 Gilpin, William *Three Essays on Picturesque Beauty; on Picturesque Travel and on Sketching Landscape: to which is added a poem, on landscape painting. Second Ed.* (London: Blamire, 1794) p. 36

25 Ibid. p. 72

26 Ibid. p. 89

27 Ibid. p. 34

28 Ibid. p. 42

29   John Thornes, *John Constable's Skies* (Birmingham: University of Birmingham Press, 1999) p. 89

30   Leslie, *Memoirs of the Life of John Constable*, Esq., RA pp. 4–5

31   Ibid. p. 350

32   Ibid. p. 123

33   All annotations taken from John Thornes, *John Constable's Skies*

34   *Examiner*, 27 May 1821

35   Thornes, *John Constable's Skies*, p. 280

36   Leslie, *Memoirs of the Life of John Constable* p. 300

37   Ibid. p. 350

38   Edmund Burke, *A Philosophical Inquiry into the Origin of our Ideas of the Sublime and Beautiful. A new edition* (Basil: Tournisen, 1792) p. 60

39   Leslie, *Memoirs of the Life of John Constable*, p. 281

40   Mark Evans, *John Constable: Oil Sketches from the Victoria and Albert Museum* (London: Victoria & Albert Museum, 2011), p. 93

41   Thornes, *John Constable's Skies*, p. 73

Chapter 3: Rain, Wind and the Wondrous Cold

1   Robert FitzRoy, Charles Darwin and Phillip King, *Narrative of the Surveying Voyages of His Majesty's Ships Adventure and Beagle, between the Years 1826 and 1836 Vol. I* (London: Colburn, 1839) pp. 189–90

2   Ibid. pp. 189–90

3   NA. ADM 51/3053 – Captains' logs: BEAGLE/1825 September 16–1829 December 31

4   Robert FitzRoy, *The Weather Book: A manual of practical meteorology* (London: Longman, Green, Longman, Roberts & Green, 1863) p. 333

5   FitzRoy et al., *Narrative of the Surveying Voyages Vol. II*, p. 71

6   Ibid. p. 333

7   *Good Words*, 1 June 1866

8   FitzRoy's Memorandum, quoted in John Gribbin and Mary Gribbin, *FitzRoy: The Remarkable Story of Darwin's Captain and the Invention of the Weather Forecast* (London: Review, 2004) pp. 301–5

9   James Weddell, *A Voyage Towards the South Pole, Performed in the Years 1822–24* (London: Longman, Rees, Orme, Brown & Green, 1825) pp. 202–3

10   Ibid. pp. 202–3

11   Ibid. p. 2

12   Ibid. p. 141

13   Herman Melville, *Moby-Dick, or, the Whale* (Boston: St Botolph Society, 1922) p. 434

14   Weddell, *A Voyage Towards the South Pole*, p. 44

15   Ibid. p. 55

16   FitzRoy et al., *Narrative of the Surveying Voyages*, p. 217

17   Ibid. p. 217

18   Ibid. p. 218

19   Ibid. p. 222

20   Ibid. p. 225

21   Ibid. p. 223

22   Ibid. p. 232

23   Ibid. p. 50

24   Ibid. p. 50

25   Ibid. p. 230

26   Ibid. p. 234

27   Henry Norton Sulivan, *Life and Letters of the Late Admiral Sir Bartholomew James Sulivan KCB 1810–1890* (London: John Murray, 1896) p. 15

28   NA. ADM 51/3053 – Captains' logs: BEAGLE/1825 September 16–1829 December 31

29   FitzRoy et al., *Narrative of the Surveying Voyages*, pp. 582–4

30   Jeffery Dennis, *Ample Instructions for the Barometer and Thermometer. Third Ed.* (London: Dennis, 1825) p. 2

31   Ibid. p. 2

32   Weddell, *A Voyage Towards the South Pole*, p. 37

33   Thomas Forster, *The Pocket Encyclopaedia of Natural Phenomena* (London: Nicholls, 1827) pp. 7–8

34   *Quarterly Journal of Science, Literature and Art, January to June 1829*, p. 425

35   George, *Treatise on Meteorology*, p. 4

36   John Claridge, *The Shepherd of Banbury's Rules to Judge the Changes of the Weather* (London: Chance and Hurst, 1827) p. iv

37   FitzRoy et al., *Narrative of the Surveying Voyages*, p. 178

38   Ibid. p. 179

39   Ibid. p. 153

40   Ibid. p. 361

41   Ibid. p. 421

42   Ibid. p. 427

43   Ibid. p. 432

44   Weddell, *A Voyage Towards the South Pole*, p. 251

45   HL. 20 May 1817, Richard Lovell Edgeworth to Francis Beaufort, Francis Beaufort Collection

46   HL. FB 748 FB to Fanny Edgeworth, 22 June 1817

47   Barrow, *Autobiographical Memoir*, p. 395

48   HL. FB 17 Diary entry for 12 May 1829

49   Harriet Martineau, *Biographical Sketches* (London: Macmillan, 1869) p. 227

50   Ibid. p. 214

51   Francis Darwin (ed.), *The Life and Letters of Charles Darwin Vol. I* (London: John Murray, 1887) p. 168

52   Frederick Burkhardt and Sydney Smith, *The Correspondence of Charles Darwin. Vol. I* (Cambridge: Cambridge University Press, 1985) pp. 135–6

53   Francis Darwin (ed.), *The Life and Letters of Charles Darwin Vol. I*, pp. 60–1

54   FitzRoy et al., *Narrative of the Surveying Voyages*, p. 37

55   R. D. Keynes (ed.), *Charles Darwin's Beagle Diary* (Cambridge: Cambridge University Press, 1988) p. 11

Chapter 4: Detectives

1   *New York Evening Post*, 3 September 1831

2   From the *Barbados Globe*, reprinted in the *Ithaca Journal*, 14 September 1831

3   Anon, *Account of the Fatal Hurricane by which Barbados Suffered in August 1831* (Bridgetown: Hyde, 1831) p. 56

4   HC Deb. 29 February 1832, Vol. 10, cc. 971–5, 971

5   Anon, *The Seaman's Practical Guide for Barbados and the Leeward Islands* (London: Smith, Elder, 1832) p. 17

6   John Poyer, *The History of Barbados from the First Discovery of the Island* (London: Mawman, 1808) p. 102

7   Ibid. p. 446

8   From the *Barbados Globe*, reprinted in the *Ithaca Journal*, 14 September 1830

9   *United Service Magazine*, Vol. 30, p. 8

10   Psalm 29. *The Holy Bible: Containing the Old and New Testaments, by the Special Command of King James I of England* (London: Whipple, 1815)

11   Elspeth Whitney, *Medieval Science and Technology* (Westport, Conn.: Greenwood Press, 2004) p. 152

12   William Faulke, *A Goodly Gallerye* (1563; Philadelphia: American Philosophical Society, 1979) pp. 28–9

13   James Shapiro, *1599: A Year in the Life of William Shakespeare* (London: Faber & Faber, 2005) pp.117–8

14   Phil, Mundt *A Scientific Search for Religious Truth* (Brisbane: Bridgeway Books, 2006) p. 49

15   'Part of a letter from John Fuller of Sussex, Esq, concerning a Strange Effect of the Late Great Storm in that County', *Philosophical Transactions of the Royal Society 1704–5*, 1 January 1704

16 Vladimir Janković, *Reading the Skies: A Cultural History of English Weather 1650–1820* (London: University of Chicago Press, 2000) p. 62

17 Defoe, *The Storm*, p. 7

18 Ibid. p. 15

19 Ibid. p. 17

20 Matthew Tindal, *Christianity as Old as the Creation* Vol. I (London, 1730) p. 6

21 John Goad, *Astro-Meteorologica, or Aphorisms and Large Significant Discourses on the Natures and Influences of the Celestial Bodies* (1686; London: Sprint, 1699) jacket quote

22 Ibid. pp. 16–17

23 Ibid. p. 25

24 Ibid. p. 27

25 Ibid. p. 39

26 Golinski, *British Weather and the Climate of Enlightenment*, p. 101

27 Weddell, *A Voyage Towards the South Pole*, p. 238

28 FitzRoy et al., *Narrative of the Surveying Voyages*, pp. 465–6

29 Harvey, *Treatise on Meteorology*, p. 3

30 *Edinburgh New Philosophical Journal*, October 1838–April 1839, p. 120

31 *Sketches of Sermons, Preached to Congregations in Various Parts of the United Kingdom and on the European Continent* Vol. 5 (New York: Bangs and Emasy, 1827) p. 47

32 William Cowper quoted in I. C. Garbett, *Morning Dew; or, Daily readings for the people of God* (1773; Bath: Binns & Goodwin, 1864) p. 222

33 Poyer, *History of Barbados*, p. 67

34 Ibid. p. 33

35 Ibid. pp. 33–4

36 Ibid. pp. 54 & 61

37 Reid, *An Attempt to Develop the Law of Storms*, pp. 1–2

38 Ibid. p. 27

39 Ibid. p. 26

40 Denison Olmstead, *Address on the Scientific Life and Labors of William C. Redfield AM* (New Haven: E. Hayes, 1857) p. 13

41 Ibid. p. 8

42 *American Journal of Science*, Vol. 20, p. 19

43 Ibid. p. 21

44 Ibid. p. 45

45 Ibid. pp. 47–8

46 Reid, *An Attempt to Develop the Law of Storms*, p. 3

47 *Edinburgh Review*, January 1839

48 *Manchester Times and Gazette*, 9 July 1836

Chapter 5: Trembling Air, Whirling Winds

1   Albert Barnes, *An Address before the Association of the Alumni of Hamilton College, Delivered 27 July 1836* (Utica: Bennett & Bright, 1836) p. 11

2   Ralph Waldo Emerson, *Miscellanies: Embracing Nature, Addresses and Lectures* (Boston: Phillips, Sampson and Company, 1856) p. 106

3   'Sketch of J. P. Espy.' Reprint from *Popular Science Monthly*, 1889

4   James P. Espy, 'Circular in relation to Meteorological Observations', *Journal of the Franklin Institute*, Vol. XIII (Philadelphia: Franklin Institute, 1834) p. 383

5   James P. Espy, *The Philosophy of Storms* (Boston: Charles Little & James Brown, 1841) p. iii

6   Ibid. p. iv

7   Harvey, *Treatise on Meteorology*, p. 149

8   Ibid. p. 109

9   W. E. Knowles Middleton, *A History of the Theories of Rain* (London: Oldborne, 1965) p. 151

10   John Blackwell, 'Observations and Experiments, made with a view to ascertain the Means by which the Spiders that produce Gossamer effect their aerial excursions', *Transactions of the Linnean Society of London*, Vol. XV (London: Longman, Rees, Orme, Brown & Green, 1832) p. 449

11   Espy, *Philosophy of Storms*, p. 167

12   Knowles Middleton, *History of the Theories of Rain*, pp. 58–62

13   L. M. Morehead, *A Few Incidents in the Life of Professor James P. Espy* (Cincinnati: R. Clarke, 1888)

14   *Journal of the Franklin Institute*, Vol. XVII, 1836, p. 240

15   *Journal of the Franklin Institute*, Vol. XV, 1835, p. 373

16   Ibid. p. 373

17   *Journal of the Franklin Institute*, Vol. XVIII. p. 106

18   Ibid. p. 107

19   Espy, *The Philosophy of Storms*, p. 489

20   *New Bedford Mercury*, 11 November 1836

21   James Rodger Fleming, *Meteorology in America, 1800–1870* (Baltimore: Johns Hopkins University Press, 1990) p. 45

22   BL. Portsmouth, 1 February 1838, W. C. Redfield Correspondence, 1822–57, 3 Vols. Microfilm, GEN MSS

23   BL. New York, 9 April 1838, ibid. GEN MSS 1078

24   Ibid.

25   *Athenaeum*, 25 August 1838

26   *Storms*, The Museum of Foreign Literature, Science, and Art, Vols 35–36 (Philadelphia: Littell, 1839) p. 242

27   *Edinburgh Review*, January 1839, p. 431

28 RS. Letter from Lt-Col. William Reid, Royal Engineers, to J.F.W. Herschel, 3 January 1839, DM/3/117

29 Ruskin quoted in *Transactions of the Meteorological Society Instituted in the Year 1823 Vol. I*, p. 59

30 RS. Archive, EC/1839/12

31 *Journal of the Franklin Institute*, Vol. XXIII, p. 371

32 BL. New York, 17 April 1839, Redfield to Reid, W. C. Redfield Correspondence, 1822–57, 3 Vols. Microfilm, GEN MSS 1078

33 Fleming, *Meteorology in America*, p. 40

34 *Journal of the Franklin Institute*, Vol. XXIII, p. 325

35 *The Knickerbocker* (New York: Clark & Edson, 1839) p. 379

36 Espy, *The Philosophy of Storms*, p. 495

37 *Rhode Island Republican*, 9 January 1839

38 *New Hampshire Sentinel*, 13 February 1839

39 *Times-Picayune*, 12 May 1839

40 *Times-Picayune*, 22 August 1840

41 Fleming, *Meteorology in America*, p. 40

42 *New Bedford Mercury*, 22 March 1839

43 BL. New York, 25 June 1839, Redfield to Reid, W. C. Redfield Correspondence, 1822–57, 3 Vols. Microfilm, GEN MSS 1078

44 The British Association Tenth Meeting, *Literary Gazette and Journal of the Belles Lettres, Arts, Sciences*, London, 10 October 1840

45 Fleming, *Meteorology in America*, p. 50

46 Ibid.

47 Espy, *The Philosophy of Storms*, p. v

48 Fleming, *Meteorology in America*, p. 53

49 Ibid. p. 67

Chapter 6: Liquid Lightning

1 *American Quarterly Register*, Vol. 12

2 *Journal of the Franklin Institute*, Vol. XXII, p. 165

3 *Transactions of the American Philosophical Society*, Vol. 7 (1841) p. 125

4 Ibid. p. 145

5 Ibid. p. 148

6 Elias Loomis, *On Certain Storms in Europe and America, December 1836* (Washington: Smithsonian, 1859) p. 1

7 *Proceedings of the American Philosophical Society*, Vol. 3 (1843) p. 55

8 Ibid. p. 56

9 Samuel F. B. Morse, *Samuel F.B. Morse: His Letters and Journals; edited and supplemented by Edward Lind Morse Vol. II* (New York: Kraus, 1972) p. 211

10   Ibid p. 216

11   Ibid. p. 107

12   Ibid. p. 5

13   Ibid. p. 6

14   Ibid. p. 41

15   Amos Kendall, *Morse's Patent. Full Exposure of Dr Chas T. Jackson's Pretensions to the Invention of the Electro-Magnetic Telegraph* (Washington: Towers, 1852) p. 57

16   James D. Reid, *The Telegraph in America and Morse Memorial* (New York: Polhemus, 1886) pp. 48–9

17   Ibid. p. 44

18   Morse, *Samuel F. B. Morse: His Letters and Journals Vol. II*, p. 17

19   Ibid. p. 18

20   Ibid. p. 38

21   Kenneth Silverman, *Lightning Man: The Accursed Life of Samuel F. B. Morse* (Boston: Da Capo, 2004)

22   Ibid.

23   Kendall, *Morse's Patent*, p. 11

24   Ibid. p. 46

25   Alfred Vail, *The American Electro Magnetic Telegraph* (Philadelphia: Lea & Blanchard, 1845) pp. 74–5

26   Kendall, *Morse's Patent*, p. 48

27   Ibid. pp. 49–51

28   Ibid. p. 19

29   Ibid. p. 54

30   Ibid. p. 58

31   Morse, *Samuel F. B. Morse: His Letters and Journals Vol. II*, p. 70

32   Ibid. p. 73

33   Ibid. p. 75

34   Ibid. p. 81

35   Ibid. p. 172

36   Ibid. p. 222

37   Ibid. p. 225

38   *Pittsfield Sun*, 6 June 1844

39   *Berkshire County Whig*, 20 June 1844

40   *Barre Gazette*, 28 June 1844

41   Henry David Thoreau, *Walden; or Life in the Woods* (1854; Wilder Publications, 2008)

42   Vail, *American Electro Magnetic Telegraph*, p. viii

43   Ibid. p. 52

44   George Brown Goode, *The Smithsonian Institution 1846–1896: The History of its First Half Century* (Washington: Smithsonian, 1897) p. 656

Chapter 7: Steady Eyes, Delicate Skies

1   HL. FB Minute Book 1846
2   Ibid.
3   Robert FitzRoy, *Good Words*, 1 June 1866
4   Darwin Correspondence Database, http://www.darwinproject.ac.uk /entry-1002, accessed on 13 September 2014
5   *The Life Boat*, 2 October 1865
6   Captain Robert FitzRoy, *Captain Fitz Roy's Statement (of Circumstances which led to a Personal Collision between Mr Sheppard and Captain Fitz Roy). August 1841* (London, 1841)
7   FitzRoy, *Good Words*, 1 June 1866
8   FitzRoy, *The Weather Book*, p. 155
9   Ibid. p. 334
10 Ibid. p. 335
11 FitzRoy, *Good Words*, 1 June 1866
12 http://www.darwinproject.ac.uk/entry-1002, accessed on 13 September 2014
13 Howard, *Seven Lectures on Meteorology*, p. 40
14 Ibid. p. 54
15 HL. Francis Beaufort Pocket Book, August 1846
16 *Illustrated London News*, 8 August 1846
17 Jonathan D. C. Webb, 'The Hailstones of 1 August 1846 in Central and Eastern England', *Weather*, Vol. 51, issue 12 (December 1996) pp. 413–19
18 'On the Amount of Radiation of Heat, at Night, from the Earth, and from Various Bodies Placed on or Near the Surface of the Earth', *Philosophical Transactions of the Royal Society*, January 1847
19 George Biddell Airy, *Autobiography* (Cambridge: Cambridge University Press, 1896) p. 2
20 James Glaisher, Camille Flammarion, W. De Fonvielle and Gaston Tissandier, *Travels in the Air* (London: Bentley & Son, 1871) p. 29
21 Ibid. p. 29
22 *Illustrated London News*, 16 March 1844
23 Ibid.
24 *Edinburgh Review*, July 1838
25 James A. Secord, *Visions of Science: Books and Readers at the Dawn of the Victorian Age* (Oxford: Oxford University Press, 2014) p. 112
26 John Frederic Daniell, *Meteorological Essays and Observations* (London: Underwood, 1823) p. xi
27 *Illustrated London News*, 11 January 1845
28 'The Game of Chess Played between London and Portsmouth', *Illustrated London News*, 12 April 1845

29   William Marriott, 'The Earliest Telegraphic Daily Meteorological Reports and Weather Maps', *Quarterly Journal of the Royal Meteorological Society*, 29 (1903) p. 123

30   Ibid. p. 130

31   RS. EC/1849/07

32   RS. James Glaisher letter to Council, 15 January 1850, MM/21/70

33   *Jackson's Oxford Journal*, 20 April 1850

34   Marriott, 'Earliest Telegraphic Daily Meteorological Reports and Weather Maps'

35   *Illustrated London News*, 3 May 1851

36   *Monthly Notices of the Royal Astronomical Society*, Vol. 64, 1904, p. 280

Chapter 8: Beginnings

1   RS. EC/1851/07

2   FitzRoy's Memorandum, quoted in Gribbin and Gribbin, *FitzRoy*, pp. 301–5

3   Frederick Burkhardt (ed.), *Origins: Selected Letters of Charles Darwin, 1822–1859*. Anniversary Edition, p. 45

4   Darwin Correspondence Database, http://www.darwinproject.ac.uk /entry-1014, accessed on 13 September 2014

5   Frederick Burkhardt and Sydney Smith, *The Correspondence of Charles Darwin: 1821–1836*, Vol. 1 (Cambridge: Cambridge University Press, 1985) p. 226

6   http://www.darwinproject.ac.uk/entry-424, accessed on 13 September 2014

7   GL. Robert FitzRoy to Sir Thomas Gladstone, 14 April 1852

8   http://www.darwinproject.ac.uk/entry-1554A, accessed on 13 September 2014

9   NA. BJ 7/2 – Maury's plan for synoptic charts: memorandum

10   NA. BJ 7/109 – Letter from G. B. Airy to Henry James regarding decision on which government department the work of digesting meteorological observations should be attached to

11   NA. BJ 7/113 – Memorandum by Robert FitzRoy 'with reference to the proposition of Lieutenant Maury'

12   NA. BJ 7/123 – Memorandum by Robert FitzRoy on the establishment of a Meteorological Office, its function and staffing

13   Robert FitzRoy, 'On British Storms', *Report of the Meeting of the British Association* (London: John Murray, 1860) p. 42

14   HC Deb. 30 June 1854, Vol. 134, cc. 959–1008

15   Lequeux, *Le Verrier*, p. 278

16 Ibid. p. 278

17 *Nautical Magazine*, Vol. 31 (1862) p. 364

18 M. Dickens and Georgina Hogarth, *The Letters of Charles Dickens*, two vols (London, 1880) p. 345

19 *Illustrated London News*, 5 January 1850

20 NA. BJ 7/108 – Letter from George Biddell Airy, Astronomer Royal, Royal Observatory, Greenwich, to Henry James regarding printing of meteorological observations and supporting objective of another meteorological conference

21 *Manchester Guardian*, 1 January 1855

22 James Glaisher, 'Snow Crystals in 1855', *Transactions of the Microscopical Society of London*, Vol. III, p. 179

23 Ibid. p. 180

24 Ibid. p. 181

25 Ibid. p. 181

26 Ibid. p. 183

27 Ibid. p. 184

28 NA. BJ 7/133 – Draft circular by Robert FitzRoy on the value of meteorology to shipping

29 NA. BJ 7/8 – Office arrangements

30 NA. BJ 7/77 – Letter from Robert FitzRoy to Matthew Maury

31 NA. BJ 7/544 – Copy of Robert FitzRoy's memorandum 'The Routine of the Meteorological Office' listing duties of himself and his staff, Lieutenant Simpkinson, Assistant, William Pattrickson, Chief Clerk and Draughtsman, J. H. Babington, Mr Townsend and Mr Harding

32 NA. BJ 7/153 – Correspondence between Henry James and Robert FitzRoy regarding FitzRoy's proposals for a new meteorological register and differing opinions as to the form of log

33 *The London, Edinburgh and Dublin Philosophical Magazine and Journal of Science*, Vol. X, July–December 1855, p. 377

34 RS. 25 Lowndes Street, 20 February 1841

35 RS. To Sir John Herschel, 4 May 1858

36 Ibid.

37 FitzRoy *Barometer and Weather Guide. Second Edition* (London: Eyre and Spottiswoode, 1859) p. 5

38 Ibid. p. 14

39 Ibid. p. 19

40 NA. BJ 7/95 – Maury to FitzRoy

41 Nicolas Courtney, *Gale Force 10: The Life and Legacy of Admiral Beaufort* (London: Review, 2002) p. 302

42 NA. BJ 7/707 – Death of Sir Francis Beaufort and the Beaufort Testimonial Fund: Correspondence and papers

43   BL. W. C. Redfield Correspondence, 1822–1857, 3 Vols. Microfilm, GEN MSS 1078

44   Ibid.

45   Howard, *Seven Lectures on Meteorology*, p. 23

46   *New York Times*, 19 August 1858

47   *Colburn's United Service Magazine*, 1859, Part III, p. 572

Chapter 9: Dangerous Paths

1   FitzRoy, *The Weather Book*, p. 311

2   Ibid. p. 312

3   Alexander McKee, *The Golden Wreck: the tragedy of the Royal Charter* (Bebbington: Avid Publications, 2000) p. 31

4   Ibid. p. 37

5   FitzRoy, *The Weather Book*, p. 316

6   Ibid. p. 306

7   Ibid. p. 320

8   McKee, *The Golden Wreck*, p. 67

9   Ibid. p. 104

10   W. F. Peacock, *A Ramble to the Wreck of the Royal Charter* (Manchester: Coles, 1860) p. 4

11   'The Shipwreck', in M. Slater and J. Drew (eds), *Dickens' Journalism, 'The Uncommercial Traveller' and Other Papers* (London: Dent, 2000) pp. 30–1

12   *Illustrated London News*, 6 November 1859

13   FitzRoy, *The Weather Book*, p. 420

14   *Philosophical Magazine*, Vol. XX, Fourth Series, p. 66

15   FitzRoy, *The Weather Book*, p. 103

16   Ibid. p. 311

17   FitzRoy, 'On British Storms', p. 42

18   FitzRoy, *The Weather Book*, p. 320

19   *Liverpool Mercury*, 28 September 1861

20   FitzRoy, 'On British Storms', p. 43

21   Darwin Correspondence Database, http://www.darwinproject.ac.uk /entry-2567, accessed on 11 September 2014

22   Nicolls, *Evolution's Captain*, p. 318

23   Howard, *Seven Lectures on Meteorology*, p. 11

24   Thomas Forster, *Annals of Some Remarkable Aerial and Alpine Voyages* (London: Keating & Brown, 1832) p. 76

25   Ibid. p. 78

26   Glaisher et al., *Travels in the Air*, p. 30

27 Ibid. p. 43

28 Ibid. p. 43

29 *Transactions of the Meteorological Society Instituted in the Year 1823 Vol. One,* p. 57

30 Glaisher et al., *Travels in the Air,* p. 44

Chapter 10: Dazzling Bright

1 Glaisher et al., *Travels in the Air,* p. 49

2 Ibid. p. 51

3 NA. BJ 7/723

4 Glaisher et al., *Travels in the Air,* p. 53

5 Ibid. p. 54

6 Ibid. p. 54

7 Ibid. p. 54

8 Glaisher et al., *Travels in the Air,* p. 21

9 *The Times,* 11 September 1862

10 NA. BJ 7/723

11 NA. BJ 7/725

12 *On the System of Forecasting the Weather pursued in Holland, Report of the 33rd Meeting of the British Association for the Advancement of Science, Aug. and September, 1863*

13 'A Visit to Admiral FitzRoy's Weather Office', *United Service Magazine,* July 1866

14 FitzRoy, *The Weather Book,* p. 178

15 *Colburn's United Service Magazine,* 1865, Part II, p. 551

16 FitzRoy, *The Weather Book,* p. 190

17 *Once a Week,* 23 February 1863

18 Ibid.

19 *Morning Post,* 31 March 1862

20 FitzRoy, *The Weather Book,* p. 169

21 Ibid. p. 218

22 *The Times,* 11 April 1862

23 FitzRoy, *The Weather Book,* p. 190

24 Darwin Correspondence Database, http://www.darwinproject.ac.uk /entry-3836, accessed on 14 September 2014

25 RS. 16 March 1863

26 FitzRoy, *The Weather Book,* p. 7

27 Ibid. p. 331

28 RS. MS/743/1/57

29 'Admiral FitzRoy on the Weather', *Eclectic Magazine,* December 1863

30   'The Weather Book: A Manual of Practical Meteorology by Rear Admiral
       FitzRoy', Athenaeum, 17 January 1863
31   Ibid.

                          Chapter 11: Endings

1   Westminster Review, Vol. 79–80, 1863, p. 261
2   Martin Brookes, Extreme Measures: The Dark Visions and Bright Ideas of
       Francis Galton (London: Bloomsbury, 2004) p. 18
3   RS. EC/1860/10
4   Brookes, Extreme Measures, p. 128
5   Ibid. p. 129
6   Francis Galton, 'A Development of the Theory of Cyclones' – accessed
       September 2014 at http://galton.org/essays/1860–1869/galton-1863
       -proc-royal-soc-cyclones.pdf
7   Francis Galton, Meteorographica, or Methods of Mapping the Weather
       (London: Macmillan, 1863) – accessed September 2014 at http://galton
       .org/books/meteorographica
8   The Reader, 19 December 1863
9   The Times, 27 January 1863
10   [Various], The Science of the Weather in a series of letters and essays
       (Glasgow: Laidlow, 1866) p. 26
11   Ibid. p. 26
12   Ibid. p. 192
13   FitzRoy, The Weather Book, p. 244
14   Ibid. p. 247
15   RS. 24 December 1862
16   RS. 16 March 1863
17   RS. 20 March 1863
18   The Times, 14 May 1864
19   The Times, 16 May 1864
20   Liverpool Mercury, 21 October 1863
21   FitzRoy, The Weather Book, p. 231
22   Darwin Correspondence Database, http://www.darwinproject.ac.uk
       /entry-2575, accessed on 14 September 2014
23   James R. Moore, The Post-Darwinian Controversies: A study of the Protestant
       struggle to come to terms with Darwin in Great Britain and America 1870–1900
       (Cambridge: Cambridge University Press, 1981) p. 91
24   Brookes, Extreme Measures, p. 142
25   Katharine Anderson, Predicting the Weather: Victorians and the Science of
       Meteorology (Chicago: University of Chicago Press, 2005) p. 163

26 Francis Galton, 'Statistical Inquiries into the Efficacy of Prayer', *Fortnightly Review* – accessed September 2014 at http://galton.org /essays/1870–1879/galton-1872-fortnightly-review-efficacy-prayer.html

27 Ibid.

28 *Punch*, 17 October 1863

29 Glaisher et al., *Travels in the Air*, p. 92

30 *Punch*, 20 January 1864

31 *Punch*, June 1863

32 *The Times*, 18 June 1864

33 Ibid.

34 *Colburn's United Service Magazine*, 1866, Part II, p. 354

35 FitzRoy, *Good Words*, 1 June 1866

36 RS. 5 October 1861

37 Lovell Reeve, *Portraits of Men of Eminence Vol. III* (London: Lovell Reeve, 1863) p. 56

38 FitzRoy, *Good Words*, 1 June 1866

39 Ibid.

40 H. E. L. Mellersh, *FitzRoy of the Beagle* (London: Maison & Lipscomb, 1968) pp. 281–4

41 Ibid. pp. 281–4

42 *Leeds Mercury*, 2 May 1865

Chapter 12: Truth Telling

1 J. D. Hooker, 2 May 1865. Darwin Correspondence Database, http://www.darwinproject.ac.uk/entry-4826, accessed 14 September 2014

2 Darwin Correspondence Database, http://www.darwinproject.ac.uk /entry-4827, accessed on 14 September 2014

3 'The Suicide of Admiral Robert FitzRoy', *Nottinghamshire Guardian*, 5 May 1865

4 GL. GG/519, 14 April 1852

5 http://www.darwinproject.ac.uk/entry-4831, accessed on 14 September 2014

6 *Journal of the Royal Geographical Society Vol. 35* (London: Murray, 1865) p. cxxxi

7 *The Times*, 26 June 1865

8 http://www.darwinproject.ac.uk/entry-4921, accessed on 14 September 2014

9 Mellersh, *FitzRoy of the Beagle*, p. 286

10 *Sporting Gazette*, 8 July 1865

11   RS. Memorandum on the Meteorological Office by Edward Sabine, 1865, MM/14/74

12   Ibid.

13   Brookes, *Extreme Measures*, p. 137

14   Dr Leone Levi, *Annals of British Legislation: Being a Digest of the Parliamentary Blue Books Vol. III* (London: Smith, Elder, 1866) p. 453

15   Ibid. p. 454

16   Ibid. p. 456

17   Ibid. p. 460

18   NA. BJ 7/960 – Milner Gibson to Thomas Farrer, 17 May 1866

19   Reeve, *Portraits of Men of Eminence Vol. I*, p. 56

20   *The Works of Lord Bacon Vol. I* (London: Bohn, 1850) p. liii

21   http://www.darwinproject.ac.uk/entry-2122, accessed on 14 September 2014

22   Charles Dickens, *The Pickwick Papers* (1836; London: Wordsworth Classics, 1992)

23   Helen Vendler, *Emily Dickinson: Selected Poems and Commentaries* (Cambridge, Mass.: Belknap, 2010) p. 431

24   Joseph Conrad, *Typhoon* (1902/1903; Ware: Wordsworth Classics, 1998) p. 31

25   Revd Francis Redford to the Board of Trade, Parliamentary Papers (1867) LXVI, pp. 185–203

26   Christopher Cooke, *Admiral FitzRoy: His Facts and Failures: a letter to the Marquis of Tweeddale* (London: Hall, 1867) p. 12

27   *Symons's Monthly Meteorological Magazine* (London: Stanford, 1867) p. 101

28   Ibid. p. 101

29   Ibid. p. 104

30   Ibid. p. 104

31   *Filey and the gales of 1860, 1867, 1869 AND 1880*, http://www.scarboroughs maritimeheritage.org.uk/afileygales.php, accessed March 2014

32   Glaisher et al., *Travels in the Air*, p. 95

33   Ibid. pp. 85–6

34   Alexander von Humboldt, *Kosmos Vol. I* (London: Baillière, 1845) p. 338

35   Howard, *Seven Lectures on Meteorology*, p. 17

36   J. L. Hunt, *James Glaisher FRS (1809–1903) Astronomer, Meteorologist and Pioneer of Weather Forecasting: 'A Venturesome Victorian'* (Royal Astronomical Society, 1996) p. 340

37   H. P. Hollis and Rev. Tucker, *Glaisher, James J., Oxford Dictionary of National Biography*

Dusk

1   M. Minnaert, *The Nature of Light & Colour in the Open Air*, (New York: Dover, 1954) p. 271

West Winds

1   *Warnings over storm due to hit England and Wales*: BBC News http:// www.bbc.co.uk/news/uk-24674537, accessed July 2014

2   *How accurate are our public forecasts?* http://www.metoffice.gov.uk /about-us/who/accuracy/forecasts, accessed July 2014

3   *The Public Weather Service's Contribution to the UK Economy*, http:// www.metoffice.gov.uk/media/pdf/h/o/P WSCG_benefits_report.pdf, accessed July 2014

4   Dame Julia Slingo, interview with the author, 23 June 2014

5   John Tyndall, 'On the Absorption and Radiation of Heat by Gasses and Vapours', *Philosophical Transactions of the Royal Society of London*, Vol. 151, 1861, p. 2

6   Ibid. p. 27

7   James Rodger Fleming, *Historical Perceptions on Climate Change* (New York: Oxford University Press, 1998)

8   Rupert Darwall, *The Age of Global Warming, A History* (London: Quartet, 2013)

9   *Summary for Policy Makers*: IPCC Report 2013, https://www.ipcc .ch/pdf/assessment report/ar4/wg1/ar4-wg1-spm.pdf, accessed July 2014

10   Climate change 'threatens our only home', warns IPCC: BBC News, http://www.bbc.co.uk/news/science-environment-24299664, accessed July 2014

11   Darwall, *The Age of Global Warming*

12   Lord Nigel Lawson, *The Trouble with Climate Change* (London: Global Warming Policy Foundation, 2014) p. 18

13   Dame Julia Slingo, interview with the author, 23 June 2014

14   Ibid.

15   Ibid.

16   Charles Moore, '*The game is up for climate change believers*', *Daily Telegraph*, http://www.telegraph.co.uk/culture/books/non_fictionreviews/10748667 /The-game-is-up-for-climate-change-believers.html

17   Ibid.

18   Dame Julia Slingo, *The Life Scientific*, BBC Radio 4, 8 April 2014

19   Dame Julia Slingo, interview with the author, 23 June 2014
20   Darwall, *The Age of Global Warming*
21   Sir Brian Hoskins, Radio 4 *Today* Programme, 13 February 2014
22   Conrad, *Typhoon*, p. 23

Select Bibliography

Newspapers, journals and parliamentary papers

Albany Journal
American Journal of Science
American Quarterly Register
Annual Register
Athenaeum
Barre Gazette
Berkshire County Whig
Boston Evening Mercantile Journal
Boston Paper
Bulletin météorologique
Colburn's United Service Magazine
Cowe's Meterological Register
Daily News
Eclectic Magazine
Edinburgh Journal of Science
Edinburgh New Philosophical Journal
Edinburgh Review
Era
Examiner
Fortnightly Review
Freeman's Journal
Good Words
Guardian
Harper's New Monthly Magazine
Illustrated London Almanac
Illustrated London News
Ithaca Journal

Jackson's Oxford Journal
Journal of Commerce
Journal of Natural Philosophy, Chemistry and the Arts
Journal of the Franklin Institute
Journal of the Royal Geographical Society
Journal of the Statistical Society
Knickerbocker
La Patrie
Leeds Mercury
Life Boat
Literary Gazette
Liverpool Mercury
London, Edinburgh and Dublin Philosophical Magazine and Journal of Science
London Intellectual Observer
Manchester Times and Gazette
Medical Times
Monthly Review or, Literary Journal
Morning Chronicle
Morning Post
Nautical Magazine
New Bedford Mercury
New Hampshire Sentinel
New Monthly Magazine
New York Journal of Commerce
New York Observer
New York Register
New York Times
Nicholson's Journal
Nottinghamshire Guardian
Once a Week
Pamphleteer
Park Lane Express
Philosophical Magazine
Philosophical Transactions of the Royal Society
Pittsfield Sun
Proceedings of the American Philosophical Society
Punch
Putnam's Monthly Magazine of American Literature, Science and Art

Quarterly Journal of Science, Literature and Art
Reader
Rhode Island Republican
Scots Magazine and Edinburgh Literary Miscellany
Sporting Gazette
Symons's Monthly Meteorological Magazine
Telegraph
The Thunderer
The Times
Times-Picayune
Transactions of the Geological Society of Pennsylvania
Transactions of the Linnean Society
Transactions of the Meteorological Society
Universal Magazine
Westminster Review

Archives

Beineke Rare Book & Manuscript Library, Yale University

W. C. Redfield correspondence, 1822–57, 3 vols. Microfilm, GEN MSS
1078

Gladstone's Library

The Gladstone/Glynne Papers

Huntington Library, San Marino, California

The Francis Beaufort Collection: private and sundry correspondence,
diaries, journals and memorabilia

National Archives, Kew

BJ 7 – FitzRoy Meteorological Department Papers
Admiralty papers, ships' logs, letters from captains, wills

National Library of Ireland, Dublin

Edgeworth and Beaufort Papers

National Meteorological Library and Archive, Exeter

Beaufort's weather diaries
Private Weather Diary: Diary of Admiral Beaufort box 1 HMS Latona,
Aquilon and Phaeton MET/2/1/2/3/539

Royal Society Archives, London

Herschel, Reid, FitzRoy, Glaisher, Beaufort and Edgeworth Papers

Other Primary Sources

*A Portable Cyclopaedia or, Compendious Dictionary of Arts and Sciences
including the Latest Discoveries* (London: Phillips, 1810)
Airy, George Biddell, *Magnetic and Meteorological Observations made at
the Royal Observatory, Greenwich in the Year 1842* (London: Palmer &
Clayton, 1844)
Airy, George Biddell, *Autobiography*, edited by Wilfrid Airy (Cambridge:
Cambridge University Press, 1896)
*Annual of Scientific Discovery of Year-Book of Facts in Science and Art for
1855* (Boston: Gould & Lincoln, 1855)
*The Annual Register, or a View to the History, Politics, and Literature for
the Year 1794* (London: Auld, 1799)
Anon. *Account of the Fatal Hurricane by which Barbados Suffered in August
1831* (Bridgetown: Hyde, 1831)
Anon. *The Seaman's Practical Guide for Barbados and the Leeward Islands*
(London: Smith, Elder, 1832)
Arago, François, *Meteorological Essays,* translated by Colonel Edward
Sabine (London: Longman, Brown, Green and Longmans, 1855)
Aristotle, *Meteorographica*, translated by H. D. P. Lee (Cambridge, Mass.:
Loeb Classical Library, 1952)
Barnes, Albert, *An Address before the Association of the Alumni of Hamilton
College, Delivered 27 July 1836* (Utica: Bennett & Bright, 1836)
Barrow, Sir John, *Autobiographical Memoir of Sir John Barrow, Bart, Late
of the Admiralty* (London: John Murray, 1847)
Beaufort, Daniel Augustus, *Memoir of a Map of Ireland* (London: Faden,
1792)
Beaufort, Francis, *Karamania, or a Brief Description of the South Coast
of Asia Minor* (London, Hunter, 1817)
Beckett, Ronald Brymer (ed.), *John Constable's Correspondence. Vols. 1–8*
(Ipswich: Suffolk Records Society, 1962–70)

Burke, Edmund, *A Philosophical Inquiry into the Origin of our Ideas of the Sublime and Beautiful. A new edition* (Basil: Tournisen, 1792)

Chambers' Information for the People, A Popular Encyclopaedia Vol. II (Philadelphia: Smith, 1855)

Claridge, John, *The Shepherd of Banbury's Rules to Judge the Changes of the Weather* (London: Hurst and Chance, 1827)

Conrad, Joseph, *Typhoon* (1902/03; Ware: Wordsworth Classics, 1998)

Cooke, Christopher, *Admiral FitzRoy, His Facts and Failures: a letter to the Marquis of Tweeddale* (London: Hall & Co, 1867)

Daniell, John Frederic, *Meteorological Essays and Observations* (London: Underwood, 1823)

Daniell, John Frederic, *Elements of Meteorology*, two vols (London: Parker, 1845)

Darwin, Charles, *On the Origin of Species* (London: John Murray, 1859)

Darwin, Erasmus, *The Botanic Garden. A Poem in Two Parts* (New York: Swords, 1798)

Darwin, Francis (ed.), *The Life and Letters of Charles Darwin Vol. I* (London: John Murray, 1887)

Davis, G., *Frostiana; or A History of the River Thames in a Frozen State, with an Account of the Late Severe Frost* (London: G. Davis, 1814)

Defoe, Daniel, *The Storm* (1704; London: Penguin, 2005)

Dennis, Jeffery, *Ample Instructions for the Barometer and Thermometer. Third Ed.* (London: Dennis, 1825)

Dibdin, Charles, *The Professional Life of Mr Dibdin, written by himself, together with the words of Six Hundred Songs Vol. III* (London: Dibdin, 1803)

Dickens, Charles, *The Pickwick Papers* (1836; London: Wordsworth Classics, 1992)

Dove, Heinrich, *The Law of Storms* (London: Board of Trade, 1858)

Edgeworth, Richard Lovell, *Memoirs of Richard Lovell Edgeworth, begun by himself and concluded by his daughter Maria Edgeworth, two vols.* (London: Bentley, 1844)

Espy, James P., *The Philosophy of Storms* (Boston: Charles Little & James Brown, 1841)

Faulke, William, *A Goodly Gallerye* (1563; Philadelphia: American Philosophical Society, 1979)

FitzRoy, Robert, Charles Darwin and Phillip King, *Narrative of the Surveying Voyages of His Majesty's Ships Adventure and Beagle, between the years 1826 and 1836. Four Vols* (London: Colburn, 1839)

FitzRoy, Captain Robert, *Captain Fitz Roy's statement (of circumstances which led to a personal collision between Mr. Sheppard and Captain Fitz Roy). August 1841* (London, 1841)

FitzRoy, Rear Admiral Robert, *Barometer and Weather Guide. Second Edition* (London: 1859)

FitzRoy, Rear Admiral Robert, *The Weather Book: A Manual of Practical Meteorology* (London: Longman, Green, Longman, Roberts, & Green, 1863)

Forster, Thomas, *Researches About Atmospheric Phaenomena. Second Ed.* (London: Baldwin, 1815)

Forster, Thomas, *The Pocket Encyclopaedia of Natural Phenomena* (London: Nicholls, 1827)

Forster, Thomas, *Annals of Some Remarkable Aerial and Alpine Voyages* (London: Keating & Brown, 1832)

Galton, Francis, *Meteorographica, or Methods of Mapping the Weather* (London: Macmillan, 1863)

Galton, Francis, 'A Development of the Theory of Cyclones', paper read at the Royal Society, 8 January 1863 – accessed from galton.org

Galton, Francis, 'Statistical Inquiries into the Efficacy of Prayer', *Fortnightly Review*, Vol XII, 1872 – accessed from galton.org

Garbett, I. C., *Morning Dew; or, Daily readings for the people of God* (Bath: Binns & Goodwin, 1864)

Gilpin, William, *Three Essays on Picturesque Beauty; on Picturesque Travel and on Sketching Landscape: to which is added a poem, on landscape painting. Second Ed.* (London: Blamire, 1794)

Glaisher, James, 'Philosophical instruments and processes as represented in the Great Exhibition in Royal Society for the Encouragement of Arts, Manufactures and Commerce. Lectures on the results of the Great Exhibition, etc.', Ser. 1, 1852

Glaisher, James, Camille Flammarion, W. De Fonvielle and Gaston Tissandier, *Travels in the Air* (London: Bentley & Son, 1871)

Goad, John, *Astro-Meteorologica or Aphorisms and Large Significant Discourses on the Natures and Influences of the Celestial Bodies* (1686; London: Sprint, 1699)

Goode, George Brown, *The Smithsonian Institution 1846–1896: The History of its First Half Century* (Washington: Smithsonian, 1897)

Harvey, George, *A Treatise on Meteorology* (London, 1834)

Howard, Luke, *The Climate of London, Two Vols* (London: Phillips, 1818)

Howard, Luke, *A Cycle of Eighteen Years in the Seasons of Britain* (London: Ridgeway, 1842)

Howard, Luke, *Essay on the Modifications of Clouds. Third Ed.* (London: Churchill, 1865)

Howard, Luke, *Seven Lectures on Meteorology* (1837; Cambridge: Cambridge University Press, 2011)

Johnson, Samuel, *A Dictionary of the English Language Vol. 3* (1755; London: Longman, Hurst, Rees, Orme & Brown, 1818)

Kaemtz, L. F., *A Complete Course of Meteorology, translated by C.V. Walker* (London: Baillière, 1845)

Kendall, Amos, *Morse's Patent. Full Exposure of Dr Chas T. Jackson's Pretensions to the Invention of the Electro-Magnetic Telegraph* (Washington: Towers, 1852)

Leslie, C. R., *Memoirs of the Life of John Constable, Esq., RA: composed chiefly of his letters. Second Ed.* (London: Longman, Brown, Green & Longmans, 1845)

Levi, Dr Leone, *Annals of British Legislation: Being a Digest of the Parliamentary Blue Books Vol. III* (London: Smith, Elder, 1866)

Loomis, Elias, *On Certain Storms in Europe and America, December, 1836* (Washington: Smithsonian, 1860)

Martineau, Harriet, *Biographical Sketches* (London: Macmillan, 1869)

Melville, Herman, *Moby-Dick, or, the Whale* (1851; Boston: St Botolph Society, 1922)

Methven, Captain Robert, *Narratives Written by Sea Commanders illustrative of the Law of Storms* (London: Weale, 1851)

Morehead, L. M., *A Few Incidents in the Life of Professor James P. Espy* (Cincinnati: R. Clarke, 1888)

Morse, Samuel Finley Breese, *Samuel F.B. Morse: His Letters and Journals; edited and supplemented by . . . Edward Lind Morse . . . with Notes and Diagrams Bearing on the Invention of the Telegraph. Two Vols* (1915; New York: Kraus, 1972)

Murphy, Patrick, *Meteorology considered in its connexion with Astronomy, Climate and the Geographical Distribution of Animals and Plants* (London: Ballière, 1836)

The New Encyclopaedia or, Universal Dictionary of Arts and Sciences. Vol. XIV (London: Vernor, Hood & Sharpe, 1807)

Newton, H. A., *Memoir of Elias Loomis 1811–1889* (Washington: Government Printing Office, 1891)

Olmstead, Denison, *Address on the Scientific Life and Labors of William C. Redfield AM* (New Haven: E. Hayes, 1857)

Oxford Dictionary of National Biography (Oxford: Oxford University Press, 2004; online edition, 2008)

Park, John James, *The Topography and Natural History of Hampstead* (London: White, Cochrane, 1814)

Pasley, C. W., *Description of the Universal Telegraph for Day and Night Signals* (London: Egerton, 1823)

Peacock, W. F., *A Ramble to the Wreck of the Royal Charter* (Manchester: Coles, 1860)

Piddington, Henry, *The Sailor's Horn-Book for the Law of Storms* (New York: John Wiley, 1848)

Poyer, John, *The History of Barbados from the First Discovery of the Island* (London: Mawman, 1808)

Reeve, Lovell, *Portraits of Men of Eminence Vols. I–III* (London: Lovell Reeve, 1863)

Reid, James D., *The Telegraph in America, and Morse Memorial* (New York: Polhemus, 1886)

Reid, Lieut.-Col. William, *An Attempt to Develop the Law of Storms* (London: Weale, 1838)

Report of a Committee appointed to consider certain questions relating to the Meteorological Department of the Board of Trade (presented to both Houses of Parliament) (London: Eyre and Spottiswoode, for H.M. Stationery Office, 1866)

Shaffner, Tal. P., *The Telegraph Manual* (New York: Pudney & Russell, 1859)

Steinmetz, Andrew, *A Manual of Weathercasts: Comprising Storm Prognostics on Land and Sea* (London: Routledge & Son, 1866)

Sulivan, Henry Norton, *Life and Letters of the Late Admiral Sir Bartholomew James Sulivan KCB 1810–1890* (London: John Murray, 1896)

Taylor, Joseph, *The Complete Weather Guide: A Collection of Practical Observations for Prognosticating the Weather* (London: Harding, 1812)

Taylor, Richard (ed.) *Scientific Memoirs Vol. III* (London: Taylor, 1843)

The Book of Common Prayer (London: Rivingtons, 1864)

The Holy Bible: Containing the Old and New Testaments, by the Special Command of King James I of England (London: Whipple, 1815)

Thomson, Thomas, *History of the Royal Society, from its Institution to the End of the Eighteenth Century* (London: Thomson, 1812)

Thoreau, Henry David, *Walden; or, Life in the Woods* (1854; Wilder Publications, 2008)

Tindal, Matthew, *Christianity as Old as the Creation Vol. I* (London, 1730)

Turnbull, Lawrence, *The Electro-Magnetic Telegraph with a Historical Account of its Rise, Progress and Present Condition* (Philadelphia: Hart, 1853)

Vail, Alfred, *The American Electro Magnetic Telegraph* (Philadelphia: Lea & Blanchard, 1845)

[Various] *The Science of the Weather in a Series of Letters and Essays* (Glasgow: Laidlow, 1866)

Vendler, Helen, *Emily Dickinson: Selected Poems and Commentaries* (Cambridge, Mass.: Belknap, 2010)

Von Humboldt, Alexander, *Kosmos Vol. I* (London: Baillière, 1845)

Weddell, James, *A Voyage Towards the South Pole, Performed in the Years 1822–24* (London: Longman, Rees, Orme, Brown & Green, 1825)

Wells, William Charles, *An Essay on Dew and Several Appearances Connected with it* (London: Longman, Green, Reader & Dyer, 1866)

Wilkins, John, *Mercury: or The Secret and Swift Messenger* (London: Baldwin, 1694)

Young, Thomas, *A Course of Lectures on Natural Philosophy and the Mechanical Arts*, two vols (London: Johnson, 1807)

Secondary Material

Anderson, Katharine, *Predicting the Weather: Victorians and the Science of Meteorology* (Chicago: University of Chicago Press, 2005)

Badt, Kurt, *John Constable's Clouds*, translated from the German by Stanley Godman (London: Routledge & Kegan Paul, 1950)

Barlow, Derek, *Origins of Meteorology: An Analytical Catalogue of the Correspondence and Papers of the First Government Meteorological Office, under Rear Admiral Robert FitzRoy, 1854–1865, and Thomas Henry Babington, 1965–1866; of the Successor Meteorological Office* (London: Public Record Office, 1996)

Bone, Stephen, *British Weather* (London: Collins, 1946)

Brookes, Martin, *Extreme Measures: The Dark Visions and Bright Ideas of Francis Galton* (London: Bloomsbury, 2004)

Burton, Jim, 'Robert FitzRoy and the Early History of the

Meteorological Office', *British Journal for the History of Science,* Vol. 19, No. 2 (July 1986)

Clarke, Desmond, *The Ingenious Mr Edgeworth* (London: Oldborne, 1965)

Courtney, Nicolas, *Gale Force 10: The Life and Legacy of Admiral Beaufort* (London: Review, 2002)

Cox, John D., *Storm Watchers: The Turbulent History of Weather Prediction from Franklin's Kite to El Niño* (Hoboken: John Wiley & Sons, 2002)

Darwall, Rupert, *The Age of Global Warming: A History* (London: Quartet, 2013)

Davis, John L., 'Weather Forecasting and the Development of Meteorological Theory at the Paris Observatory 1853–1878', *Annals of Science,* 41 (1984)

Desmond, Adrian and James Moore, *Darwin* (London: Penguin, 1992)

DeYoung, Donald, *Weather and Bible : 100 Questions and Answers* (Grand Rapids: Baker Books, 1992)

Evans, Mark, *John Constable: Oil Sketches from the Victoria and Albert Museum* (London: Victoria & Albert Museum, 2011)

Fleming, James Rodger, *Meteorology in America, 1800–1870* (Baltimore: Johns Hopkins University Press, 1990)

Fleming, James Rodger, *Historical Perceptions on Climate Change* (Oxford, New York: Oxford University Press, 1998)

Friendly, Alfred, *Beaufort of the Admiralty: The Life of Sir Francis Beaufort* (London: Hutchinson, 1977)

Golinski, Jan, *British Weather and the Climate of Enlightenment* (Chicago: University of Chicago Press, 2007)

Gribbin, John and Mary Gribbin, *FitzRoy: The Remarkable Story of Darwin's Captain and the Invention of the Weather Forecast* (London: Review, 2004)

Halford, Pauline, *Storm Warning* (Stroud: Sutton, 2004)

Hamblyn, Richard, *The Invention of Clouds: How an Amateur Meteorologist Forged the Language of the Skies* (London: Picador, 2001)

Holmes, Richard, *The Age of Wonder: How the Romantic Generation Discovered the Beauty and Terror of Science* (London: Harper Press, 2008)

Holmes, Richard, *Falling Upwards* (London: Collins, 2013)

Hunt, J. L., *James Glaisher FRS (1809–1903) Astronomer, Meteorologist and Pioneer of Weather Forecasting: 'A Venturesome Victorian'* (London: Royal Astronomical Society, 1996)

Jankovic, Vladimir, *Reading the Skies: A Cultural History of English Weather 1650–1820* (Chicago, London: University of Chicago Press, 2000)

Kemp, Peter, *The Oxford Companion to Ships and the Sea* (St Albans: Granada Publishing, 1979)

Keynes, R. D. (ed.), *Charles Darwin's Beagle Diary* (Cambridge: Cambridge University Press, 1988)

Knowles Middleton, W. E., *A History of the Theories of Rain* (London: Oldborne, 1965)

Lawson, Lord Nigel, *The Trouble with Climate Change* (London: Global Warming Policy Foundation, 2014)

Leary, Patrick, *A Brief History of the Illustrated London News*, accessed online at www.gale.co.uk/lin

Lequeux, James, *Le Verrier – Magnificent and Detestable Astronomer*, edited and with an introduction by William Sheehan; translated by Bernard Sheehan (New York: Springer, 2013)

Longshore, David, *Encyclopaedia of Hurricanes, Typhoons and Cyclones* (London: FitzRoy Dearborne, 1999)

Ludlum, David McWilliams, *Early American Hurricanes, 1492–1870* (Boston: American Meteorological Society, 1963)

Marriott, William, 'The Earliest Telegraphic Daily Meteorological Reports and Weather Maps', *Quarterly Journal of the Royal Meteorological Society*, 29 (1903)

McKee, Alexander, *The Golden Wreck: The Tragedy of the Royal Charter* (Bebbington: Avid Publications, 2000)

Mellersh, H. E. L., *FitzRoy of the Beagle* (London: Maison & Lipscomb, 1968)

Minnaert, M., *The Nature of Light & Colour in the Open Air* (New York: Dover, 1954)

Monmontier, Mark, *Air Apparent: How Meteorologists Learned to Map, Predict and Dramatize Weather* (Chicago: University of Chicago Press, 1999)

Moore, James R., *The Post-Darwinian Controversies: A Study of the Protestant struggle to come to terms with Darwin in Great Britain and America 1870–1900* (Cambridge: Cambridge University Press, 1981)

Nicolls, Peter, *Evolution's Captain: The Tragic Fate of Robert FitzRoy, the Man who Sailed Charles Darwin Around the World* (London: Profile Books, 2004)

Pesic, Peter, *Sky in a Bottle* (Cambridge, Mass.: MIT Press, 2005)

Secord, James A., *Visions of Science: Books and Readers at the Dawn of the Victorian Age* (Oxford: Oxford University Press, 2014)

Shirley, Andrew, *The Rainbow: A portrait of John Constable* (London: Joseph, 1949)

Silverman, Kenneth, *Lightning Man: The Accursed Life of Samuel F.B. Morse* (Boston: Da Capo, 2004)

Slater, M. and J. Drew (eds), *Dickens' Journalism, 'The Uncommercial Traveller' and Other Papers* (London: Dent, 2000)

Thompson, Robert Luther, *Wiring a Continent: The History of the Telegraph Industry in the United States 1832–1866* (New York: Arno Press, 1972)

Thornes, John, *John Constable's Skies* (Birmingham: University of Birmingham Press, 1999)

Uglow, Jenny, *Lunar Men: The Friends that Made the Future* (London: Faber & Faber, 2003)

Walker, Malcolm, *History of the Meteorological Office* (Cambridge: Cambridge University Press, 2011)

Wheeler, H. F. B. and A. M. Broadley, *Napoleon and the Invasion of England: The Story of the Great Terror* (Cirencester: Nonsuch, 2007)

Whitney, Elspeth, *Medieval Science and Technology* (Westport, Conn.: Greenwood Press, 2004)

Wood, Gillen D'Arcy, 'Constable, Clouds, Climate Change', *Wordsworth Circle*, Vol. 38, 1–2 (2007)

Index